MODERN EARTH SCIENCE

Test Generator: Test Item Listing

HOLT, RINEHART AND WINSTON

Harcourt Brace & Company

Austin • New York • Orlando • Atlanta • San Francisco • Boston • Dallas • Toronto • London

Printed in the United States of America

ISBN 0-03-050667-0

5 179   03 02 01 00

Contents

About the Test Generator: Test Item Listing v

Chapter 1 Test Items 1

Chapter 2 Test Items 20

Chapter 3 Test Items 43

Chapter 4 Test Items 62

Chapter 5 Test Items 81

Chapter 6 Test Items 99

Chapter 7 Test Items 118

Chapter 8 Test Items 140

Chapter 9 Test Items 159

Chapter 10 Test Items 173

Chapter 11 Test Items 192

Chapter 12 Test Items 213

Chapter 13 Test Items 235

Chapter 14 Test Items 253

Chapter 15 Test Items 275

Chapter 16 Test Items 297

Chapter 17 Test Items 316

Chapter 18 Test Items 337

Chapter 19 Test Items 353

Chapter 20 Test Items 372

Chapter 21 Test Items 389

Chapter 22 Test Items 408

Chapter 23 Test Items 428

Chapter 24 Test Items 448

Chapter 25 Test Items 470

Chapter 26 Test Items 487

Chapter 27 Test Items 506

Chapter 28 Test Items 528

Chapter 29 Test Items 548

Chapter 30 Test Items 568

About the Test Generator: Test Item Listing

The *Modern Earth Science Test Generator* contains over 2800 test items for *Modern Earth Science*. For easy reference, this *Test Item Listing* contains a complete printout of the test items that are stored on the test generator diskettes. The test items are briefly described below. For directions on how to install and use the computer program, please refer to the *HRW Test Generator User's Guide* in the test generator package.

Chapter Files

The test items are stored in chapter files. There is one file for each chapter of your textbook. Each test item is classified by its question type and its learning objective. This allows you to use the question types and the learning objectives as criteria for selecting items and creating a test. With the test generator, you can edit the test items that are supplied on the diskettes or write your own questions. Your user's guide more fully describes these features of the program.

Test Item Listing

Each test item in this booklet is listed as shown in the example below. You will notice that some information about the test item is included in addition to the test item itself. A description of the various information fields follows the example.

Item Type: Multiple Choice   Objective: 1   Section 2.1
Level: A: Standard   Tag: None

The part of the earth that is solid but has the ability to flow is the
a) lithosphere.   b) oceanic crust.   c) asthenosphere.   d) inner core.

Item Type

Each test item in the test generator is classified as one of the following types:
- True/False
- Multiple Choice
- Fill in the Blank
- Essay

Objective

Each test item is also classified according to the learning objectives that are listed at the beginning of each chapter section in your textbook. Each test item has a two-part objective code. The first part of the code lists the number of the objective as it appears in the textbook. The second part lists the chapter section in which the learning objective appears.

Level

Each test item is classified according to two levels of difficulty. Level A, or Standard, test items are slightly less demanding than Level B, or Enriched, test items.

Tag

Test items that refer to a graphic are tagged with a diagram number. A tag denotes that a graphic will print with the tagged test item when you generate a test. The graphics in each chapter file appear at the end of the chapter listing.

Answer

A test item's answer immediately follows the test item's listing. For easier identification, a small black box appears in the lower-right corner of the answer box.

Technical Support

The *HRW Test Generator User's Guide* should answer any questions you have about the program. However, if you need assistance, call the Holt, Rinehart and Winston Technical Support Center at 1-800-323-9239. You can also access a knowledge base of solutions to common problems and answers to common questions by calling the Technical Support Center's Fax on Demand Service at 1-800-352-1680.

In addition, you can contact the Technical Support Center through the World Wide Web at http://www.hrwtechsupport.com or by E-mail at tsc@hrwtechsupport.com.

Chapter 1

1) Item Type: True-False   Objective: 1   Section 1.1
    Level: B: Enriched   Tag: None

_____ Nearly three-fourths of the earth's surface is covered by oceans.

T	■

2) Item Type: True-False   Objective: 1   Section 1.1
    Level: A: Standard   Tag: None

_____ Meteorology is the study of the earth's oceans.

F	■

3) Item Type: True-False   Objective: 1   Section 1.1
    Level: A: Standard   Tag: None

_____ Earth science includes the study of oceans.

T	■

4) Item Type: True-False   Objective: 1   Section 1.1
    Level: A: Standard   Tag: None

_____ Astronomy is a branch of earth science.

T	■

5) Item Type: True-False   Objective: 1   Section 1.1
    Level: A: Standard   Tag: None

_____ Exploration for oil deposits would be done most likely by an ecologist.

F	■

6) Item Type: True-False   Objective: 1   Section 1.1
    Level: A: Standard   Tag: None

_____ Meteorology is a major area of study within earth science.

T	■

7) Item Type: True-False   Objective: 1   Section 1.1
    Level: A: Standard   Tag: None

_____ An astronomer studies the universe beyond the earth.

T	■

8) Item Type: True-False   Objective: 2   Section 1.1
    Level: A: Standard   Tag: None

    _____ The atmosphere is one part of the geosphere.

    ┌─────────┐
    │ F   ▄▄ │
    └─────────┘

9) Item Type: True-False   Objective: 2   Section 1.1
    Level: A: Standard   Tag: None

    _____ An ecosystem is a largely self-supporting, physically distinct system.

    ┌─────────┐
    │ T   ▄▄ │
    └─────────┘

10) Item Type: True-False   Objective: 2   Section 1.1
    Level: A: Standard   Tag: None

    _____ Most plastics are biodegradable.

    ┌─────────┐
    │ F   ▄▄ │
    └─────────┘

11) Item Type: True-False   Objective: 2   Section 1.1
    Level: A: Standard   Tag: None

    _____ The physical environment of the earth supports the biosphere.

    ┌─────────┐
    │ T   ▄▄ │
    └─────────┘

12) Item Type: True-False   Objective: 2   Section 1.1
    Level: A: Standard   Tag: None

    _____ Earth scientists study the geosphere.

    ┌─────────┐
    │ T   ▄▄ │
    └─────────┘

13) Item Type: True-False   Objective: 2   Section 1.1
    Level: A: Standard   Tag: None

    _____ Biodegradable wastes pose a major threat to the environment.

    ┌─────────┐
    │ F   ▄▄ │
    └─────────┘

14) Item Type: True-False   Objective: 2   Section 1.1
    Level: A: Standard   Tag: None

    _____ Ecology is a field of study that unites earth science and biology.

    ┌─────────┐
    │ T   ▄▄ │
    └─────────┘

15) Item Type: True-False   Objective: 2   Section 1.1
    Level: A: Standard   Tag: None

_____ Plants produce food through the process of photosynthesis.

T

16) Item Type: True-False   Objective: 3   Section 1.2
    Level: A: Standard   Tag: None

_____ A kilogram is a standard unit of measurement.

T

17) Item Type: True-False   Objective: 3   Section 1.2
    Level: A: Standard   Tag: None

_____ Scientific methods are guides to scientific problem solving.

T

18) Item Type: True-False   Objective: 3   Section 1.2
    Level: A: Standard   Tag: None

_____ Eratosthenes used shadows and simple geometry to calculate the circumference of the earth.

T

19) Item Type: True-False   Objective: 3   Section 1.2
    Level: A: Standard   Tag: None

_____ One way scientists state a problem is by asking questions based on observations.

T

20) Item Type: True-False   Objective: 3   Section 1.2
    Level: A: Standard   Tag: None

_____ A controlled experiment usually tests several variables at one time.

F

21) Item Type: True-False   Objective: 3   Section 1.2
    Level: B: Enriched   Tag: None

_____ A hypothesis may be based on facts established through experimentation.

T

22) Item Type: True-False Objective: 3 Section 1.2
    Level: B: Enriched Tag: None

_____ Proposing a hypothesis is usually the first step when following scientific methods.

F

23) Item Type: True-False Objective: 3 Section 1.2
    Level: B: Enriched Tag: None

_____ A hypothesis must be tested before any observations can be made.

F

24) Item Type: True-False Objective: 3 Section 1.2
    Level: B: Enriched Tag: None

_____ Scientific research always follows the sequential steps called scientific methods.

F

25) Item Type: True-False Objective: 3 Section 1.2
    Level: A: Standard Tag: None

_____ A scientific law is a rule for conducting scientific experiments.

F

26) Item Type: True-False Objective: 4 Section 1.2
    Level: A: Standard Tag: None

_____ Scientists have simulated climatic conditions on a computer to help explain the extinction of the dinosaurs.

T

27) Item Type: True-False Objective: 4 Section 1.2
    Level: B: Enriched Tag: None

_____ Rock layers containing deformed quartz particles provide important evidence for the big bang theory.

F

28) Item Type: True-False Objective: 4 Section 1.2
    Level: A: Standard Tag: None

_____ Dust raised by the impact of a meteorite on earth may have blocked the sun's rays for many years.

T

29) Item Type: True-False   Objective: 4   Section 1.2
     Level: A: Standard   Tag: None

_____ Scientists believe that at one time an abnormally high amount of dust may have
       formed a cloud over the earth.

T

30) Item Type: True-False   Objective: 4   Section 1.2
     Level: A: Standard   Tag: None

_____ Controlled experiments have confirmed the meteorite-impact hypothesis.

F

31) Item Type: True-False   Objective: 4   Section 1.2
     Level: A: Standard   Tag: None

_____ Iridium is a common element found in earth rocks.

F

32) Item Type: True-False   Objective: 5   Section 1.3
     Level: A: Standard   Tag: None

_____ In order for a theory to become a law, it must be proven correct every time it is
       tested.

T

33) Item Type: True-False   Objective: 5   Section 1.3
     Level: A: Standard   Tag: None

_____ Once a scientific law is well established, it becomes a theory.

F

34) Item Type: True-False   Objective: 6   Section 1.3
     Level: A: Standard   Tag: None

_____ When a chemical element is heated enough, it produces a bright-line spectrum.

T

35) Item Type: True-False   Objective: 6   Section 1.3
     Level: A: Standard   Tag: None

_____ According to the big bang theory, all galaxies in the universe are moving toward
       the earth.

F

36) Item Type: True-False   Objective: 6   Section 1.3
     Level: B: Enriched   Tag: None

    ____ The wavelength of red light is longer than that of blue light.

T

37) Item Type: True-False   Objective: 6   Section 1.3
     Level: A: Standard   Tag: None

    ____ Galaxies in the universe are moving closer together.

F

38) Item Type: True-False   Objective: 6   Section 1.3
     Level: A: Standard   Tag: None

    ____ As a light source moves away from an observer, the wavelengths of light will appear shorter to the observer.

F

39) Item Type: True-False   Objective: 7   Section 1.3
     Level: A: Standard   Tag: None

    ____ The big bang theory proposes that all the matter and energy in the universe was once compressed into an extremely small volume.

T

40) Item Type: True-False   Objective: 8   Section 1.3
     Level: A: Standard   Tag: None

    ____ Background radiation is thought to be energy left from the big bang explosion.

T

41) Item Type: Multiple Choice   Objective: 1   Section 1.1
     Level: A: Standard   Tag: None

Which of the following is a meteorologist most likely to study?

a) mountain formations   b) marine life

c) rainfall trends   d) meteorites

c

42) Item Type: Multiple Choice   Objective: 1   Section 1.1
    Level: A: Standard   Tag: None

The relationship between plants and animals in a lake would most likely be studied by

a) a meteorologist.   b) an ecologist.

c) an oceanographer.   d) a geologist.

b

43) Item Type: Multiple Choice   Objective: 1   Section 1.1
    Level: A: Standard   Tag: None

Which of the following would an astronomer be most likely to study?

a) a comet's path through the solar system

b) the height of tides during a full moon

c) an iridium-laden rock layer produced by meteorites

d) the amount of solar energy absorbed by a tree

a

44) Item Type: Multiple Choice   Objective: 2   Section 1.1
    Level: B: Enriched   Tag: None

The ozone layer in the atmosphere shields the earth from

a) fluorocarbons.   b) excessive heat from the sun.

c) pollution.   d) the sun's ultraviolet rays.

d

45) Item Type: Multiple Choice   Objective: 2   Section 1.1
    Level: A: Standard   Tag: None

Which of the following is a part of the biosphere?

a) the oceans   b) the moon

c) the earth's core   d) the sun

a

46) Item Type: Multiple Choice   Objective: 2   Section 1.1
    Level: B: Enriched   Tag: Diagram   0171

The relationships shown in this diagram would most likely be studied by

a) a geologist.   b) an ecologist.   c) a meteorologist.   d) an astronomer.

b

47) Item Type: Multiple Choice   Objective: 2   Section 1.1
    Level: A: Standard   Tag: None

The largest ecosystem is called the

a) biosphere.   b) atmosphere.   c) hydrosphere.   d) geosphere.

a

48) Item Type: Multiple Choice   Objective: 2   Section 1.1
    Level: B: Enriched   Tag: None

The ozone in the upper atmosphere is a form of

a) carbon.   b) hydrogen.   c) oxygen.   d) silicon.

c

49) Item Type: Multiple Choice   Objective: 2   Section 1.1
    Level: A: Standard   Tag: None

The ecosystem that encompasses all other ecosystems is called the

a) geosphere.   b) atmosphere.   c) biosphere.   d) hydrosphere.

c

50) Item Type: Multiple Choice   Objective: 3   Section 1.2
    Level: A: Standard   Tag: None

Which of the following is an important means of gathering information?

a) stating the problem   b) making a measurement

c) forming a hypothesis   d) stating a conclusion

b

51) Item Type: Multiple Choice   Objective: 3   Section 1.2
    Level: A: Standard   Tag: None

Eratosthenes used careful observation and simple geometry to determine the

a) distance between the earth and the moon.

b) circumference of the earth.

c) distance between the earth and the sun.

d) circumference of the sun.

52) Item Type: Multiple Choice   Objective: 3   Section 1.2
Level: A: Standard   Tag: None

A possible explanation of or solution to a problem is called

a) a law.   b) a hypothesis.   c) a conclusion.   d) an observation.

b

53) Item Type: Multiple Choice   Objective: 3   Section 1.2
Level: A: Standard   Tag: None

The first step in using scientific methods to solve a problem is to

a) state the problem.   b) form a hypothesis.

c) reach a conclusion.   d) gather information.

a

54) Item Type: Multiple Choice   Objective: 4   Section 1.2
Level: A: Standard   Tag: None

The meteorite-impact hypothesis is supported by the existence of layers of rock containing high levels of

a) quartz.   b) silicon.   c) ozone.   d) iridium.

d

55) Item Type: Multiple Choice   Objective: 4   Section 1.2
Level: A: Standard   Tag: None

The meteorite-impact hypothesis is one explanation for the discovery of

a) background radiation in some galaxies.   b) bright-line spectra in some elements.

c) a red shift in some stars.   d) high iridium levels in some rocks.

d

56) Item Type: Multiple Choice   Objective: 5   Section 1.3
Level: A: Standard   Tag: None

A hypothesis or set of hypotheses supported by the results of experimentation and observation is called a(n)

a) law.   b) theory.   c) observation.   d) fact.

b

57) Item Type: Multiple Choice   Objective: 5   Section 1.3
Level: A: Standard   Tag: None

A well-established and proven theory is most likely to become

a) a scientific law.   b) a hypothesis.   c) an observation.   d) a conclusion.

a

58) Item Type: Multiple Choice   Objective: 6   Section 1.3
Level: B: Enriched   Tag: Diagram   0173

In the diagram above, compared with the light coming from source **X,** the light coming
from source **Y** appears

a) brighter.   b) less bright.   c) more red.   d) more blue.

c

59) Item Type: Multiple Choice   Objective: 6   Section 1.3
Level: B: Enriched   Tag: Diagram   0173

As an observer moves toward sources **X** and **Y** in the diagram above, light from the two
sources will appear to become

a) red-shifted.   b) less bright.   c) redder.   d) blue-shifted.

d

60) Item Type: Multiple Choice   Objective: 6   Section 1.3
Level: B: Enriched   Tag: None

When a light source moves toward an observer, the light waves appear to be

a) moving slower.   b) longer.   c) moving faster.   d) shorter.

d

61) Item Type: Multiple Choice   Objective: 6   Section 1.3
Level: B: Enriched   Tag: Diagram   0174

Look at the diagram above. One wavelength is equal to the distance between points

a) **1** and **3.**   b) **1** and **4.**   c) **2** and **3.**   d) **2** and **4.**

a

62) Item Type: Multiple Choice   Objective: 7   Section 1.3
Level: A: Standard   Tag: None

A spectroscope is most likely to be used to determine a star's

a) distance from earth.   b) diameter.   c) chemical makeup.   d) brightness.

c

63) Item Type: Multiple Choice   Objective: 7   Section 1.3
Level: A: Standard   Tag: None

It is thought that before the big bang all the matter and energy in the universe was in the form of one

a) extremely small volume.

b) solar system.

c) expanding cloud.

d) galaxy.

64) Item Type: Multiple Choice   Objective: 8   Section 1.3
Level: B: Enriched   Tag: None

The amount of red shift in the light emitted by a galaxy is most likely to be used to determine

a) the temperature of the stars in the galaxy.

b) the galaxy's diameter.

c) the speed at which the galaxy is traveling.

d) the galaxy's age.

65) Item Type: Multiple Choice   Objective: 8   Section 1.3
Level: A: Standard   Tag: None

The red shift observed in a distant galaxy's spectra is evidence for the

a) meteorite-impact hypothesis.

b) presence of background radiation.

c) destruction of the ozone.

d) expansion of the universe.

66) Item Type: Multiple Choice   Objective: 8   Section 1.3
Level: A: Standard   Tag: None

The discovery of background radiation distributed evenly throughout the universe provided support for the

a) Doppler effect.

b) meteorite-impact hypothesis.

c) big bang theory.

d) law of gravitation.

67) Item Type: Multiple Choice   Objective: 8   Section 1.3
     Level: A: Standard   Tag: None

A spectroscope is most likely to be used to determine a star's

a) position.   b) composition.   c) brightness.   d) origin.

b

68) Item Type: Multiple Choice   Objective: 8   Section 1.3
     Level: A: Standard   Tag: None

Background radiation distributed evenly throughout the universe probably resulted from

a) the Doppler effect.   b) moving.

c) the big bang.   d) starlight spectra.

c

69) Item Type: Fill in the Blank   Objective: 3   Section 1.2
     Level: B: Enriched   Tag: None

The earth's circumference was first determined accurately by _____.

Eratosthenes

70) Item Type: Fill in the Blank   Objective: 3   Section 1.2
     Level: B: Enriched   Tag: None

A controlled experiment is designed to test a single factor called a(n)

_____.

variable

71) Item Type: Fill in the Blank   Objective: 3   Section 1.2
     Level: A: Standard   Tag: Diagram   0172

The temperature indicated by this thermometer is _____.

14°C

72) Item Type: Fill in the Blank   Objective: 4   Section 1.2
     Level: A: Standard   Tag: None

Scientists believe that the iridium found in some rock layers may originally have come

from a(n) _____.

meteorite

73) Item Type: Fill in the Blank   Objective: 4   Section 1.2
Level: A: Standard   Tag: None

Scientists proposed the meteorite-impact hypothesis to explain the extinction of the

_____.

dinosaurs ▪

74) Item Type: Fill in the Blank   Objective: 5   Section 1.3
Level: A: Standard   Tag: None

A scientist's possible explanation of or solution to a problem is called a(n)

_____.

hypothesis ▪

75) Item Type: Fill in the Blank   Objective: 5   Section 1.3
Level: A: Standard   Tag: None

A theory that is well-established through research and experimentation is most likely to

become a scientific _____.

law ▪

76) Item Type: Fill in the Blank   Objective: 5   Section 1.3
Level: A: Standard   Tag: None

A rule that consistently describes a natural phenomenon is known as a scientific

_____.

law ▪

77) Item Type: Fill in the Blank   Objective: 6   Section 1.3
Level: A: Standard   Tag: None

The apparent shift in the wavelengths of energy emitted by an energy source moving away

from or toward an observer is known as the _____.

Doppler effect ▪

78) Item Type: Fill in the Blank   Objective: 6   Section 1.3
    Level: A: Standard   Tag: None

When heated, each element produces a series of thin colored lines called a(n)

_____.

bright-line spectrum

79) Item Type: Fill in the Blank   Objective: 6   Section 1.3
    Level: A: Standard   Tag: None

When light passes through a glass prism, it produces a band of colors called the

_____.

spectrum

80) Item Type: Fill in the Blank   Objective: 6   Section 1.3
    Level: A: Standard   Tag: None

Sunlight is made up of a mixture of different-colored light called the

_____.

spectrum

81) Item Type: Fill in the Blank   Objective: 6   Section 1.3
    Level: B: Enriched   Tag: None

When a light source is moving toward an observer, the wavelengths of light appear

_____.

shorter

82) Item Type: Fill in the Blank   Objective: 6   Section 1.3
    Level: A: Standard   Tag: None

A bright-line spectrum uniquely identifies each _____.

element

83) Item Type: Fill in the Blank   Objective: 6   Section 1.3
    Level: B: Enriched   Tag: None

Scientists determine the speed at which galaxies travel by examining the degree of red

_____.

shift

84) Item Type: Fill in the Blank   Objective: 6   Section 1.3
    Level: A: Standard   Tag: None

Scientists studying starlight determine what elements are present in the stars by using a(n)

_____.

| spectroscope | ■ |

85) Item Type: Fill in the Blank   Objective: 7   Section 1.3
    Level: A: Standard   Tag: None

The explosion that scientists think created the universe is called the

_____.

| big bang | ■ |

86) Item Type: Fill in the Blank   Objective: 8   Section 1.3
    Level: B: Enriched   Tag: None

The discovery of red shift in the spectra of galaxies provided evidence that the universe

is _____.

| expanding | ■ |

87) Item Type: Essay   Objective: 2   Section 1.1
    Level: B: Enriched   Tag: None

Describe the relationships between plants, animals, and microorganisms in a tropical rain forest.

_____

_____

_____

_____

_____

Plants in the rain forest use sunlight to produce food. The plants are then eaten by animals, which are in turn eaten by other animals. When the plants and animals die, microorganisms convert the bodies of the dead organisms into chemicals that enter the soil and nourish other plants and animals.

Chapter 1

88) Item Type: Essay   Objective: 4   Section 1.2
    Level: A: Standard   Tag: None

What does the meteorite-impact hypothesis state?

_____

_____

_____

_____

_____

> This hypothesis states that millions of years ago a giant meteorite crashed into the earth. The impact of the collision raised dust that blocked the sun's rays and probably caused the earth to become colder. Plant life began to die, and many animal species, including the dinosaurs, became extinct.

89) Item Type: Essay   Objective: 6   Section 1.3
    Level: B: Enriched   Tag: None

How might travelers in space use the Doppler effect to determine whether they are moving toward or away from a distant galaxy?

_____

_____

_____

_____

_____

> Space travelers could monitor the bright-line spectrum emitted by the galaxy. If the spectrum is shifted toward the red end, the travelers are moving away from the galaxy. If the spectrum is shifted toward the blue end, the travelers are moving toward the galaxy.

90) Item Type: Essay   Objective: 7   Section 1.3
Level: A: Standard   Tag: None

Describe the big bang theory.

_____

_____

_____

_____

_____

> Billions of years ago, all matter and energy in the universe was compressed into a hot, dense sphere. This sphere exploded, sending the matter and energy hurtling outward in a giant cloud. As the cloud expanded, some of the matter condensed into galaxies. Today the universe is still expanding.

91) Item Type: Essay   Objective: 8   Section 1.3
Level: B: Enriched   Tag: None

Describe what changes in the spectra of galaxies tell scientists about the universe.

_____

_____

_____

_____

_____

> Scientists have found that the line of spectrum of every galaxy tested is shifted toward the red end. No galaxy showed a shift toward the blue end. The red shift indicates that all galaxies in the universe are moving away from the earth, and that the universe is expanding.

1)   Tag Name: Diagram 0171

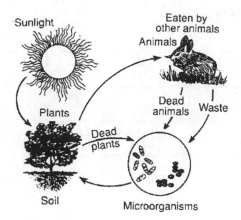

2)   Tag Name: Diagram 0172

3)   Tag Name: Diagram 0173

X

|600 nm|

Y

|700 nm|

4)   Tag Name: Diagram 0174

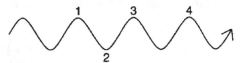

Traveling Light Wave

1) Item Type: True-False   Objective: 1   Section 2.1
   Level: A: Standard   Tag: None

   _____ The lithosphere is made up of the solid part of the upper mantle and the crust.

   T

2) Item Type: True-False   Objective: 1   Section 2.1
   Level: A: Standard   Tag: None

   _____ The outer core is the only zone of the earth that is thought to be liquid.

   T

3) Item Type: True-False   Objective: 1   Section 2.1
   Level: A: Standard   Tag: None

   _____ The solid rock of the asthenosphere has the ability to flow.

   T

4) Item Type: True-False   Objective: 1   Section 2.1
   Level: A: Standard   Tag: None

   _____ The core makes up about two-thirds of the earth's mass.

   F

5) Item Type: True-False   Objective: 2   Section 2.1
   Level: A: Standard   Tag: None

   _____ Studies of seismic waves suggest that the lithosphere is probably more rigid than
   the asthenosphere.

   T

6) Item Type: True-False   Objective: 2   Section 2.1
   Level: A: Standard   Tag: None

   _____ When an earthquake occurs, S waves can be detected on the side of the earth
   opposite the earthquake.

   F

7) Item Type: True-False   Objective: 2   Section 2.1
   Level: A: Standard   Tag: None

   _____ P waves travel more slowly than S waves.

   F

8) Item Type: True-False   Objective: 2   Section 2.1
     Level: A: Standard   Tag: None

_____ P waves travel faster through liquids than through solids.

F

9) Item Type: True-False   Objective: 3   Section 2.1
     Level: A: Standard   Tag: None

_____ Because the sun contains large amounts of iron, it has a magnetic field.

F

10) Item Type: True-False   Objective: 3   Section 2.1
     Level: A: Standard   Tag: None

_____ The earth's magnetic field affects an area that extends beyond the atmosphere.

T

11) Item Type: True-False   Objective: 4   Section 2.1
     Level: A: Standard   Tag: None

_____ The weight of an object on the earth depends on its mass and its distance from the earth's center.

T

12) Item Type: True-False   Objective: 4   Section 2.1
     Level: B: Enriched   Tag: None

_____ According to Newton's law of gravitation, the force of attraction between any two objects depends on their masses and the distance between them.

T

13) Item Type: True-False   Objective: 4   Section 2.1
     Level: B: Enriched   Tag: None

_____ According to Newton's law of gravitation, the force of attraction between any two objects depends on the weight and shape of the objects.

F

14) Item Type: True-False   Objective: 4   Section 2.1
     Level: A: Standard   Tag: None

_____ Based on Newton's law of gravitation, an object weighs less at the equator than at the North Pole.

T

15) Item Type: True-False   Objective: 5   Section 2.2
    Level: A: Standard   Tag: None

_____ Rays of the sun that strike the earth at about 90° produce the greatest amount of heat on the earth's surface.

T

16) Item Type: True-False   Objective: 5   Section 2.2
    Level: A: Standard   Tag: None

_____ At its closest point to the sun, the earth is said to be at aphelion.

F

17) Item Type: True-False   Objective: 5   Section 2.2
    Level: A: Standard   Tag: None

_____ As the earth orbits the sun, the direction of the tilt of the earth's axis changes.

F

18) Item Type: True-False   Objective: 5   Section 2.2
    Level: A: Standard   Tag: None

_____ Each rotation of the earth takes about one year.

F

19) Item Type: True-False   Objective: 6   Section 2.2
    Level: A: Standard   Tag: None

_____ When the Southern Hemisphere experiences summer, the Northern Hemisphere experiences winter.

T

20) Item Type: True-False   Objective: 6   Section 2.2
    Level: A: Standard   Tag: None

_____ Precession is the wobbling of the earth's axis as it turns in space.

T

21) Item Type: True-False   Objective: 6   Section 2.2
    Level: A: Standard   Tag: None

_____ During the summer solstice, the sun's vertical rays strike the earth along the Tropic of Cancer.

T

22) Item Type: True-False   Objective: 6   Section 2.2
    Level: A: Standard   Tag: None

_____ On the winter solstice there are 24 hours of daylight north of the Arctic Circle.

F

23) Item Type: True-False   Objective: 7   Section 2.2
    Level: A: Standard   Tag: None

_____ At noon in a standard time zone, the sun is highest over the center of the zone.

T

24) Item Type: True-False   Objective: 7   Section 2.2
    Level: A: Standard   Tag: None

_____ Circadian rhythms occur in 24-hour cycles.

T

25) Item Type: True-False   Objective: 7   Section 2.2
    Level: B: Enriched   Tag: None

_____ When it is 5:00 P.M. Friday immediately west of the international date line, it is 5:00 P.M. Thursday immediately east of the line.

T

26) Item Type: True-False   Objective: 7   Section 2.2
    Level: A: Standard   Tag: None

_____ Circadian rhythms regulate patterns of sleeping and waking.

T

27) Item Type: True-False   Objective: 7   Section 2.2
    Level: A: Standard   Tag: None

_____ There are exactly 24 standard time zones around the world.

T

28) Item Type: True-False   Objective: 8   Section 2.3
    Level: B: Enriched   Tag: Diagram 0274

_____ The satellite in this diagram is traveling in a polar orbit.

F

29) Item Type: True-False   Objective: 8   Section 2.3
    Level: A: Standard   Tag: None

_____ A satellite in polar orbit passes over a different portion of the earth's surface during each orbit.

T

30) Item Type: True-False   Objective: 8   Section 2.3
    Level: A: Standard   Tag: None

_____ Any object in orbit around another body with a smaller mass is a satellite.

F

31) Item Type: True-False   Objective: 8   Section 2.3
    Level: A: Standard   Tag: None

_____ A satellite will stay in orbit only if there is an exact balance between its mass and its weight.

F

32) Item Type: True-False   Objective: 8   Section 2.3
    Level: A: Standard   Tag: None

_____ A satellite encounters more air resistance in higher-altitude orbits.

F

33) Item Type: True-False   Objective: 8   Section 2.3
    Level: A: Standard   Tag: None

_____ A satellite in orbit travels slowest at perigee and fastest at apogee.

F

34) Item Type: True-False   Objective: 8   Section 2.3
    Level: A: Standard   Tag: None

_____ Satellites in geosynchronous orbit revolve around the earth once every 24 hours.

T

35) Item Type: True-False   Objective: 9   Section 2.3
    Level: A: Standard   Tag: None

_____ *Landsats* can locate forested areas on the earth's surface.

T

36) Item Type: True-False   Objective: 9   Section 2.3
Level: B: Enriched   Tag: None

_____ The orbit of *Landsats* around the earth is geosynchronous.

F

37) Item Type: True-False   Objective: 9   Section 2.3
Level: A: Standard   Tag: None

_____ The space shuttle is a satellite that can return to the earth's surface.

T

38) Item Type: Multiple Choice   Objective: 1   Section 2.1
Level: A: Standard   Tag: None

Which of the following best describes the material that makes up the earth's asthenosphere?

a) a rigid solid   b) a solid that is able to flow

c) a liquid at high temperature   d) a gas under great pressure

b

39) Item Type: Multiple Choice   Objective: 1   Section 2.1
Level: B: Enriched   Tag: None

What percentage of the earth's hydrosphere is made up of salt water from the ocean?

a) 3%   b) 45%   c) 70%   d) 97%

d

40) Item Type: Multiple Choice   Objective: 1   Section 2.1
Level: A: Standard   Tag: None

The lithosphere is made up of the upper mantle and the

a) crust.   b) asthenosphere.   c) hydrosphere.   d) core.

a

41) Item Type: Multiple Choice   Objective: 1   Section 2.1
Level: A: Standard   Tag: None

The part of the earth that is solid but has the ability to flow is the

a) lithosphere.   b) oceanic crust.   c) asthenosphere.   d) inner core.

c

42) Item Type: Multiple Choice   Objective: 1   Section 2.1
Level: A: Standard   Tag: None

The outermost zone of the earth is made up of

a) shadow zones and the Moho.   b) the core and the magnetosphere.

c) oceanic crust and continental crust.   d) the asthenosphere and the lithosphere.

c

43) Item Type: Multiple Choice   Objective: 1   Section 2.1
Level: A: Standard   Tag: None

The earth's atmosphere is made up of mostly oxygen and

a) nitrogen.   b) water vapor.   c) hydrogen.   d) carbon dioxide.

a

44) Item Type: Multiple Choice   Objective: 1   Section 2.1
Level: A: Standard   Tag: None

Which zone of the earth is made up of liquid iron?

a) the asthenosphere   b) the outer core   c) the upper mantle   d) the inner core

b

45) Item Type: Multiple Choice   Objective: 2   Section 2.1
Level: B: Enriched   Tag: Diagram 0271

At which location in the diagram is the Moho found?

a) *1*   b) *2*   c) *3*   d) *4*

c

46) Item Type: Multiple Choice   Objective: 2   Section 2.1
Level: B: Enriched   Tag: None

S waves are most easily blocked by the earth's

a) asthenosphere.   b) lithosphere.   c) inner core.   d) outer core.

d

47) Item Type: Multiple Choice   Objective: 2   Section 2.1
Level: A: Standard   Tag: None

Seismic waves increase in speed when they enter

a) the hydrosphere.   b) a strong magnetic field.

c) the outer core.   d) a more rigid material.

d

48) Item Type: Multiple Choice   Objective: 2   Section 2.1
Level: A: Standard   Tag: None

Because certain seismic waves are not detected in shadow zones, it is hypothesized that the earth's interior is

a) strongly magnetic.   b) entirely solid.   c) partially liquid.   d) uniformly rigid.

c

49) Item Type: Multiple Choice   Objective: 2   Section 2.1
Level: A: Standard   Tag: None

Where do most seismic waves come from?

a) the sun's rays   b) within the core of the earth

c) the magnetosphere   d) earthquakes and explosions

d

50) Item Type: Multiple Choice   Objective: 2   Section 2.1
Level: A: Standard   Tag: None

One characteristic of S waves is that they travel

a) faster than P waves.   b) only through solid material.

c) faster through less rigid material.   d) only through liquid material.

b

51) Item Type: Multiple Choice   Objective: 2   Section 2.1
Level: A: Standard   Tag: None

The boundary between the earth's crust and mantle where the speed of seismic waves changes is called the

a) Moho.   b) shadow zone.   c) magnetosphere.   d) hydrosphere.

a

52) Item Type: Multiple Choice   Objective: 3   Section 2.1
Level: A: Standard   Tag: None

Geologists believe the source of the earth's magnetic field may be in the

a) core.   b) magnetosphere.   c) crust.   d) magnetic pole.

a

53) Item Type: Multiple Choice   Objective: 3   Section 2.1
     Level: A: Standard   Tag: None

Which of the following best describes the magnetosphere?

a) the region of liquid iron in the earth's outer core

b) the region of space affected by the earth's magnetic field

c) the region of solid iron in the earth's inner core

d) the part of the atmosphere affected by the moon's magnetic field

b

54) Item Type: Multiple Choice   Objective: 4   Section 2.1
     Level: A: Standard   Tag: None

The mass of an object is defined by

a) how much the object weighs on earth.

b) the amount of water the object displaces.

c) the strength of the pull of gravity on the object.

d) the amount of matter in the object.

d

55) Item Type: Multiple Choice   Objective: 4   Section 2.1
     Level: B: Enriched   Tag: None

What is the weight at the earth's surface of an object with a mass of 5 kg?

a) 10 N   b) 50 N   c) 100 N   d) 500 N

b

56) Item Type: Multiple Choice   Objective: 4   Section 2.1
     Level: A: Standard   Tag: None

The force of gravity between two massive objects will increase if the objects are

a) at rest.   b) closer together.   c) in motion.   d) farther apart.

b

57) Item Type: Multiple Choice   Objective: 4   Section 2.1
Level: A: Standard   Tag: None

At which of the following locations would an object weigh 0.3 percent more than it does at the equator?

a) the North Pole

b) the Tropic of Cancer

c) 2,500 km above the earth

d) the Tropic of Capricorn

a

58) Item Type: Multiple Choice   Objective: 5   Section 2.2
Level: B: Enriched   Tag: None

Which of the following is most likely to result from an increase in the earth's rotation rate?

a) an increase in the length of a day

b) an increase in the length of a year

c) a decrease in the length of a day

d) a decrease in the length of a year

c

59) Item Type: Multiple Choice   Objective: 5   Section 2.2
Level: A: Standard   Tag: None

In which month does the earth reach perihelion?

a) January   b) March   c) July   d) September

a

60) Item Type: Multiple Choice   Objective: 5   Section 2.2
Level: A: Standard   Tag: None

When the earth is at the farthest point in its orbit from the sun, it is said to be at

a) the elliptical.   b) perihelion.   c) the equinox.   d) aphelion.

d

61) Item Type: Multiple Choice   Objective: 5   Section 2.2
    Level: B: Enriched   Tag: None

Which one of the following phrases is true of the tilt of the earth's axis?

a) It remains constant throughout the year.

b) It remains parallel to the plane of its orbit.

c) It changes with the seasons.

d) It is perpendicular to the plane of its orbit.

a

62) Item Type: Multiple Choice   Objective: 5   Section 2.2
    Level: A: Standard   Tag: None

Which of the following best defines the term aphelion?

a) the day with the fewest hours of daylight

b) the closest point of the earth's orbit to the sun

c) the day with the most hours of daylight

d) the farthest point of the earth's orbit from the sun

d

63) Item Type: Multiple Choice   Objective: 6   Section 2.2
    Level: A: Standard   Tag: None

When the South Pole tilts toward the sun, the Northern Hemisphere experiences

a) fall.   b) winter.   c) spring.   d) summer.

b

64) Item Type: Multiple Choice   Objective: 6   Section 2.2
    Level: A: Standard   Tag: None

Each year, the area south of the Antarctic Circle experiences 24 hours of darkness on

a) December 21 or 22.   b) March 21 or 22.

c) June 21 or 22.   d) September 22 or 23.

c

Chapter 2

65) Item Type: Multiple Choice Objective: 6 Section 2.2
Level: A: Standard Tag: None

On the vernal equinox, the sun's vertical rays strike the earth along the

a) Arctic Circle. b) Tropic of Cancer.

c) Antarctic Circle. d) Tropic of Capricorn.

c

66) Item Type: Multiple Choice Objective: 6 Section 2.2
Level: B: Enriched Tag: None

On the summer solstice, the sun's vertical rays strike the earth along the

a) Antarctic Circle. b) Tropic of Capricorn.

c) Arctic Circle. d) Tropic of Cancer.

d

67) Item Type: Multiple Choice Objective: 6 Section 2.2
Level: B: Enriched Tag: Diagram 0272

Which number on the diagram represents the earth's position on the autumnal equinox?

a) *1* b) *2* c) *3* d) *4*

d

68) Item Type: Multiple Choice Objective: 6 Section 2.2
Level: B: Enriched Tag: None

Which of the following occurs each year on September 22 or 23?

a) The number of hours of daylight and of darkness are equal everywhere on the earth.

b) The North Pole tilts directly toward the sun.

c) The sun's vertical rays strike along the Tropic of Cancer.

d) The moon is at perihelion in its orbit around the earth.

a

69) Item Type: Multiple Choice   Objective: 6   Section 2.2
Level: A: Standard   Tag: None

There are 12 hours of daylight and 12 hours of darkness south of the Antarctic Circle on the day

a) of the summer solstice.

b) that marks the beginning of daylight saving time.

c) of the vernal equinox.

d) that marks the end of standard time.

70) Item Type: Multiple Choice   Objective: 7   Section 2.2
Level: A: Standard   Tag: Diagram 0273

According to the diagram above, what is the correct time in Denver when it is noon in Boston?

a) 9 A.M.   b) 10 A.M.   c) 1 P.M.   d) 2 P.M.

71) Item Type: Multiple Choice   Objective: 7   Section 2.2
Level: A: Standard   Tag: None

Approximately how many degrees of the earth's total circumference does each time zone cover?

a) 1°   b) 15°   c) 45°   d) 90°

72) Item Type: Multiple Choice   Objective: 7   Section 2.2
Level: B: Enriched   Tag: None

If it is 3:00 P.M. Tuesday in the time zone bordering the east side of the international date line, in the time zone bordering the west side it is

a) 3:00 A.M. Wednesday.   b) 3:00 A.M. Tuesday.

c) 3:00 P.M. Wednesday.   d) 3:00 P.M. Monday.

73) Item Type: Multiple Choice   Objective: 7   Section 2.2
Level: A: Standard   Tag: None

The sun is highest over the center of each standard time zone at

a) 10:30 A.M.   b) 12:00 noon   c) 1:30 P.M.   d) 3:00 P.M.

Chapter 2

74) Item Type: Multiple Choice   Objective: 7   Section 2.2
    Level: A: Standard   Tag: None

How many standard time zones are there around the earth?

a) 4   b) 12   c) 24   d) 48

c

75) Item Type: Multiple Choice   Objective: 7   Section 2.2
    Level: B: Enriched   Tag: Diagram 0273

According to the diagram above, the cities of St. Louis and Los Angeles are separated by approximately how many degrees of longitude?

a) 2°   b) 15°   c) 30°   d) 90°

c

76) Item Type: Multiple Choice   Objective: 8   Section 2.3
    Level: B: Enriched   Tag: None

Which of the following statements is true of a satellite in a polar orbit?

a) It remains at the same point above the equator.

b) It moves in the same direction as the earth's rotation.

c) It completes one revolution every 24 hours.

d) It passes over a different portion of the earth during each revolution.

d

77) Item Type: Multiple Choice   Objective: 8   Section 2.3
    Level: A: Standard   Tag: None

The speed necessary to keep a satellite in orbit around the earth is determined primarily by the satellite's

a) orbit type.   b) altitude.   c) size.   d) weight.

b

78) Item Type: Multiple Choice   Objective: 8   Section 2.3
    Level: A: Standard   Tag: None

A satellite that passes over a different portion of the earth during each revolution is traveling

a) at apogee.   b) in a geosynchronous orbit.

c) at perigee.   d) in a polar orbit.

d

79) Item Type: Multiple Choice   Objective: 8   Section 2.3
Level: A: Standard   Tag: None

A satellite will remain in orbit if its speed is adequate for its

a) mass.

b) distance from apogee.

c) altitude.

d) distance from perigee.

80) Item Type: Multiple Choice   Objective: 8   Section 2.3
Level: A: Standard   Tag: None

Navigation satellites are used to

a) transmit weather information.

b) survey the ocean floor.

c) study the distant regions of space.

d) send out radio signals to ships and aircraft.

81) Item Type: Multiple Choice   Objective: 9   Section 2.3
Level: B: Enriched   Tag: None

Landsat satellites are most frequently used to

a) identify features on the earth's surface.

b) locate ships that are lost or in trouble.

c) transmit communication signals across continents.

d) analyze the atmosphere of planets.

82) Item Type: Multiple Choice   Objective: 9   Section 2.3
Level: B: Enriched   Tag: None

Satellites in geosynchronous orbits are most often used for

a) communications.

b) surveying the earth's surface.

c) weather information.

d) surveying the ocean floor.

83) Item Type: Multiple Choice   Objective: 9   Section 2.3
    Level: B: Enriched   Tag: None

Which of the following has been especially useful in identifying features on the earth's surface?

a) communications satellites   b) Landsat

c) meteorological satellites   d) GPS

b

84) Item Type: Multiple Choice   Objective: 9   Section 2.3
    Level: B: Enriched   Tag: None

The space shuttle is most useful for which of the following purposes?

a) relaying telephone signals   b) locating aircraft

c) predicting weather patterns   d) carrying cargo

d

85) Item Type: Multiple Choice   Objective: 9   Section 2.3
    Level: A: Standard   Tag: None

Which satellite network is used for the accurate navigation of ships and aircraft?

a) the space shuttle   b) GIS   c) Landsat   d) GPS

d

86) Item Type: Multiple Choice   Objective: 9   Section 2.3
    Level: A: Standard   Tag: None

A satellite in a geosynchronous orbit would probably be more useful than one in a polar orbit for

a) relaying television signals across the Atlantic.

b) studying global weather patterns.

c) mapping the earth's major population centers.

d) locating ships lost at sea.

a

87) Item Type: Fill in the Blank   Objective: 1   Section 2.1
    Level: A: Standard   Tag: None

Together, the earth's crust and upper mantle are called the _____.

lithosphere

Chapter 2

88) Item Type: Fill in the Blank   Objective: 1   Section 2.1
    Level: B: Enriched   Tag: None

The zone of the earth that consists of a solid part and a liquid part is the

_____.

core

89) Item Type: Fill in the Blank   Objective: 2   Section 2.1
    Level: A: Standard   Tag: None

The boundary between the earth's crust and mantle is called the _____.

Moho

90) Item Type: Fill in the Blank   Objective: 2   Section 2.1
    Level: A: Standard   Tag: None

A vibration that travels through the earth is called a(n) _____.

seismic wave

91) Item Type: Fill in the Blank   Objective: 3   Section 2.1
    Level: A: Standard   Tag: None

The region of space that is affected by the earth's magnetic field is the

_____.

magnetosphere

92) Item Type: Fill in the Blank   Objective: 4   Section 2.1
    Level: A: Standard   Tag: None

The force exerted on an object by the pull of gravity is called the object's

_____.

weight

93) Item Type: Fill in the Blank   Objective: 5   Section 2.2
    Level: A: Standard   Tag: None

The point at which a planet is closest to the sun is called _____.

perihelion

Chapter 2

94) Item Type: Fill in the Blank   Objective: 6   Section 2.2
    Level: A: Standard   Tag: None

    The wobbling of the earth's axis as it turns in space is called _____.

    | precession |

95) Item Type: Fill in the Blank   Objective: 6   Section 2.2
    Level: B: Enriched   Tag: None

    The sun's rays strike the earth at a 90° angle along the Tropic of Capricorn on the winter

    _____.

    | solstice |

96) Item Type: Fill in the Blank   Objective: 6   Section 2.2
    Level: A: Standard   Tag: None

    The day in March when the number of hours of daylight and of darkness are equal is

    called the _____.

    | vernal equinox |

97) Item Type: Fill in the Blank   Objective: 7   Section 2.2
    Level: A: Standard   Tag: None

    The system under which clocks are set one hour ahead in April is called

    _____.

    | daylight saving time |

98) Item Type: Fill in the Blank   Objective: 7   Section 2.2
    Level: A: Standard   Tag: None

    Jet lag is caused by an upset in the body's biological _____.

    | clock |

99)   Item Type: Essay   Objective: 1   Section 2.1
      Level: B: Enriched   Tag: None

Describe the three major zones of the earth.

_____

_____

_____

_____

_____

The thin, solid, outermost layer covering the earth is the crust. The crust is composed of oceanic and continental crust. Beneath the crust is the mantle. The uppermost part of the mantle is solid; the rest of the mantle is a solid that can flow. Below the mantle is the core, which contains iron. The outer core is liquid; the inner core is solid.

100) Item Type: Essay   Objective: 4   Section 2.1
      Level: A: Standard   Tag: None

Explain the relationship between weight and location on the earth's surface.

_____

_____

_____

_____

The motion of the earth's rotation causes the earth to bulge slightly at the equator and flatten at the poles. The distance between the surface and the earth's center is, therefore, greatest at the equator and least at the poles. Since the force of gravity decreases as the distance from the earth's center increases, an object will weigh more at the poles than at the equator.

101) Item Type: Essay Objective: 5 Section 2.2
Level: A: Standard Tag: None

Describe how day length is affected by the tilt of the earth's axis.

_____

_____

_____

_____

_____

When the North Pole points toward the sun, the Northern Hemisphere has longer periods of daylight than the Southern Hemisphere. When the North Pole points away from the sun, the Southern Hemisphere has longer periods of daylight. On the autumnal equinox and on the vernal equinox, the North Pole points neither toward nor away from the sun, and on those days, hours of daylight and darkness are equal everywhere on earth.

102) Item Type: Essay Objective: 6 Section 2.2
Level: B: Enriched Tag: None

Describe the relative position of the earth and the sun on the winter solstice.

_____

_____

_____

_____

By December, the earth is halfway through its orbit and the North Pole points away from the sun. On the day of the winter solstice (December 21 or 22), the sun's rays strike the earth at a 90° angle along the Tropic of Capricorn.

103) Item Type: Essay   Objective: 7   Section 2.2
Level: B: Enriched   Tag: None

When it is 4:00 P.M. on Thursday in Sydney, Australia, it is 6:00 P.M. on Wednesday in Nome, Alaska. Describe the relative positions of these cities in terms of time zones and the international date line.

_____

_____

_____

_____

Because Sydney is one day ahead of Nome, Nome must be on the east side of the international date line and Sydney must be on the west side. Also, Sydney must be two time zones west of Nome because the time in Sydney is two hours earlier.

1) Tag Name: Diagram 0271

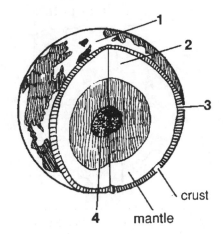

2) Tag Name: Diagram 0272

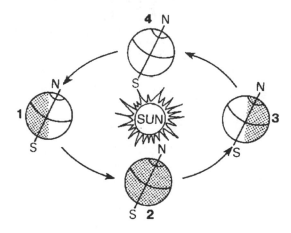

HRW material copyrighted under notice appearing earlier in this work.

TEST ITEM LISTING **41**

3)   Tag Name: Diagram 0273

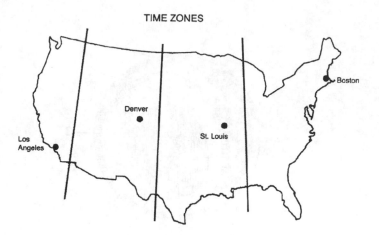

4)   Tag Name: Diagram 0274

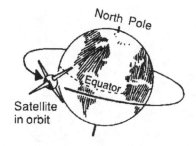

1) Item Type: True-False   Objective: 1   Section 3.1
   Level: B: Enriched   Tag: None

_____ The distance covered by one degree of longitude is the same as the distance covered by one degree of latitude.

F

2) Item Type: True-False   Objective: 1   Section 3.1
   Level: A: Standard   Tag: None

_____ Meridians are circles around the earth parallel to the equator.

F

3) Item Type: True-False   Objective: 1   Section 3.1
   Level: A: Standard   Tag: None

_____ A meridian is an imaginary semicircle that runs from pole to pole.

T

4) Item Type: True-False   Objective: 1   Section 3.1
   Level: A: Standard   Tag: None

_____ The latitude of the equator is 0°.

T

5) Item Type: True-False   Objective: 1   Section 3.1
   Level: A: Standard   Tag: None

_____ The distance represented by a degree of longitude is greatest at the equator.

T

6) Item Type: True-False   Objective: 2   Section 3.1
   Level: B: Enriched   Tag: None

_____ The earth's axis of rotation intersects the earth's surface at the geographic poles.

T

7) Item Type: True-False   Objective: 2   Section 3.1
   Level: A: Standard   Tag: None

_____ Each degree of latitude is equal to 60 seconds of latitude.

F

Chapter 3

8) Item Type: True-False   Objective: 2   Section 3.1
    Level: A: Standard   Tag: None

    _____ The distance covered by a degree of latitude depends upon where the degree is measured.

    F

9) Item Type: True-False   Objective: 2   Section 3.1
    Level: A: Standard   Tag: None

    _____ True north is another name for the geomagnetic North Pole.

    F

10) Item Type: True-False   Objective: 3   Section 3.1
    Level: A: Standard   Tag: None

    _____ Most maps are drawn with south at the top.

    F

11) Item Type: True-False   Objective: 3   Section 3.1
    Level: A: Standard   Tag: None

    _____ Directions on a map should be determined relative to the meridians and parallels.

    T

12) Item Type: True-False   Objective: 4   Section 3.2
    Level: A: Standard   Tag: None

    _____ On a gnomonic projection, lines of longitude are parallel.

    F

13) Item Type: True-False   Objective: 4   Section 3.2
    Level: A: Standard   Tag: None

    _____ On a Mercator projection, only the areas near 0° latitude are shown with relative accuracy.

    T

14) Item Type: True-False   Objective: 4   Section 3.2
    Level: A: Standard   Tag: None

    _____ On a polyconic projection, the relative size and shape of small areas are nearly the same as those on a globe.

    T

15) Item Type: True-False   Objective: 4   Section 3.2
     Level: A: Standard   Tag: None

_____ The globe is a type of map projection.

F

16) Item Type: True-False   Objective: 5   Section 3.2
     Level: B: Enriched   Tag: None

_____ The ratio of 1:10,000 can also be expressed as 1 cm equals 1 km.

F

17) Item Type: True-False   Objective: 5   Section 3.2
     Level: B: Enriched   Tag: None

_____ On a map, a fractional scale intended for use with metric measurements must be converted if it is to be used with standard measurements.

F

18) Item Type: True-False   Objective: 5   Section 3.2
     Level: A: Standard   Tag: Diagram   0373

_____ According to the map above, the distance between Osaka and Tokyo is approximately 500 km.

T

19) Item Type: True-False   Objective: 5   Section 3.2
     Level: A: Standard   Tag: Diagram   0373

_____ According to the map above, Osaka is a capital city.

F

20) Item Type: True-False   Objective: 6   Section 3.3
     Level: A: Standard   Tag: None

_____ Mean sea level is the average level of high tide.

F

21) Item Type: True-False   Objective: 6   Section 3.3
     Level: A: Standard   Tag: None

_____ On a map, relief is the difference in elevation between one contour line and the next.

F

22) Item Type: True-False   Objective: 6   Section 3.3
    Level: A: Standard   Tag: None

_____ On a topographic map, elevation is measured as the distance above or below mean sea level.

T

23) Item Type: True-False   Objective: 6   Section 3.3
    Level: A: Standard   Tag: None

_____ On a topographic map, elevation is represented by contour lines.

T

24) Item Type: True-False   Objective: 6   Section 3.3
    Level: A: Standard   Tag: None

_____ On topographic maps, contour lines connect points having the same elevation.

T

25) Item Type: True-False   Objective: 6   Section 3.3
    Level: A: Standard   Tag: None

_____ Topographic maps show only natural features of an area.

F

26) Item Type: True-False   Objective: 6   Section 3.3
    Level: B: Enriched   Tag: None

_____ Topographic maps of plains and other relatively flat areas have large contour intervals.

F

27) Item Type: True-False   Objective: 6   Section 3.3
    Level: A: Standard   Tag: None

_____ Eastern and western boundaries of U.S.G.S. maps are indicated by meridians of longitude.

T

28) Item Type: True-False   Objective: 7   Section 3.3
    Level: A: Standard   Tag: None

_____ Index contours are contour lines that are printed bolder in order to make map reading easier.

T

29) Item Type: True-False   Objective: 7   Section 3.3
     Level: A: Standard   Tag: None

_____ Landsat satellites are used by cartographers.

T

30) Item Type: Multiple Choice   Objective: 1   Section 3.1
     Level: A: Standard   Tag: Diagram 0371

As shown in the diagram above, the earth's axis of rotation intersects the earth's surface at the

a) magnetic poles.   b) geographic poles.   c) prime meridian.   d) equator.

b

31) Item Type: Multiple Choice   Objective: 1   Section 3.1
     Level: A: Standard   Tag: Diagram 0371

The vertical lines in the diagram above represent

a) relief.   b) parallels.   c) seconds.   d) meridians.

d

32) Item Type: Multiple Choice   Objective: 1   Section 3.1
     Level: A: Standard   Tag: None

Circular lines that run perpendicular to the equator are called

a) depression contours.   b) parallels.   c) great circles.   d) meridians.

d

33) Item Type: Multiple Choice   Objective: 1   Section 3.1
     Level: A: Standard   Tag: None

The prime meridian passes through which of the following countries?

a) United States   b) England   c) Russia   d) Brazil

b

34) Item Type: Multiple Choice   Objective: 2   Section 3.1
     Level: A: Standard   Tag: None

The North Pole has a latitude of

a) 0° N.   b) 45° N.   c) 90° N.   d) 180° N.

c

35) Item Type: Multiple Choice   Objective: 2   Section 3.1
Level: B: Enriched   Tag: None

What is the location of a point that is 85° south of the equator and 350° east of the prime meridian?

a) 85° S, 350° E   b) 65° S, 350° W   c) 85° S, 10° E   d) 85° S, 10° W

d

36) Item Type: Multiple Choice   Objective: 2   Section 3.1
Level: A: Standard   Tag: None

The shortest distance between any two points on the globe can be found by drawing a

a) parallel.   b) great circle.   c) prime meridian.   d) conic projection.

b

37) Item Type: Multiple Choice   Objective: 2   Section 3.1
Level: B: Enriched   Tag: None

A plane leaves Greenwich, England, flying in a westerly direction. After flying three-fourths of the way around the earth, the plane's longitude would be

a) 90° E.   b) 90° W.   c) 270° E.   d) 270° W.

a

38) Item Type: Multiple Choice   Objective: 2   Section 3.1
Level: B: Enriched   Tag: None

Which line of latitude is found 11,100 km north of the South Pole?

a) 10° N   b) 10° S   c) 100° N   d) 100° S

a

39) Item Type: Multiple Choice   Objective: 2   Section 3.1
Level: A: Standard   Tag: None

A city's location on the earth with respect to the prime meridian is given by the city's

a) elevation.   b) declination.   c) latitude.   d) longitude.

d

40) Item Type: Multiple Choice   Objective: 2   Section 3.1
Level: B: Enriched   Tag: None

The circumference of the earth is approximately

a) 40 km.   b) 4,000 km.   c) 40,000 km.   d) 400,000 km.

c

HRW material copyrighted under notice appearing earlier in this work.

Chapter 3

41) Item Type: Multiple Choice   Objective: 3   Section 3.1
    Level: A: Standard   Tag: None

The latitude of the equator is

a) 0°.   b) 90°.   c) 180°.   d) 360°.

```
a  ▄▄
```

42) Item Type: Multiple Choice   Objective: 3   Section 3.1
    Level: A: Standard   Tag: None

In the Northern Hemisphere, magnetic declination is measured in degrees east or west of

a) true north.   b) the prime meridian.

c) geomagnetic north.   d) the great circle.

```
a  ▄▄
```

43) Item Type: Multiple Choice   Objective: 4   Section 3.2
    Level: B: Enriched   Tag: None

Meridians and parallels are shown as perpendicular lines on a

a) gnomonic projection.   b) Mercator projection.

c) polyconic projection.   d) conic projection.

```
b  ▄▄
```

44) Item Type: Multiple Choice   Objective: 4   Section 3.2
    Level: B: Enriched   Tag: None

Which of the following are shown as straight lines on a Mercator projection?

a) compass directions b) coastlines   c) contour lines   d) great circles

```
a  ▄▄
```

45) Item Type: Multiple Choice   Objective: 4   Section 3.2
    Level: A: Standard   Tag: Diagram 0372

According to the information above, a city having a population of 15,000 should be
indicated on a map with which symbol?

a) ○   b) ⊗   c) ◉   d) ☐

```
b  ▄▄
```

46) Item Type: Multiple Choice   Objective: 4   Section 3.2
Level: A: Standard   Tag: None

On a map, a list of symbols and their meanings is called

a) a legend.   b) a scale.   c) an index contour.   d) a map projection.

a

47) Item Type: Multiple Choice   Objective: 4   Section 3.2
Level: B: Enriched   Tag: None

Which type of map is produced by fitting together a series of map projections?

a) Mercator   b) conic   c) polyconic   d) gnomonic

c

48) Item Type: Multiple Choice   Objective: 4   Section 3.2
Level: A: Standard   Tag: None

On a Mercator projection, the greatest distortion is produced

a) at the equator.   b) at the poles.

c) along the prime meridian.   d) midway between the poles.

b

49) Item Type: Multiple Choice   Objective: 4   Section 3.2
Level: A: Standard   Tag: None

The type of projection formed by placing a sheet of paper so that it touches the globe at only one point is called a

a) polyconic projection.   b) Mercator projection.

c) gnomonic projection.   d) conic projection.

c

50) Item Type: Multiple Choice   Objective: 4   Section 3.2
Level: A: Standard   Tag: None

Great circles are shown as straight lines on which type of map projection?

a) gnomonic   b) Mercator   c) polyconic   d) conic

a

51) Item Type: Multiple Choice   Objective: 4   Section 3.2
Level: A: Standard   Tag: None

Which of the following is true of a Mercator projection?

a) Coastline shapes are accurate.

b) Spacing between parallels is equal.

c) Compass directions are curved.

d) Equatorial regions are greatly distorted.

a

52) Item Type: Multiple Choice   Objective: 5   Section 3.2
Level: B: Enriched   Tag: None

On a map with a scale of 1:50,000, a measured distance of 1 cm on the map is equal to what distance on the earth's surface?

a) 50,000 m   b) 5,000 m   c) 500 m   d) 50 m

c

53) Item Type: Multiple Choice   Objective: 5   Section 3.2
Level: A: Standard   Tag: None

To find the actual distance between two points on the earth, the distance between the two points on the map must be measured and compared to the

a) map projection.   b) scale.   c) contour interval.   d) legend.

b

54) Item Type: Multiple Choice   Objective: 6   Section 3.3
Level: B: Enriched   Tag: None

On a topographic map, contour lines that bend to form a V-shape indicate a

a) hilltop.   b) road.   c) cliff.   d) valley.

d

55) Item Type: Multiple Choice   Objective: 6   Section 3.3
Level: A: Standard   Tag: None

A contour line that is bold and labeled is called

a) an index contour.   b) a depression contour.

c) a contour interval.   d) a map scale.

a

56) Item Type: Multiple Choice   Objective: 6   Section 3.3
    Level: A: Standard   Tag: None

The difference in elevation between the highest and lowest parts of an area being mapped is called the

a) altitude.   b) relief.   c) topography.   d) contour interval.

b

57) Item Type: Multiple Choice   Objective: 6   Section 3.3
    Level: B: Enriched   Tag: None

Contour intervals are most likely to be smallest on maps of

a) hilly areas.   b) flat areas.

c) high-altitude areas.   d) low-altitude areas.

b

58) Item Type: Multiple Choice   Objective: 6   Section 3.3
    Level: A: Standard   Tag: None

The elevation difference between two contour lines on a map is called the

a) map scale.   b) contour interval.   c) map projection.   d) index contour.

b

59) Item Type: Multiple Choice   Objective: 7   Section 3.3
    Level: A: Standard   Tag: None

On a topographic map, rapid increases in elevation are represented by

a) depression contours.

b) V-shaped contour lines.

c) contour lines forming closed loops.

d) closely spaced contour lines.

d

60) Item Type: Multiple Choice   Objective: 7   Section 3.3
    Level: A: Standard   Tag: None

On topographic maps, closely spaced contour lines indicate

a) valleys.   b) depressions.   c) mountain peaks.   d) steep slopes.

d

61) Item Type: Multiple Choice   Objective: 7   Section 3.3
Level: A: Standard   Tag: None

On topographic maps, contour lines that form closed loops indicate

a) hilltops.   b) cliffs.   c) lakes.   d) islands.

a

62) Item Type: Multiple Choice   Objective: 7   Section 3.3
Level: A: Standard   Tag: None

On a topographic map, green areas indicate

a) woodlands.   b) bodies of water.   c) towns.   d) mountainous regions.

a

63) Item Type: Multiple Choice   Objective: 7   Section 3.3
Level: A: Standard   Tag: None

Closely spaced contour lines crossing a river indicate a

a) gentle course.   b) bridge.   c) steep course.   d) valley.

c

64) Item Type: Multiple Choice   Objective: 7   Section 3.3
Level: A: Standard   Tag: Diagram 0374

In the diagram above, what type of feature is indicated by the arrow?

a) peak   b) valley   c) stream   d) depression

d

65) Item Type: Multiple Choice   Objective: 7   Section 3.3
Level: A: Standard   Tag: Diagram 0374

What is the elevation of the point labeled *X* on the map above?

a) 400 m   b) 450 m   c) 500 m   d) 600 m

c

66) Item Type: Multiple Choice   Objective: 7   Section 3.3
Level: A: Standard   Tag: Diagram 0375

What is the elevation of the point labeled *A* in the map above?

a) 1,310 feet   b) 1,320 feet

c) 1,330 feet   d) 1,340 feet

c

Chapter 3

67) Item Type: Multiple Choice   Objective: 7   Section 3.3
     Level: B: Enriched   Tag: Diagram 0375

According to the map above, what is the approximate elevation of the point where Fuller Brook crosses Fuller Road?

a) 1,275 feet   b) 1,295 feet

c) 1,305 feet   d) 1,325 feet

b

68) Item Type: Fill in the Blank   Objective: 1   Section 3.1
     Level: A: Standard   Tag: None

The line of longitude that passes through Greenwich, England, is called the

_____.

prime meridian

69) Item Type: Fill in the Blank   Objective: 1   Section 3.1
     Level: A: Standard   Tag: None

One degree of latitude consists of 60 equal parts called _____.

minutes

70) Item Type: Fill in the Blank   Objective: 1   Section 3.1
     Level: A: Standard   Tag: None

The angular distance north or south of the equator is called _____.

latitude

71) Item Type: Fill in the Blank   Objective: 2   Section 3.1
     Level: A: Standard   Tag: None

A circle that divides the earth into equal halves is called a(n) _____.

great circle

72) Item Type: Fill in the Blank   Objective: 2   Section 3.1
     Level: A: Standard   Tag: None

The line of longitude selected to be 0° is called the _____.

prime meridian

73) Item Type: Fill in the Blank   Objective: 2   Section 3.1
     Level: B: Enriched   Tag: None

The angular distance between the equator and the North Pole is _____.

90°

Chapter 3

74) Item Type: Fill in the Blank   Objective: 2   Section 3.1
Level: A: Standard   Tag: None

The only line of latitude that is a great circle is _____.

> the equator

75) Item Type: Fill in the Blank   Objective: 3   Section 3.1
Level: A: Standard   Tag: None

Magnetic declination is the angle between the direction in which the compass needle points

and the direction of the _____.

> geographic pole

76) Item Type: Fill in the Blank   Objective: 3   Section 3.1
Level: A: Standard   Tag: None

True north is in the direction of the geographic _____.

> North Pole

77) Item Type: Fill in the Blank   Objective: 3   Section 3.1
Level: A: Standard   Tag: None

The angle between the geographic North Pole and the direction in which a compass needle

points is called magnetic _____.

> declination

78) Item Type: Fill in the Blank   Objective: 3   Section 3.1
Level: A: Standard   Tag: None

Magnetic declination is the angle between geographic north and _____.

> magnetic north

79) Item Type: Fill in the Blank   Objective: 4   Section 3.2
Level: A: Standard   Tag: None

Maps made by combining several conic projections are called _____.

> polyconic projections

80) Item Type: Fill in the Blank   Objective: 5   Section 3.2
Level: B: Enriched   Tag: None

A map on which 1 cm represents 100 m has a fractional scale of _____.

> 1:10,000

81) Item Type: Fill in the Blank   Objective: 5   Section 3.2
Level: A: Standard   Tag: None

The map feature that indicates the relationship between distance as it is drawn on the map

and actual distance is called the _____.

scale

82) Item Type: Fill in the Blank   Objective: 6   Section 3.3
Level: A: Standard   Tag: None

On a topographic map, the difference in elevation between one contour line and the next is

the _____.

contour interval

83) Item Type: Fill in the Blank   Objective: 7   Section 3.3
Level: A: Standard   Tag: None

On a topographic map, contour lines that are printed bolder and that have their elevations

labeled are called _____.

index contours

84) Item Type: Essay   Objective: 2   Section 3.1
Level: B: Enriched   Tag: None

Why is a degree of longitude at the equator not equal to the same distance as a degree of
longitude at 45° N latitude?

_____

_____

_____

_____

_____

All lines of longitude intersect at the poles. As these lines approach the equator, they are
farther and farther apart. Therefore, the distance between two adjacent lines of longitude,
which represents one degree, is greater at the equator than at 45° N latitude.

85) Item Type: Essay   Objective: 2   Section 3.1
     Level: B: Enriched   Tag: None

Interpret the location of a city that is located at 20° S, 50° W.

_____

_____

_____

_____

_____

The city has a latitude of 20° S. Therefore, it is located south of the equator (i.e., in the Southern Hemisphere). The city's longitude is 50° W. Therefore, it is also located 50° W of the prime meridian.

86) Item Type: Essay   Objective: 3   Section 3.1
     Level: A: Standard   Tag: None

Explain the term magnetic declination.

_____

_____

_____

_____

_____

Because of the tilt of the imaginary magnet inside the earth, the geomagnetic poles and the geographic poles are located in different places. A compass needle points to the geomagnetic North Pole. The angle between the direction of the geographic pole and the direction in which the compass needle points is called magnetic declination.

87) Item Type: Essay   Objective: 3   Section 3.1
    Level: A: Standard   Tag: None

How can a magnetic compass be used to find geographic north?

_____

_____

_____

_____

_____

> Magnetic compasses point toward the geomagnetic North Pole. In order to determine true geographic north using a magnetic compass, the magnetic declination must be known. By correcting the compass reading to account for the declination, geographic north may be determined.

88) Item Type: Essay   Objective: 4   Section 3.2
    Level: B: Enriched   Tag: None

Why are gnomonic projections useful to navigators in plotting air travel routes?

_____

_____

_____

_____

_____

> Navigators can use great circles to show the shortest distance between any two points on the earth. On gnomonic projections, great circles appear as straight lines.

89) Item Type: Essay   Objective: 4   Section 3.2
     Level: B: Enriched   Tag: None

Describe the distortion produced by gnomonic and conic projections.

_____

_____

_____

_____

_____

On a gnomonic projection, the unequal spacing between parallels causes great distortion in both direction and distance. The distortion increases as the distance from the point of contact used to make the projection increases. Areas on a conic projection along the parallel of latitude where the cone and globe are in contact are distorted least.

1)   Tag Name: Diagram 0371

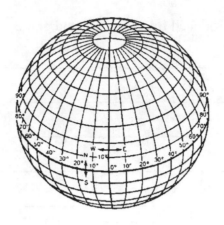

2)   Tag Name: Diagram 0372

POPULATION

⊛ Capital Cities   ⊗ 10,000 to 25,000   ⊡ 100,000 to 500,000

○ Under 5,000   ⊙ 25,000 to 50,000   ◉ 500,000 and over

◉ 5,000 to 10,000   ● 50,000 to 100,000

3)   Tag Name: Diagram 0373

4) Tag Name: Diagram 0374

5) Tag Name: Diagram 0375

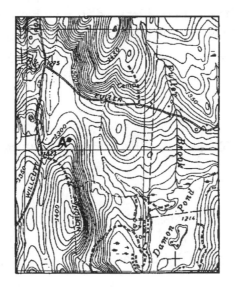

1) Item Type: True-False   Objective: 1   Section 4.1
   Level: A: Standard   Tag: None

_____ The theory of plate tectonics explains how some coastal mountain ranges may have been formed.

> T

2) Item Type: True-False   Objective: 1   Section 4.1
   Level: A: Standard   Tag: None

_____ Wegener based the theory of continental drift on the phenomenon of seafloor spreading.

> F

3) Item Type: True-False   Objective: 1   Section 4.1
   Level: A: Standard   Tag: None

_____ Wegener speculated that continental drift may have crumpled the crust, producing mountain ranges such as the Andes.

> T

4) Item Type: True-False   Objective: 1   Section 4.1
   Level: B: Enriched   Tag: None

_____ Alfred Wegener hypothesized that continents are fixed and unable to move.

> F

5) Item Type: True-False   Objective: 1   Section 4.1
   Level: A: Standard   Tag: None

_____ The theory of continental drift states that present-day continents were once a single landmass.

> T

6) Item Type: True-False   Objective: 2   Section 4.1
   Level: A: Standard   Tag: None

_____ Fossil records provide a clue to earlier positions of lithospheric plates.

> T

7) Item Type: True-False   Objective: 2   Section 4.1
   Level: B: Enriched   Tag: None

_____ The magnetic orientation of molten rock from a mid-ocean ridge will always reflect a reversed polarity of the earth's magnetic field at that time.

> F

8) **Item Type:** True-False **Objective:** 2 **Section** 4.1
   Level: B: Enriched **Tag:** None

   _____ Island arcs are evidence of continental drift.

   F

9) **Item Type:** True-False **Objective:** 2 **Section** 4.1
   Level: A: Standard **Tag:** None

   _____ Identical fossil remains of Mesosaurus in eastern South America and western Africa support the theory of continental drift.

   T

10) **Item Type:** True-False **Objective:** 2 **Section** 4.1
    Level: B: Enriched **Tag:** None

    _____ The similar age and types of rocks from western Africa and eastern Brazil support the theory of continental drift.

    T

11) **Item Type:** True-False **Objective:** 3 **Section** 4.1
    Level: A: Standard **Tag:** None

    _____ The magnetic orientation of a rock is determined when the rock solidifies.

    T

12) **Item Type:** True-False **Objective:** 3 **Section** 4.1
    Level: A: Standard **Tag:** None

    _____ All Mid-Atlantic Ridge rock samples examined have proven to be older than the oldest-known continental rocks.

    F

13) **Item Type:** True-False **Objective:** 3 **Section** 4.1
    Level: A: Standard **Tag:** None

    _____ Magma is the molten rock that wells up through fissures in the earth's crust.

    T

14) **Item Type:** True-False **Objective:** 3 **Section** 4.1
    Level: A: Standard **Tag:** None

    _____ Movement of the earth's crust away from a mid-ocean ridge is called seafloor spreading.

    T

15) Item Type: True-False　　Objective: 3　　Section 4.1
Level: A: Standard　　Tag: None

_____ Paleomagnetism provided conclusive evidence of seafloor spreading.

T

16) Item Type: True-False　　Objective: 3　　Section 4.1
Level: B: Enriched　　Tag: None

_____ Bands of alternately older and younger rocks occur on each side of rifts in mid-ocean ridges.

F

17) Item Type: True-False　　Objective: 3　　Section 4.1
Level: A: Standard　　Tag: None

_____ Rock along the Mid-Atlantic Ridge is generally younger than continental rock.

T

18) Item Type: True-False　　Objective: 3　　Section 4.1
Level: B: Enriched　　Tag: None

_____ The earth's oldest rocks are found along the Mid-Atlantic Ridge.

F

19) Item Type: True-False　　Objective: 3　　Section 4.1
Level: A: Standard　　Tag: None

_____ The molten rock that rises through rifts at seafloor-spreading locations is called magma.

T

20) Item Type: True-False　　Objective: 4　　Section 4.2
Level: A: Standard　　Tag: None

_____ The theory of plate tectonics states that the earth's crust is made of a single massive landmass.

F

21) Item Type: True-False　　Objective: 4　　Section 4.2
Level: A: Standard　　Tag: None

_____ Plate tectonics is a theory that describes the earth's surface as being in motion, with broken plates of the crust drifting slowly over the material of the asthenosphere.

T

22) Item Type: True-False   Objective: 4   Section 4.2
     Level: A: Standard   Tag: None

_____ According to plate tectonics, the asthenosphere rides on lithospheric plates.

F

23) Item Type: True-False   Objective: 5   Section 4.2
     Level: A: Standard   Tag: None

_____ The Mid-Atlantic Ridge is an example of a divergent plate boundary.

T

24) Item Type: True-False   Objective: 5   Section 4.2
     Level: A: Standard   Tag: None

_____ A plate boundary at which one plate slides under another is called a transform fault boundary.

F

25) Item Type: True-False   Objective: 5   Section 4.2
     Level: A: Standard   Tag: None

_____ Island arcs form along convergent plate boundaries.

T

26) Item Type: True-False   Objective: 5   Section 4.2
     Level: A: Standard   Tag: None

_____ Transform faults occur at divergent plate boundaries.

T

27) Item Type: True-False   Objective: 5   Section 4.2
     Level: A: Standard   Tag: None

_____ A transform fault occurs where two plates collide directly.

F

28) Item Type: True-False   Objective: 5   Section 4.2
     Level: A: Standard   Tag: None

_____ Most divergent boundary locations are marked by mid-ocean ridges.

T

29) Item Type: True-False   Objective: 5   Section 4.2
     Level: A: Standard   Tag: None

_____ The San Andreas Fault is an example of a transform fault boundary.

T

30) Item Type: True-False   Objective: 6   Section 4.2
     Level: A: Standard   Tag: None

_____ Many scientists believe that lithospheric plate movement is caused by temperature differences in the mantle.

T

31) Item Type: True-False   Objective: 6   Section 4.2
     Level: A: Standard   Tag: None

_____ Convection currents are thought to originate within the continental crust.

F

32) Item Type: True-False   Objective: 7   Section 4.2
     Level: A: Standard   Tag: None

_____ A rift valley is an example of a suspect terrane.

F

33) Item Type: True-False   Objective: 7   Section 4.2
     Level: A: Standard   Tag: None

_____ An island arc is an example of a suspect terrane.

F

34) Item Type: True-False   Objective: 7   Section 4.2
     Level: A: Standard   Tag: None

_____ Suspect terranes form primarily along divergent plate boundaries.

F

35) Item Type: True-False   Objective: 7   Section 4.2
     Level: B: Enriched   Tag: None

_____ Panthalassa is an ancient suspect terrane.

F

36) Item Type: Multiple Choice   Objective: 1   Section 4.1
Level: B: Enriched   Tag: None

Panthalassa was the name Wegener chose to describe

a) the continent of South America.

b) fossil remains on the coast of Africa.

c) a huge, ancient ocean.

d) the Mid-Atlantic Ridge.

c

37) Item Type: Multiple Choice   Objective: 1   Section 4.1
Level: B: Enriched   Tag: None

According to Wegener, Pangaea began breaking up approximately how many years ago?

a) 100 million   b) 200 million   c) 300 million   d) 400 million

b

38) Item Type: Multiple Choice   Objective: 1   Section 4.1
Level: A: Standard   Tag: None

Which scientist first proposed that the continents were once joined in a single landmass called Pangaea?

a) Hess   b) Suess   c) Dietz   d) Wegener

d

39) Item Type: Multiple Choice   Objective: 1   Section 4.1
Level: A: Standard   Tag: None

Pangaea was the term Wegener used to describe

a) a huge, ancient ocean.   b) the earth.

c) the Eurasian continent.   d) an ancient single landmass.

d

40) Item Type: Multiple Choice   Objective: 2   Section 4.1
Level: B: Enriched   Tag: None

What evidence found in tropical regions of southern Africa and South America most strongly supports the theory that the continents were once joined?

a) coal deposits   b) mountain chains   c) glacial debris   d) land bridges

c

41) Item Type: Multiple Choice   Objective: 2   Section 4.1
    Level: B: Enriched   Tag: None

What geologic discovery in the 1960's provided the scientific evidence to verify the occurrence of continental drift?

a) fossils of the same plants and animals on continental coasts

b) eroded land bridges that once connected continents

c) coal deposits in the United States, Europe, and Siberia

d) reversed magnetic polarities in rocks on land and ocean floors

42) Item Type: Multiple Choice   Objective: 2   Section 4.1
    Level: B: Enriched   Tag: Diagram 0471

Which theory is most directly supported by the discovery of rocks similar in age and type in the areas shaded on the map?

a) relativity   b) suspect terranes   c) plate tectonics   d) seafloor spreading

43) Item Type: Multiple Choice   Objective: 2   Section 4.1
    Level: A: Standard   Tag: None

Which of the following was used as evidence for the theory of continental drift?

a) convection currents

b) magma intrusions

c) similarity in continental margins

d) magnitudes of past earthquakes

44) Item Type: Multiple Choice   Objective: 3   Section 4.1
    Level: B: Enriched   Tag: None

The island of Haimaey off the south coast of Iceland is a dramatic example of the effects of

a) seafloor spreading.   b) convection currents.

c) suspect terranes.   d) paleomagnetism.

a

45) Item Type: Multiple Choice   Objective: 3   Section 4.1
Level: A: Standard   Tag: None

As rock moves away from a mid-ocean ridge, it is replaced by

a) continental crust.   b) molten rock.   c) suspect terranes. d) older rock.

b

46) Item Type: Multiple Choice   Objective: 3   Section 4.1
Level: B: Enriched   Tag: None

What discovery did scientists make in the 1940's when they compared continental rocks to rocks near centers of seafloor spreading?

a) The continental rocks were younger.

b) The rocks near spreading centers had unique magnetic properties.

c) The rocks at spreading centers were younger.

d) The continental rocks had unique mineral properties.

c

47) Item Type: Multiple Choice   Objective: 3   Section 4.1
Level: B: Enriched   Tag: None

Paleomagnetism of the ocean floor is used to identify the

a) number of plate boundaries.

b) approximate time the rock cooled and hardened.

c) relative strength of the earth's magnetic field.

d) kind of rock that forms the earth's crust.

b

48) Item Type: Multiple Choice   Objective: 3   Section 4.1
Level: A: Standard   Tag: None

Seafloor spreading occurs at

a) subduction zones.   b) terrane boundaries.

c) transform faults.   d) divergent boundaries.

d

49) Item Type: Multiple Choice   Objective: 3   Section 4.1
    Level: A: Standard   Tag: None

When examining rocks from both sides of the Mid-Atlantic Ridge, scientists found evidence for the phenomenon of

a) suspect terranes.   b) magnetic reversal.

c) volcano formation.   d) convection currents.

b

50) Item Type: Multiple Choice   Objective: 4   Section 4.2
    Level: A: Standard   Tag: None

The earth's layer of solid rock that flows under pressure is called the

a) crust.   b) asthenosphere.   c) lithosphere.   d) hydrosphere.

b

51) Item Type: Multiple Choice   Objective: 4   Section 4.2
    Level: A: Standard   Tag: Diagram 0472

In the diagram above, which of the following has formed at **A**?

a) island arc   b) ocean trench

c) mid-ocean ridge   d) magnetic band sequence

b

52) Item Type: Multiple Choice   Objective: 4   Section 4.2
    Level: A: Standard   Tag: Diagram 0472

According to the diagram above, what is occurring at point **B**?

a) seafloor spreading   b) volcanic eruption

c) magnetic reversal   d) trench development

b

53) Item Type: Multiple Choice   Objective: 4   Section 4.2
    Level: A: Standard   Tag: None

Seafloor spreading occurs at which of the following plate boundaries?

a) divergent   b) transform fault   c) convergent   d) subduction

a

54) Item Type: Multiple Choice   Objective: 4   Section 4.2
Level: A: Standard   Tag: None

The theory of plate tectonics is most directly based on which of the following interactions?

a) continental plates moving over oceanic plates

b) the Panthalassa plate being subducted under the Pangaea plate

c) suspect terrane plates moving over magma

d) lithospheric plates riding on the asthenosphere

d

55) Item Type: Multiple Choice   Objective: 5   Section 4.2
Level: B: Enriched   Tag: None

The rift valley of the Red Sea is an example of a

a) convergent boundary.   b) subduction zone.

c) transform fault.   d) divergent boundary.

d

56) Item Type: Multiple Choice   Objective: 5   Section 4.2
Level: A: Standard   Tag: None

The type of collision that occurs when two lithospheric plates converge is determined primarily by plate

a) density.   b) temperature.   c) size.   d) magnetism.

a

57) Item Type: Multiple Choice   Objective: 5   Section 4.2
Level: A: Standard   Tag: None

Which of the following is located at a convergent plate boundary?

a) eastern coast of South America   b) Himalayan Mountains

c) San Francisco Bay area   d) Mid-Atlantic Ridge

b

58) Item Type: Multiple Choice   Objective: 5   Section 4.2
Level: A: Standard   Tag: None

Which of the following may result from the collision of one plate with another?

a) a convergent boundary   b) a divergent boundary

c) a rift valley   d) a transform fault boundary

a

59) Item Type: Multiple Choice   Objective: 6   Section 4.2
      Level: A: Standard   Tag: None

Where in the earth do convection currents occur?

a) in the lithosphere   b) along a subduction zone

c) in the asthenosphere   d) along a rift valley

c

60) Item Type: Multiple Choice   Objective: 6   Section 4.2
      Level: B: Enriched   Tag: None

Scientific evidence indicating that convection currents move the plates has been found by studying

a) transform faults.   b) mountain chains.   c) coral atolls.   d) rift valleys.

d

61) Item Type: Multiple Choice   Objective: 6   Section 4.2
      Level: A: Standard   Tag: None

The transfer of heat through the movement of heated fluid material in the earth's crust is called

a) tectonics.   b) convection.   c) paleomagnetism.   d) subduction.

b

62) Item Type: Multiple Choice   Objective: 6   Section 4.2
      Level: A: Standard   Tag: None

Many scientists believe that the motion of lithospheric plates is due to

a) subduction.   b) reversal of magnetic polarity.

c) convection.   d) seafloor spreading.

c

63) Item Type: Multiple Choice   Objective: 7   Section 4.2
      Level: A: Standard   Tag: None

The Palo Alto Hills are probably an example of

a) a landmass bridge.   b) an island arc.

c) a glacial landform.   d) a suspect terrane.

d

64) Item Type: Multiple Choice   Objective: 7   Section 4.2
     Level: A: Standard   Tag: None

Which of the following areas is considered a terrane?

a) the Andes Mountains in South America

b) the west coast of Africa

c) the Palo Alto Hills of California

d) the south coast of Iceland

c

65) Item Type: Multiple Choice   Objective: 7   Section 4.2
     Level: A: Standard   Tag: None

Pieces of land bounded by faults that have different geologic features from those of neighboring land are most likely to be

a) island arcs.   b) hot spots.   c) suspect terranes.   d) spreading centers.

c

66) Item Type: Multiple Choice   Objective: 7   Section 4.2
     Level: B: Enriched   Tag: None

The theory of suspect terranes provides an explanation of how

a) continents form.   b) magma spreads.

c) convection currents occur.   d) faulting occurs.

a

67) Item Type: Fill in the Blank   Objective: 1   Section 4.1
     Level: A: Standard   Tag: None

The theory proposing that continents were once joined in a single landmass is called the

theory of _____.

continental drift

68) Item Type: Fill in the Blank   Objective: 1   Section 4.1
     Level: B: Enriched   Tag: None

According to the theory of continental drift, the eastern coast of South America was once

joined to the continent of _____.

Africa

69) Item Type: Fill in the Blank   Objective: 1   Section 4.1
Level: B: Enriched   Tag: None

The name of the huge landmass that Wegener believed covered much of the earth

200 million years ago is _____.

| Pangaea |

70) Item Type: Fill in the Blank   Objective: 1   Section 4.1
Level: A: Standard   Tag: None

The theory that first hypothesized movement of the continents is called the theory of

_____.

| continental drift |

71) Item Type: Fill in the Blank   Objective: 3   Section 4.1
Level: A: Standard   Tag: None

Which property of iron-bearing rocks provided evidence for the hypothesis of seafloor

spreading? _____

| magnetic orientation |

72) Item Type: Fill in the Blank   Objective: 3   Section 4.1
Level: A: Standard   Tag: None

An undersea mountain range with a narrow valley running down its center is called a(n)

_____.

| mid-ocean ridge |

73) Item Type: Fill in the Blank   Objective: 4   Section 4.2
Level: A: Standard   Tag: None

The two types of crust that make up the surface of the earth are _____.

| oceanic crust and continental crust |

74) Item Type: Fill in the Blank   Objective: 4   Section 4.2
Level: B: Enriched   Tag: Diagram 0473

The layer labeled **Y** in the diagram above is the part of the lithosphere known as

_____.

| continental crust |

75) Item Type: Fill in the Blank   Objective: 5   Section 4.2
     Level: B: Enriched   Tag: Diagram 0473

The term for the type of plate boundary shown at the point labeled **X** in this diagram is

_____.

divergent	

76) Item Type: Fill in the Blank   Objective: 5   Section 4.2
     Level: B: Enriched   Tag: None

The type of plate boundary that causes extensive volcanic activity in Iceland is called a(n)

_____.

divergent plate boundary	

77) Item Type: Fill in the Blank   Objective: 5   Section 4.2
     Level: A: Standard   Tag: None

A fault formed at the point where two plates slide past each other is called a(n)

_____.

transform fault	

78) Item Type: Fill in the Blank   Objective: 6   Section 4.2
     Level: B: Enriched   Tag: None

Many scientists attribute the movement of lithospheric plates to the process of heat transfer

called _____.

convection	

79) Item Type: Fill in the Blank   Objective: 6   Section 4.2
     Level: B: Enriched   Tag: None

The layer of the earth in which convection currents occur is the _____.

asthenosphere	

80) Item Type: Fill in the Blank   Objective: 6   Section 4.2
     Level: A: Standard   Tag: None

Many scientists think that the movement of lithospheric plates is caused by

_____.

convection currents	

81) Item Type: Fill in the Blank   Objective: 7   Section 4.2
     Level: A: Standard   Tag: None

A small piece of land that is scraped off descending ocean floor at a subduction zone is

called a(n) _____.

| terrane ■ |

82) Item Type: Fill in the Blank   Objective: 7   Section 4.2
     Level: A: Standard   Tag: None

Pieces of land bordered by faults that have geologic features different from those of

neighboring pieces of land are called _____.

| terranes ■ |

83) Item Type: Fill in the Blank   Objective: 7   Section 4.2
     Level: B: Enriched   Tag: None

New discoveries concerning the ways in which the continents grow along convergent

boundaries provide the basis for the theory of _____.

| suspect terranes ■ |

84) Item Type: Fill in the Blank   Objective: 7   Section 4.2
     Level: A: Standard   Tag: None

The theory referring to new land that is added to continental margins at subduction zones

is called the _____.

| theory of suspect terranes ■ |

85) Item Type: Essay   Objective: 2   Section 4.1
     Level: A: Standard   Tag: None

What hypothesis is Alfred Wegener known for, and what evidence supported his
hypothesis?

_____

_____

_____

_____

_____

Wegener first proposed the theory of continental drift. There is evidence to support his
hypothesis, such as the similarities in the coastlines of continents, the fossil remains in
South America and Africa, the ages and types of rocks found on different continents,
similarities between mountain chains on different continents, and the climatic histories of
continents.

86) Item Type: Essay   Objective: 5   Section 4.2
     Level: B: Enriched   Tag: None

On the west coast of South America, an oceanic plate is colliding with a continental plate.
According to the theory of plate tectonics, what type of geologic activity is occurring at
this plate boundary? Describe the activity.

_____

_____

_____

_____

The oceanic plate, which is denser, is forced under the continental crust, forming a
subduction zone. As it is subducted, material from the oceanic plate melts and then rises to
the surface on the continental crust, forming volcanic mountains (Andes).

87) Item Type: Essay   Objective: 4   Section 4.2
Level: B: Enriched   Tag: None

Describe the relationship between the lithosphere and the asthenosphere in terms of plate tectonics.

_____

_____

_____

_____

_____

The oceanic and continental crust of the earth and part of the earth's upper mantle make up the lithosphere—the solid, relatively rigid outer shell of the earth. The lithosphere is broken into separate plates. These plates float on the denser asthenosphere beneath, a layer of solid rock that flows slowly when under pressure. The continents are carried along on the moving lithospheric plates.

88) Item Type: Essay   Objective: 5   Section 4.2
Level: B: Enriched   Tag: None

What happens when a plate with oceanic crust collides with a plate with continental crust?

_____

_____

_____

_____

_____

The oceanic plate, which is denser, is forced under the continental crust, forming a subduction zone. As it is subducted, material from the oceanic plate melts and then rises to the surface on the continental crust, forming volcanic mountains.

89) Item Type: Essay   Objective: 6   Section 4.2
Level: B: Enriched   Tag: None

What are convection currents, and how do they affect the earth's asthenosphere?

_____

_____

_____

_____

_____

Convection currents occur beneath the earth's surface in the asthenosphere. Hot material deep within the earth's asthenosphere rises to the base of the lithosphere. Once it reaches the lithosphere, the hot material cools. New rising hot material pushes the cooler material aside. This cooler, more dense material sinks back down. This process results in a continuous circulation of material in the asthenosphere.

90) Item Type: Essay   Objective: 7   Section 4.2
Level: B: Enriched   Tag: None

What are the three identifying characteristics of a terrane?

_____

_____

_____

_____

First, a terrane contains rock and fossils that differ from the rock and fossils of neighboring terranes. Second, there are major faults at the boundaries of a terrane. Finally, the magnetic properties of a terrane do not match those of neighboring terranes.

1)   Tag Name: Diagram 0471

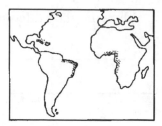

2)   Tag Name: Diagram 0472

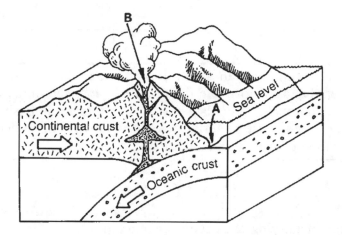

3)   Tag Name: Diagram 0473

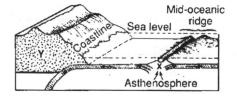

1) Item Type: True-False   Objective: 1   Section 5.1
     Level: A: Standard   Tag: None

_____ Isostatic adjustments can occur as a result of the erosion of mountain ranges.

T

2) Item Type: True-False   Objective: 1   Section 5.1
     Level: A: Standard   Tag: None

_____ When land is covered by a glacier, the crust beneath it will sink lower into the mantle.

T

3) Item Type: True-False   Objective: 1   Section 5.1
     Level: A: Standard   Tag: None

_____ A very thick deposit of material on the ocean floor is called a hot spot.

F

4) Item Type: True-False   Objective: 1   Section 5.1
     Level: B: Enriched   Tag: None

_____ When a glacier retreats from an area, the part of the crust that was formerly ice-covered actually sinks deeper into the mantle.

F

5) Item Type: True-False   Objective: 1   Section 5.1
     Level: A: Standard   Tag: None

_____ Isostatic adjustments are constantly occurring in areas of the earth's crust with mountain ranges.

T

6) Item Type: True-False   Objective: 2   Section 5.1
     Level: A: Standard   Tag: None

_____ The major cause of crustal deformation is the movement of the earth's lithospheric plates.

T

7) Item Type: True-False   Objective: 2   Section 5.1
     Level: A: Standard   Tag: None

_____ Compression is the force that pulls rocks apart.

F

8) Item Type: True-False   Objective: 2   Section 5.1
    Level: A: Standard   Tag: None

_____ Stress in the crust can be caused by isostatic adjustments.

| T | ▪ |

9) Item Type: True-False   Objective: 2   Section 5.1
    Level: A: Standard   Tag: None

_____ Forces that cause deformation of the crust are usually the result of a change in the volume of the mantle.

| F | ▪ |

10) Item Type: True-False   Objective: 2   Section 5.1
    Level: A: Standard   Tag: None

_____ Compression causes crustal rocks to be squeezed together.

| T | ▪ |

11) Item Type: True-False   Objective: 2   Section 5.1
    Level: A: Standard   Tag: None

_____ A change in the weight of part of the crust can result in crustal deformation.

| T | ▪ |

12) Item Type: True-False   Objective: 2   Section 5.1
    Level: A: Standard   Tag: None

_____ Stress is a force that applies pressure to the rocks in the earth's crust.

| T | ▪ |

13) Item Type: True-False   Objective: 3   Section 5.2
    Level: A: Standard   Tag: None

_____ Strain causes the folding and faulting of crustal rocks.

| F | ▪ |

14) Item Type: True-False   Objective: 3   Section 5.2
    Level: A: Standard   Tag: None

_____ Anticlines generally form ridges in response to folding.

| T | ▪ |

15) Item Type: True-False   Objective: 4   Section 5.2
     Level: A: Standard   Tag: None

_____ Normal faults form in regions where the crust is diverging.

T

16) Item Type: True-False   Objective: 4   Section 5.2
     Level: A: Standard   Tag: None

_____ Along a strike-slip fault, the rock on either side of the fault planes moves vertically.

F

17) Item Type: True-False   Objective: 4   Section 5.2
     Level: A: Standard   Tag: None

_____ A thrust fault is a type of normal fault.

F

18) Item Type: True-False   Objective: 4   Section 5.2
     Level: A: Standard   Tag: None

_____ When motion occurs along a break in rock, the break is called a fracture.

F

19) Item Type: True-False   Objective: 4   Section 5.2
     Level: B: Enriched   Tag: None

_____ A reverse fault occurs when compression causes a hanging wall to move up relative to a footwall.

T

20) Item Type: True-False   Objective: 4   Section 5.2
     Level: A: Standard   Tag: None

_____ A thrust fault has no hanging wall.

F

21) Item Type: True-False   Objective: 5   Section 5.3
     Level: A: Standard   Tag: None

_____ Folded mountains are commonly found where continents have collided.

T

22) Item Type: True-False   Objective: 5   Section 5.3
      Level: A: Standard   Tag: None

    _____ Plate collisions occur along converging plate boundaries.

T

23) Item Type: True-False   Objective: 5   Section 5.3
      Level: A: Standard   Tag: None

    _____ The circum-Pacific belt is primarily an area of converging plate boundaries.

T

24) Item Type: True-False   Objective: 5   Section 5.3
      Level: A: Standard   Tag: None

    _____ According to the theory of plate tectonics, India was once a separate continent riding on the Indian plate.

T

25) Item Type: True-False   Objective: 5   Section 5.3
      Level: A: Standard   Tag: None

    _____ Collisions between oceanic and continental crust often result in the formation of volcanic mountains.

T

26) Item Type: True-False   Objective: 6   Section 5.3
      Level: A: Standard   Tag: None

    _____ Mountains that are formed when molten rock erupts violently onto the earth's surface are called dome mountains.

F

27) Item Type: True-False   Objective: 6   Section 5.3
      Level: A: Standard   Tag: None

    _____ The Himalayas are dome mountains.

F

28) Item Type: True-False   Objective: 6   Section 5.3
      Level: A: Standard   Tag: None

    _____ The Appalachian Mountains are fault-block mountains.

F

29) Item Type: True-False   Objective: 6   Section 5.3
     Level: A: Standard   Tag: None

_____ Most plateaus are formed when thick, horizontal layers of rock are slowly uplifted.

T

30) Item Type: True-False   Objective: 6   Section 5.3
     Level: A: Standard   Tag: None

_____ Mid-ocean ridges are huge folded mountain chains.

F

31) Item Type: Multiple Choice   Objective: 1   Section 5.1
     Level: A: Standard   Tag: None

As a volcanic mountain range is built, isostatic adjustment will cause the crust beneath it to

a) break.   b) sink.   c) rise.   d) fold.

b

32) Item Type: Multiple Choice   Objective: 1   Section 5.1
     Level: A: Standard   Tag: None

Up-and-down motions of the crust are called

a) thrust faulting.   b) strain movement.

c) isostatic adjustments.   d) compressional stress.

c

33) Item Type: Multiple Choice   Objective: 1   Section 5.1
     Level: A: Standard   Tag: None

Crust sinks deeper into the mantle as a result of increased crustal

a) temperature.   b) thickness.   c) drift.   d) tension.

b

34) Item Type: Multiple Choice   Objective: 1   Section 5.1
     Level: A: Standard   Tag: None

Isostatic adjustments constantly occur in areas of the crust that are characterized by

a) mountains.

b) faults.

c) grabens.

d) folds.

a

35) Item Type: Multiple Choice   Objective: 1   Section 5.1
    Level: B: Enriched   Tag: None

Rising of the crust would most likely be caused by the crust being

a) compressed.   b) thinned.   c) folded.   d) fractured.

b

36) Item Type: Multiple Choice   Objective: 1   Section 5.1
    Level: B: Enriched   Tag: None

As rivers flow into the ocean and deposit thick layers of sediment on the ocean floor, what will the crust under the sediment do?

a) rise   b) fracture   c) sink   d) erupt

c

37) Item Type: Multiple Choice   Objective: 1   Section 5.1
    Level: B: Enriched   Tag: None

Isostatic adjustments continue until the forces pressing the crust up and down are

a) balanced.   b) doubled.   c) decreased.   d) removed.

a

38) Item Type: Multiple Choice   Objective: 1   Section 5.1
    Level: A: Standard   Tag: None

The balance between the forces pushing the crust down and the forces pushing it up is called

a) strain.   b) isostasy.   c) convergence.   d) tectonics.

b

39) Item Type: Multiple Choice   Objective: 1   Section 5.1
    Level: A: Standard   Tag: None

Which of the following is the major cause of deformation of the earth's crust?

a) magma formation   b) hot spots

c) plate tectonics   d) divergent margins

c

Chapter 5

40) Item Type: Multiple Choice   Objective: 2   Section 5.1
    Level: A: Standard   Tag: None

A very thick accumulation of materials deposited near shore on the ocean floor causes the ocean floor to

a) rise.   b) fracture.   c) erupt.   d) sink.

d

41) Item Type: Multiple Choice   Objective: 2   Section 5.1
    Level: B: Enriched   Tag: Diagram 0571

What type of force has acted on the rocks in this diagram?

a) shearing   b) tension   c) compression   d) strain

b

42) Item Type: Multiple Choice   Objective: 2   Section 5.1
    Level: A: Standard   Tag: None

A change in the shape or volume of crustal rock due to stress is called

a) shearing.   b) isostasy.   c) compression.   d) strain.

d

43) Item Type: Multiple Choice   Objective: 3   Section 5.2
    Level: A: Standard   Tag: None

Folding of rocks is most likely to happen when rocks undergo

a) tension.   b) shearing.   c) compression.   d) cooling.

c

44) Item Type: Multiple Choice   Objective: 3   Section 5.2
    Level: B: Enriched   Tag: None

Rocks deeper in the earth's crust are less likely to deform by breaking than rocks closer to the earth's surface because of

a) lower stresses.   b) higher temperature.   c) higher strains.   d) lower altitude.

b

45) Item Type: Multiple Choice   Objective: 3   Section 5.2
    Level: A: Standard   Tag: None

The permanent deformation of a rock without the rock breaking is called

a) collision.   b) folding.   c) faulting.   d) fracturing.

b

46) Item Type: Multiple Choice   Objective: 3   Section 5.2
Level: A: Standard   Tag: None

Where are rocks most likely to break?

a) in the upper mantle

b) near the earth's surface

c) deep in the crust

d) in large monoclines

b

47) Item Type: Multiple Choice   Objective: 3   Section 5.2
Level: B: Enriched   Tag: None

A deformed rock may return to its original shape if the deforming force was applied

a) unevenly.   b) slowly.   c) uniformly.   d) quickly.

b

48) Item Type: Multiple Choice   Objective: 3   Section 5.2
Level: A: Standard   Tag: None

In the Appalachian Mountains, synclines most often form

a) lakes.   b) valleys.   c) ridges.   d) plateaus.

c

49) Item Type: Multiple Choice   Objective: 4   Section 5.2
Level: A: Standard   Tag: None

Which of the following causes reverse faults?

a) compression   b) shearing   c) tension   d) folding

a

50) Item Type: Multiple Choice   Objective: 4   Section 5.2
Level: B: Enriched   Tag: Diagram 0573

What type of fault is pictured in the diagram above?

a) normal   b) abnormal   c) thrust   d) strike-slip

c

51) Item Type: Multiple Choice   Objective: 4   Section 5.2
Level: B: Enriched   Tag: Diagram 0573

In the diagram above, the rocks in **Block 1** form

a) an anticline.   b) a footwall.   c) a syncline.   d) a hanging wall.

d

52) Item Type: Multiple Choice   Objective: 4   Section 5.2
    Level: B: Enriched   Tag: Diagram 0573

In the diagram above, line **X** represents a

a)   fault plane.   b)   collision line.   c)   subduction zone.   d)   fold peak.

a

53) Item Type: Multiple Choice   Objective: 4   Section 5.2
    Level: B: Enriched   Tag: None

Where do strike-slip faults often occur?

a)   at mid-ocean ridges   b)   in the lower half of the crust

c)   between grabens   d)   along transform fault boundaries

d

54) Item Type: Multiple Choice   Objective: 4   Section 5.2
    Level: B: Enriched   Tag: None

Large-scale normal faulting formed the

a)   Grand Canyon.

b)   Great Lakes of the northern United States.

c)   Alps.

d)   great rift valleys of eastern Africa.

d

55) Item Type: Multiple Choice   Objective: 4   Section 5.2
    Level: A: Standard   Tag: None

Normal faults usually occur in a series of parallel fault lines producing landforms that are

a)   rounded.   b)   flat.   c)   steplike.   d)   ramplike.

c

56) Item Type: Multiple Choice   Objective: 4   Section 5.2
    Level: A: Standard   Tag: None

The footwall consists of rocks located

a)   at the base of a cliff.   b)   below the fault plane.

c)   under a plate boundary.   d)   along a transform fault.

b

57) Item Type: Multiple Choice   Objective: 4   Section 5.2
Level: A: Standard   Tag: None

A thrust fault is a special type of

a) hanging wall.   b) footwall.   c) reverse fault.   d) normal fault.

c

58) Item Type: Multiple Choice   Objective: 5   Section 5.3
Level: A: Standard   Tag: None

A collision between a continental plate and an oceanic plate is most likely to produce

a) hot spots.   b) mountain ranges.   c) dome mountains.   d) volcanic islands.

b

59) Item Type: Multiple Choice   Objective: 5   Section 5.3
Level: A: Standard   Tag: None

Which of the following usually form along convergent plate boundaries?

a) mountains   b) normal faults   c) grabens   d) ocean basins

a

60) Item Type: Multiple Choice   Objective: 5   Section 5.3
Level: B: Enriched   Tag: None

Scientists from several countries are using laser beams and the satellite called *Lageos* to

a) measure elevation.   b) track plate movement.

c) predict earthquakes.   d) locate fault lines.

b

61) Item Type: Multiple Choice   Objective: 5   Section 5.3
Level: B: Enriched   Tag: None

The northern ocean crust of the African plate is subducting the continental crust of Eurasia, causing

a) the Himalayas to become higher.

b) the Mediterranean Sea to become larger.

c) the Himalayas to erode.

d) the Mediterranean Sea to become smaller.

d

62) Item Type: Multiple Choice   Objective: 5   Section 5.3
    Level: B: Enriched   Tag: None

A collision between continental crust and oceanic crust formed the

a) Mid-Atlantic Ridge.   b) Cascade Mountains.

c) Himalaya Mountains.   d) Adirondack Mountains.

b

63) Item Type: Multiple Choice   Objective: 5   Section 5.3
    Level: A: Standard   Tag: None

Which of the following was formed by a collision between continental and oceanic crusts?

a) Mount St. Helens   b) the Tibetan Plateau

c) Mount Everest   d) the Imperial Valley

a

64) Item Type: Multiple Choice   Objective: 5   Section 5.3
    Level: B: Enriched   Tag: None

A collision between oceanic crust and oceanic crust formed the

a) Hawaiian Islands.   b) Cascades.   c) Fiji Islands.   d) Alps.

c

65) Item Type: Multiple Choice   Objective: 5   Section 5.3
    Level: A: Standard   Tag: None

During collisions between continental and oceanic crusts, magma is produced when the oceanic crust is partially

a) folded.   b) compressed.   c) faulted.   d) melted.

d

66) Item Type: Multiple Choice   Objective: 6   Section 5.3
    Level: B: Enriched   Tag: None

The oceanic crust of the Mediterranean Sea floor is

a) colliding with the lithospheric plate to the east.

b) folding due to tensional stresses.

c) subducting beneath continental crust to the north.

d) fracturing along strike-slip faults.

c

67) Item Type: Multiple Choice   Objective: 6   Section 5.3
    Level: A: Standard   Tag: None

The structures that form where parts of the earth's crust have been broken by faults are called

a) grabens.   b) plateaus.   c) synclines.   d) monoclines.

a

68) Item Type: Multiple Choice   Objective: 6   Section 5.3
    Level: A: Standard   Tag: None

Most plateaus are found

a) near mountain ranges.   b) bordering ocean basins.

c) beneath grabens.   d) near diverging boundaries.

a

69) Item Type: Multiple Choice   Objective: 6   Section 5.3
    Level: A: Standard   Tag: None

What type of mountains are commonly located where continents have collided?

a) folded   b) volcanic   c) fault-block   d) dome

a

70) Item Type: Multiple Choice   Objective: 6   Section 5.3
    Level: A: Standard   Tag: None

A group of adjacent mountain ranges makes up a mountain

a) belt.   b) system.   c) boundary.   d) ridge.

b

71) Item Type: Multiple Choice   Objective: 6   Section 5.3
    Level: A: Standard   Tag: None

Mount Everest is a part of which mountain range?

a) Alps   b) Andes   c) Urals   d) Himalayas

d

72) Item Type: Multiple Choice   Objective: 6   Section 5.3
    Level: B: Enriched   Tag: Diagram 0574

What type of mountains are shown in this diagram?

a) folded   b) volcanic   c) dome   d) fault-block

d

73) Item Type: Multiple Choice   Objective: 6   Section 5.3
Level: B: Enriched   Tag: None

What are the largest groups of mountains on the earth called?

a) ridges   b) ranges   c) belts   d) systems

c

74) Item Type: Multiple Choice   Objective: 6   Section 5.3
Level: B: Enriched   Tag: None

Which of the following is an example of a graben?

a) Fiji Islands   b) Azores   c) Black Hills   d) Imperial Valley

d

75) Item Type: Multiple Choice   Objective: 6   Section 5.3
Level: B: Enriched   Tag: None

What formations develop after faulting breaks the crust into blocks and a block slips downward relative to the surrounding blocks?

a) plateaus   b) fault-block mountains c) grabens   d) thrust faults

c

76) Item Type: Multiple Choice   Objective: 6   Section 5.3
Level: B: Enriched   Tag: None

What type of mountain forms when molten rock rises through the crust and pushes up the overlying rock?

a) folded   b) fault-block   c) dome   d) volcanic

c

77) Item Type: Fill in the Blank   Objective: 1   Section 5.1
Level: B: Enriched   Tag: None

A very thick accumulation of sand, gravel, and rock material deposited by a river on the

floor of the ocean causes it to _____.

sink

78) Item Type: Fill in the Blank   Objective: 1   Section 5.1
Level: A: Standard   Tag: None

The balancing of the up-and-down forces between the crust and the mantle is called

_____.

isostasy

79) Item Type: Fill in the Blank   Objective: 3   Section 5.2
    Level: B: Enriched   Tag: Diagram 0572

The term for the type of fold shown in this diagram is _____.

anticline

80) Item Type: Fill in the Blank   Objective: 4   Section 5.2
    Level: A: Standard   Tag: None

The actual surface of a break in the crustal rocks at a fault is called a fault

_____.

plane

81) Item Type: Fill in the Blank   Objective: 4   Section 5.2
    Level: A: Standard   Tag: None

The crustal rocks below the broken surface at a normal fault make up the

_____.

footwall

82) Item Type: Fill in the Blank   Objective: 4   Section 5.2
    Level: A: Standard   Tag: None

The type of fault in which the rocks on either side of the fault plane move horizontally is

called a(n) _____.

slip-strike fault

83) Item Type: Essay   Objective: 4   Section 5.2
    Level: A: Standard   Tag: None

Describe a reverse fault.

_____

_____

_____

_____

In a reverse fault, the fault plane is usually almost vertical. This type of fault forms when compression causes the hanging wall to move up relative to the footwall.

84) Item Type: Essay   Objective: 4   Section 5.2
    Level: B: Enriched   Tag: None

The San Andreas Fault in California is located at a fault boundary where the Pacific plate is moving toward the northwest past the North American plate. What type of fault is the San Andreas Fault? What type of stress is most likely acting on the rocks along this fault?

_____

_____

_____

The San Andreas Fault is a strike-slip or transform fault. These faults occur at boundaries where one plate is sliding past another and is causing the rocks on either side of the fault plane to move horizontally. Shearing is the type of stress that is most likely acting on the rocks along the fault because shearing pushes rocks in two opposite horizontal directions.

85) Item Type: Essay   Objective: 4   Section 5.2
    Level: B: Enriched   Tag: None

Describe the parts of a normal fault.

_____

_____

_____

_____

_____

A normal fault is one in which the break in the rock is almost vertical. The rocks above
the normal fault plane, called the hanging wall, move down relative to the rocks below the
fault plane, called the footwall.

86) Item Type: Essay   Objective: 5   Section 5.3
    Level: B: Enriched   Tag: None

Describe a collision between two plates of oceanic crust.

_____

_____

_____

_____

_____

The oceanic crust of one plate subducts the oceanic crust of the other plate. As the
underlying oceanic crust plunges deeper into the mantle, the intense heat melts its crustal
material, forming magma. The magma rises and breaks through the oceanic crust of the
overriding plate and forms volcanic mountains on the ocean floor.

1) Tag Name: Diagram 0571

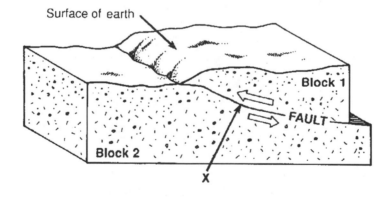

2) Tag Name: Diagram 0572

3) Tag Name: Diagram 0573

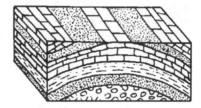

4)   Tag Name: Diagram 0574

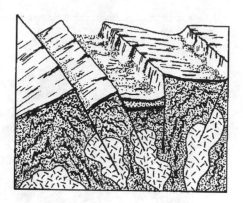

1) Item Type: True-False   Objective: 1   Section 6.1
    Level: A: Standard   Tag: None

_____ The earthquakes that cause the most damage usually have a shallow focus.

T

2) Item Type: True-False   Objective: 2   Section 6.1
    Level: A: Standard   Tag: None

_____ A fault zone is a group of interconnected faults.

T

3) Item Type: True-False   Objective: 2   Section 6.1
    Level: A: Standard   Tag: None

_____ The New Madrid, Missouri, earthquake of 1812 resulted from movement along the San Andreas fault.

F

4) Item Type: True-False   Objective: 2   Section 6.1
    Level: B: Enriched   Tag: None

_____ Plates are scraping past each other at the Eurasian-Melanesian mountain belt.

F

5) Item Type: True-False   Objective: 2   Section 6.1
    Level: A: Standard   Tag: None

_____ Most earthquakes occur near plate boundaries.

T

6) Item Type: True-False   Objective: 3   Section 6.2
    Level: A: Standard   Tag: None

_____ L waves are the fastest-moving seismic waves.

F

7) Item Type: True-False   Objective: 3   Section 6.2
    Level: A: Standard   Tag: None

_____ P waves move faster through liquids than through other materials.

F

8) Item Type: True-False   Objective: 3   Section 6.2
     Level: A: Standard   Tag: None

_____ P waves moving through the earth can travel through solids, liquids, and gases.

T

9) Item Type: True-False   Objective: 4   Section 6.2
     Level: B: Enriched   Tag: None

_____ Scientists analyze the difference between the arrival times of P waves and S waves to determine an earthquake's epicenter.

T

10) Item Type: True-False   Objective: 4   Section 6.2
     Level: A: Standard   Tag: None

_____ Scientists locate the epicenter of an earthquake by determining its magnitude.

F

11) Item Type: True-False   Objective: 4   Section 6.2
     Level: A: Standard   Tag: None

_____ Scientists must locate the epicenter of an earthquake before they can determine its distance from a seismograph.

F

12) Item Type: True-False   Objective: 5   Section 6.2
     Level: A: Standard   Tag: None

_____ The magnitude of an earthquake is an approximate measure of how much energy the earthquake releases.

T

13) Item Type: True-False   Objective: 5   Section 6.2
     Level: A: Standard   Tag: None

_____ Microquakes are usually strong enough to be felt by people.

F

14) Item Type: True-False   Objective: 5   Section 6.2
     Level: A: Standard   Tag: None

_____ A seismograph records energy released by an earthquake.

T

15) Item Type: True-False   Objective: 5   Section 6.2
     Level: A: Standard   Tag: None

_____ An earthquake with the highest magnitude is given a rating of 31.7 on the Richter scale.

| F |

16) Item Type: True-False   Objective: 6   Section 6.3
     Level: B: Enriched   Tag: None

_____ Movement of the ground is the direct cause of most injuries related to earthquakes.

| F |

17) Item Type: True-False   Objective: 6   Section 6.3
     Level: B: Enriched   Tag: None

_____ Duration does not affect the intensity of an earthquake.

| F |

18) Item Type: True-False   Objective: 6   Section 6.3
     Level: A: Standard   Tag: None

_____ Most buildings are not designed to withstand the swaying motion caused by earthquakes.

| T |

19) Item Type: True-False   Objective: 7   Section 6.3
     Level: A: Standard   Tag: None

_____ Tsunamis are seismic waves that travel through the earth's core.

| F |

20) Item Type: True-False   Objective: 7   Section 6.3
     Level: A: Standard   Tag: None

_____ The epicenter of an earthquake that causes a tsunami is located on the ocean floor.

| T |

21) Item Type: True-False   Objective: 8   Section 6.3
     Level: A: Standard   Tag: None

_____ Following simple safety rules during an earthquake can help prevent death and injury.

| T |

22) Item Type: True-False   Objective: 9   Section 6.3
Level: A: Standard   Tag: None

_____ Scientists think that seismic gaps are likely locations of future earthquakes.

T

23) Item Type: True-False   Objective: 9   Section 6.3
Level: A: Standard   Tag: None

_____ Changes in the magnetic and electrical properties of rocks along a fault may be
warnings that an earthquake will occur in the near future.

T

24) Item Type: True-False   Objective: 9   Section 6.3
Level: A: Standard   Tag: None

_____ Earthquake predictions based on records of past earthquakes in the area can be off
by years.

T

25) Item Type: Multiple Choice   Objective: 1   Section 6.1
Level: A: Standard   Tag: None

When friction prevents the rocks on either side of a fault from moving past each other, the
fault is said to be

a) fractured.   b) subducting.   c) locked.   d) elastic.

c

26) Item Type: Multiple Choice   Objective: 1   Section 6.1
Level: A: Standard   Tag: None

The place where slippage first occurs is called an earthquake's

a) focus.   b) epicenter.   c) magnitude.   d) intensity.

a

27) Item Type: Multiple Choice   Objective: 1   Section 6.1
Level: A: Standard   Tag: None

Geologists use the elastic rebound theory to explain

a) the cause of tsunamis.   b) the intensity of an earthquake.

c) the magnitude of tsunamis.   d) the cause of an earthquake.

d

28) Item Type: Multiple Choice   Objective: 1   Section 6.1
Level: A: Standard   Tag: None

Which of the following best describes aftershocks?

a) a series of small tremors occurring after a major earthquake

b) seismic waves that cannot travel through liquids

c) areas along a fault where slippage and fracturing first occur

d) giant ocean waves that originate at a fault zone

a

29) Item Type: Multiple Choice   Objective: 1   Section 6.1
Level: A: Standard   Tag: None

Deep-focus earthquakes usually occur in areas in which one plate

a) slides past another.   b) remains stationary against another.

c) moves apart from another.   d) subducts under another.

d

30) Item Type: Multiple Choice   Objective: 1   Section 6.1
Level: A: Standard   Tag: None

Ninety percent of continental earthquakes have

a) the same depth.   b) a shallow focus.

c) the same type of eruptions.   d) a deep focus.

b

31) Item Type: Multiple Choice   Objective: 1   Section 6.1
Level: B: Enriched   Tag: None

How far below the earth's surface do intermediate-focus earthquakes occur?

a) 10 to 30 km   b) 30 to 70 km

c) 70 to 300 km   d) 300 to 650 km

c

32) Item Type: Multiple Choice   Objective: 1   Section 6.1
Level: B: Enriched   Tag: None

Which type of earthquake usually occurs farther inland than other earthquakes?

a) deep-focus   b) shallow-focus

c) intermediate-focus   d) microquakes

a

33) Item Type: Multiple Choice   Objective: 1   Section 6.1
     Level: B: Enriched   Tag: None

Which of the following generally causes the most damage?

   a)   shallow-focus earthquakes   b)   deep-focus earthquakes

   c)   intermediate-focus earthquakes   d)   aftershocks

34) Item Type: Multiple Choice   Objective: 1   Section 6.1
     Level: A: Standard   Tag: None

The force that can change the size and shape of rocks is called

   a)   stress.   b)   magnitude.   c)   elasticity.   d)   friction.

35) Item Type: Multiple Choice   Objective: 2   Section 6.1
     Level: B: Enriched   Tag: None

Most earthquakes occur along or near the edges of the

   a)   North American plate.   b)   earth's oceans and lakes.

   c)   Eurasian plate.   d)   earth's lithospheric plates.

36) Item Type: Multiple Choice   Objective: 2   Section 6.1
     Level: B: Enriched   Tag: None

In which major earthquake zone are earthquakes produced mainly by plates moving away from each other?

   a)   the Pacific Ring of Fire

   b)   the Mid-Atlantic Ridge

   c)   the San Andreas Fault area

   d)   the Eurasian-Melanesian mountain belt

37) Item Type: Multiple Choice   Objective: 2   Section 6.1
   Level: B: Enriched   Tag: None

The San Andreas Fault zone has formed where the edge of the Pacific plate is slipping

a) under the North American plate.

b) over the North American plate.

c) south along the North American plate.

d) north along the North American plate.

d

38) Item Type: Multiple Choice   Objective: 2   Section 6.1
   Level: B: Enriched   Tag: None

Where did the most widely-felt series of earthquakes in the United States occur?

a) Alaska   b) California   c) Hawaii   d) Missouri

d

39) Item Type: Multiple Choice   Objective: 2   Section 6.1
   Level: A: Standard   Tag: None

The Pacific Ring of Fire is an earthquake zone that forms a ring around

a) the Atlantic Ocean.   b) South America.

c) the Pacific Ocean.   d) North America.

c

40) Item Type: Multiple Choice   Objective: 3   Section 6.2
   Level: B: Enriched   Tag: None

L waves are especially destructive when traveling through

a) water.   b) solid rock.   c) a fault zone.   d) loose earth.

d

41) Item Type: Multiple Choice   Objective: 3   Section 6.2
   Level: A: Standard   Tag: None

Which type of seismic wave travels the fastest?

a) L wave   b) P wave   c) S wave   d) surface wave

b

42) Item Type: Multiple Choice Objective: 3 Section 6.2
Level: A: Standard Tag: None

Which type of seismic wave causes rock particles to move together and apart in the same direction as the wave is moving?

a) L wave b) S wave c) P wave d) surface wave

c

43) Item Type: Multiple Choice Objective: 3 Section 6.2
Level: B: Enriched Tag: None

How many separate sensing devices make up a seismograph?

a) 1 b) 2 c) 3 d) 4

c

44) Item Type: Multiple Choice Objective: 3 Section 6.2
Level: A: Standard Tag: None

S waves can travel only through

a) gases. b) solids. c) liquids. d) gases and liquids.

b

45) Item Type: Multiple Choice Objective: 3 Section 6.2
Level: A: Standard Tag: None

Which type of seismic wave causes rock particles to move at right angles to the direction in which the wave is traveling?

a) S waves b) P waves c) L waves d) primary waves

a

46) Item Type: Multiple Choice Objective: 3 Section 6.2
Level: A: Standard Tag: None

Which of the following generally causes the most damage during an earthquake?

a) aftershock b) primary wave c) secondary wave d) surface wave

d

47) Item Type: Multiple Choice Objective: 4 Section 6.2
Level: B: Enriched Tag: None

What is the minimum number of seismograph stations a scientist must have data from in order to locate the epicenter of an earthquake?

a) 1 b) 2 c) 3 d) 4

c

48) Item Type: Multiple Choice   Objective: 4   Section 6.2
    Level: B: Enriched   Tag: None

If a seismograph station detects S waves shortly after it detects P waves, then the earthquake was

a) far away.   b) nearby.   c) very weak.   d) very strong.

49) Item Type: Multiple Choice   Objective: 4   Section 6.2
    Level: A: Standard   Tag: None

To determine how far away from a seismograph station an earthquake occurred, scientists plot the difference in arrival times between

a) P and S waves.   b) S and L waves.

c) P and L waves.   d) seismic waves and tsunamis.

50) Item Type: Multiple Choice   Objective: 5   Section 6.2
    Level: B: Enriched   Tag: None

How much more energy is released by an earthquake with a magnitude of 3.5 than by an earthquake with a magnitude of 2.5?

a) 1,000 times more energy   b) 317 times more energy

c) 100 times more energy   d) 31.7 times more energy

51) Item Type: Multiple Choice   Objective: 5   Section 6.2
    Level: A: Standard   Tag: None

The Mercalli scale is used to express the

a) magnitude of an earthquake.

b) distance of an earthquake from a seismograph.

c) intensity of an earthquake.

d) depth of an earthquake's focus.

52) Item Type: Multiple Choice   Objective: 5   Section 6.2
Level: B: Enriched   Tag: None

An earthquake with a magnitude of 6.5 is classified as a

a) major earthquake.

b) moderate earthquake.

c) minor earthquake.

d) microquake.

b

53) Item Type: Multiple Choice   Objective: 5   Section 6.2
Level: A: Standard   Tag: None

The magnitude of an earthquake is a direct measure of

a) how much energy it releases.

b) how much damage it causes.

c) how many tsunamis it creates.

d) how many aftershocks it causes.

a

54) Item Type: Multiple Choice   Objective: 5   Section 6.2
Level: A: Standard   Tag: None

The largest earthquake ever recorded had a magnitude of

a) II.   b) 8.1.   c) 8.9.   d) X.

c

55) Item Type: Multiple Choice   Objective: 5   Section 6.2
Level: A: Standard   Tag: None

The Mercalli scale includes a description that expresses

a) the magnitude of a tsunami.

b) how much damage a tsunami causes.

c) the magnitude of an earthquake.

d) how much damage an earthquake causes.

d

56) Item Type: Multiple Choice   Objective: 5   Section 6.2
Level: A: Standard   Tag: None

Which of the following is used to indicate an earthquake having the highest intensity?

a) II   b) 8.9   c) X   d) 31.7

c

57) Item Type: Multiple Choice   Objective: 5   Section 6.2
    Level: A: Standard   Tag: None

What does the Mercalli scale describe?

a) the energy of an earthquake

b) the intensity of an earthquake

c) the focus of an earthquake

d) the likelihood of an earthquake

b

58) Item Type: Multiple Choice   Objective: 6   Section 6.3
    Level: A: Standard   Tag: None

During an earthquake, a person is most likely to be safe in

a) a car.   b) an open field.   c) a cellar.   d) a building.

b

59) Item Type: Multiple Choice   Objective: 7   Section 6.3
    Level: A: Standard   Tag: None

During the Alaskan earthquake of 1964, most deaths were caused by

a) the collapse of buildings.   b) a tsunami.

c) vibration of the ground.   d) cracks formed in the earth.

b

60) Item Type: Multiple Choice   Objective: 7   Section 6.3
    Level: B: Enriched   Tag: Diagram 0673

Which point in this diagram indicates a likely source of a tsunami?

a) *A*   b) *B*   c) *C*   d) *D*

a

61) Item Type: Multiple Choice   Objective: 7   Section 6.3
    Level: A: Standard   Tag: None

The Seismic Sea Wave Warning System (SSWWS) issues warnings of

a) tsunamis.   b) earthquakes.   c) aftershocks.   d) landslides.

a

62) Item Type: Multiple Choice   Objective: 7   Section 6.3
    Level: A: Standard   Tag: None

Tsunamis result from earthquakes with epicenters located

a) on the ocean floor.   b) in loose soil.

c) along coastlines.   d) in continental faults.

a

63) Item Type: Multiple Choice   Objective: 7   Section 6.3
    Level: B: Enriched   Tag: None

The seismograph stations that make up the SSWWS are located in and around the

a) Atlantic Ocean.   b) Indian Ocean.

c) Pacific Ocean.   d) Great Lakes.

c

64) Item Type: Multiple Choice   Objective: 7   Section 6.3
    Level: A: Standard   Tag: None

What is a tsunami?

a) a deep-focus earthquake   b) a shallow-focus earthquake

c) a type of seismic wave   d) a giant ocean wave

d

65) Item Type: Multiple Choice   Objective: 7   Section 6.3
    Level: A: Standard   Tag: None

Landslides on the seafloor are most likely to produce

a) tsunamis.   b) fault zones.   c) seismic gaps.   d) elastic rebound.

a

66) Item Type: Multiple Choice   Objective: 7   Section 6.3
    Level: A: Standard   Tag: None

Faulting and underwater landslides are primary causes of

a) tsunamis.   b) seismic gaps.   c) microquakes.   d) subduction.

a

67) Item Type: Multiple Choice   Objective: 8   Section 6.3
    Level: A: Standard   Tag: None

In general, where is the safest place for an individual to park when in a car during an earthquake?

a) close to water   b) in a tunnel   c) in the open   d) under a bridge

c

68) Item Type: Multiple Choice   Objective: 8   Section 6.3
    Level: A: Standard   Tag: Diagram 0674

During an earthquake, which position shown in the diagram above would provide the greatest safety from falling and flying debris?

a) position **A**   b) position **B**

c) position **C**   d) position **D**

a

69) Item Type: Multiple Choice   Objective: 8   Section 6.3
    Level: A: Standard   Tag: None

Where are earthquakes most likely to occur?

a) near inactive faults   b) in hard-packed soil

c) near active faults   d) in loose soil

c

70) Item Type: Multiple Choice   Objective: 9   Section 6.3
    Level: A: Standard   Tag: None

Tests done in Colorado showed that earthquakes were less severe after faults were injected with

a) water.   b) natural gas.   c) rock.   d) loose earth.

a

71) Item Type: Multiple Choice   Objective: 9   Section 6.3
    Level: A: Standard   Tag: None

Zones of immobile rock along faults are called

a) epicenters.   b) seismic gaps.   c) ridges.   d) subduction zones.

b

72) Item Type: Multiple Choice   Objective: 9   Section 6.3
     Level: A: Standard   Tag: None

Scientific tests conducted at Rangely, Colorado, showed that earthquakes were less severe when

a) the strain in rocks along a fault was artificially increased.

b) loose soil was deposited above a fault zone.

c) a giant magnet was placed above a fault zone.

d) water was injected along a fault.

d

73) Item Type: Multiple Choice   Objective: 9   Section 6.3
     Level: A: Standard   Tag: None

Which of the following is often monitored by scientists in an attempt to predict earthquakes more accurately?

a) peculiarities in animal behavior

b) depths of past focus locations

c) increases in gas seepage from strained rocks

d) intensities of past earthquakes

c

74) Item Type: Multiple Choice   Objective: 9   Section 6.3
     Level: A: Standard   Tag: None

Which of the following is sometimes used to predict earthquakes?

a) magnitudes of past earthquakes

b) the direction in which L waves travel

c) changes in the speed of tsunamis

d) slight tilting of the ground

d

75) Item Type: Multiple Choice   Objective: 9   Section 6.3
    Level: A: Standard   Tag: None

A seismic gap is a region in which

a) there are no seismographs.

b) tsunamis never occur.

c) a fault is locked.

d) only deep-focus earthquakes occur.

c

76) Item Type: Fill in the Blank   Objective: 1   Section 6.1
    Level: A: Standard   Tag: None

Earthquakes take place along faults in the earth's _____.

crust

77) Item Type: Fill in the Blank   Objective: 1   Section 6.1
    Level: B: Enriched   Tag: Diagram 0671

The point that indicates the epicenter of the earthquake in the diagram is

_____.

Point **B**

78) Item Type: Fill in the Blank   Objective: 1   Section 6.1
    Level: A: Standard   Tag: None

When friction keeps a fault in an immobile state, the fault is said to be

_____.

locked

79) Item Type: Fill in the Blank   Objective: 2   Section 6.1
    Level: A: Standard   Tag: None

The western coast of North America is part of a major earthquake zone called the

_____.

Pacific Ring of Fire

80) Item Type: Fill in the Blank   Objective: 2   Section 6.1
    Level: A: Standard   Tag: None

A group of interconnected faults is called a(n) _____.

| fault zone ▪ |

81) Item Type: Fill in the Blank   Objective: 2   Section 6.1
    Level: A: Standard   Tag: None

The fault extending along much of the length of California is called the

_____.

| San Andreas Fault ▪ |

82) Item Type: Fill in the Blank   Objective: 3   Section 6.2
    Level: A: Standard   Tag: None

Seismic waves that can penetrate the liquid part of the earth's core are called

_____.

| P waves ▪ |

83) Item Type: Fill in the Blank   Objective: 3   Section 6.2
    Level: A: Standard   Tag: None

The last waves to be recorded by a seismograph are the _____.

| L waves ▪ |

84) Item Type: Fill in the Blank   Objective: 3   Section 6.2
    Level: B: Enriched   Tag: Diagram 0672

The seismic waves shown in this diagram are called _____.

| L (or surface) waves ▪ |

85) Item Type: Fill in the Blank   Objective: 3   Section 6.2
    Level: B: Enriched   Tag: None

Seismic waves that cause rock particles to move at right angles to the direction in which

the waves are traveling are called _____.

| S (or secondary) waves ▪ |

Chapter 6

86) Item Type: Fill in the Blank   Objective: 4   Section 6.2
Level: A: Standard   Tag: None

Scientists plot three intersecting circles to locate an earthquake's _____.

| epicenter |

87) Item Type: Fill in the Blank   Objective: 5   Section 6.2
Level: A: Standard   Tag: None

The amount of energy released by an earthquake is also known as its

_____.

| magnitude |

88) Item Type: Fill in the Blank   Objective: 9   Section 6.3
Level: A: Standard   Tag: None

Zones of immobile rock along faults are called _____.

| seismic gaps |

89) Item Type: Essay   Objective: 7   Section 6.3
Level: B: Enriched   Tag: None

How may faulting cause a tsunami?

Faulting may cause a sudden drop or rise in the ocean floor, which may result in a similar drop or rise in mass of water. This mass of water churns up and down as it adjusts to the change in sea level and sets in motion a series of long, low waves that develop into tsunamis.

90) Item Type: Essay   Objective: 7   Section 6.3
Level: B: Enriched   Tag: None

Describe the two events that generally cause tsunamis.

_____

_____

_____

_____

_____

> Two events that may cause tsunamis are faulting and underwater landslides. Faulting may bring about a sudden rise or drop in the level of a part of the ocean floor, causing a large mass of water to churn up and down. This water movement sets into motion long, low waves that develop into tsunamis. An earthquake may trigger a severe underwater landslide, throwing water above the landslide in an up-and-down motion and thereby creating a series of tsunamis.

1) Tag Name: Diagram 0671

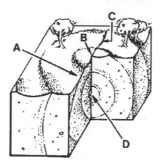

2) Tag Name: Diagram 0672

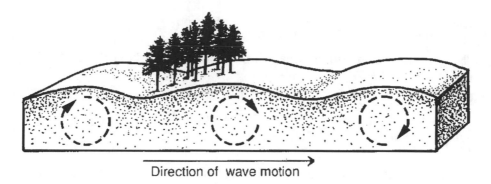

Direction of wave motion

3) Tag Name: Diagram 0673

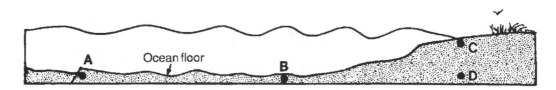

Ocean floor

4) Tag Name: Diagram 0674

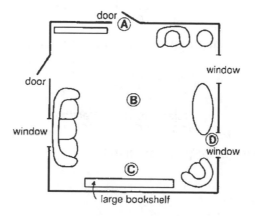

1)   Item Type: True-False   Objective: 1   Section 7.1
        Level: B: Enriched   Tag: None

      _____   Temperature and pressure generally keep the rocks in the asthenosphere below the melting point.

T

2)   Item Type: True-False   Objective: 1   Section 7.1
        Level: A: Standard   Tag: None

      _____   Solid rock located deep in the mantle is called magma.

F

3)   Item Type: True-False   Objective: 1   Section 7.1
        Level: A: Standard   Tag: None

      _____   Magma may sometimes break through to the earth's surface.

T

4)   Item Type: True-False   Objective: 2   Section 7.1
        Level: A: Standard   Tag: None

      _____   Magma is able to rise upward through the earth's crust because it is less dense than the surrounding rocks.

T

5)   Item Type: True-False   Objective: 2   Section 7.1
        Level: A: Standard   Tag: None

      _____   Magma forms at subducted plate boundaries.

T

6)   Item Type: True-False   Objective: 2   Section 7.1
        Level: A: Standard   Tag: None

      _____   The movement of magma toward or onto the earth's surface is called volcanism.

T

7)   Item Type: True-False   Objective: 3   Section 7.1
        Level: A: Standard   Tag: None

      _____   Volcanic activity is frequent in island arcs.

T

8) Item Type: True-False   Objective: 3   Section 7.1
    Level: A: Standard   Tag: None

_____ A major zone of active volcanoes encircles the Atlantic Ocean.

F

9) Item Type: True-False   Objective: 3   Section 7.1
    Level: A: Standard   Tag: None

_____ An area in which one lithospheric plate is being moved under another is called a subduction zone.

T

10) Item Type: True-False   Objective: 3   Section 7.1
    Level: A: Standard   Tag: None

_____ The islands that make up Japan are the result of a hot spot.

F

11) Item Type: True-False   Objective: 3   Section 7.1
    Level: A: Standard   Tag: None

_____ Earthquakes frequently occur near island arcs.

T

12) Item Type: True-False   Objective: 4   Section 7.2
    Level: A: Standard   Tag: None

_____ Felsic lavas usually erupt along with large amounts of water vapor.

T

13) Item Type: True-False   Objective: 5   Section 7.2
    Level: A: Standard   Tag: None

_____ Dust and ash from the destruction of Krakatau blocked some of the sun's rays for several years.

T

14) Item Type: True-False   Objective: 5   Section 7.2
    Level: A: Standard   Tag: None

_____ Tephra is usually thrown from volcanoes that eject felsic lava.

T

15) Item Type: True-False   Objective: 5   Section 7.2
    Level: A: Standard   Tag: None

_____   Volcanic dust in the atmosphere often reduces average global temperatures by as
much as 20°C.

F

16) Item Type: True-False   Objective: 6   Section 7.2
    Level: A: Standard   Tag: None

_____   Fissures are commonly found at the top of shield cones.

F

17) Item Type: True-False   Objective: 6   Section 7.2
    Level: A: Standard   Tag: None

_____   The Hawaiian Islands are examples of cinder cones.

F

18) Item Type: True-False   Objective: 6   Section 7.2
    Level: A: Standard   Tag: None

_____   Cinder cones are generally much steeper than shield cones.

T

19) Item Type: True-False   Objective: 6   Section 7.2
    Level: A: Standard   Tag: None

_____   Rock piles around volcanic vents are known as cones.

T

20) Item Type: True-False   Objective: 6   Section 7.2
    Level: A: Standard   Tag: None

_____   A caldera is the funnel-shaped pit at the top of a volcanic vent.

F

21) Item Type: True-False   Objective: 7   Section 7.2
    Level: A: Standard   Tag: None

_____   Knowledge of previous eruptions of a particular volcano is generally helpful in
predicting its future eruptions.

T

22) Item Type: True-False   Objective: 7   Section 7.2
    Level: A: Standard   Tag: None

_____ Volcanic bulges are measured with balloons.

F

23) Item Type: True-False   Objective: 7   Section 7.2
    Level: A: Standard   Tag: None

_____ Temperature changes within rocks can contribute to small earthquakes.

T

24) Item Type: True-False   Objective: 8   Section 7.3
    Level: A: Standard   Tag: None

_____ The heat produced by intense meteorite bombardment may be responsible for much of the moon's ancient volcanism.

T

25) Item Type: True-False   Objective: 8   Section 7.3
    Level: A: Standard   Tag: None

_____ Plate tectonics was probably the primary cause of volcanism on the moon.

F

26) Item Type: True-False   Objective: 8   Section 7.3
    Level: A: Standard   Tag: None

_____ Meteorite bombardment of the moon may have generated the heat necessary to form magma.

T

27) Item Type: True-False   Objective: 8   Section 7.3
    Level: A: Standard   Tag: None

_____ There is some evidence for plate tectonics on Mars.

F

28) Item Type: True-False   Objective: 8   Section 7.3
    Level: A: Standard   Tag: None

_____ Convection currents probably produced the magma on the moon.

F

29) Item Type: True-False   Objective: 8   Section 7.3
    Level: A: Standard   Tag: None

_____ Evidence of volcanism exists in several areas of our solar system.

T

30) Item Type: True-False   Objective: 9   Section 7.3
    Level: B: Enriched   Tag: None

_____ It is believed that Io's volcanoes eject several thousand metric tons of material every second.

T

31) Item Type: True-False   Objective: 9   Section 7.3
    Level: A: Standard   Tag: None

_____ The material that erupts from volcanoes on Io is felsic.

F

32) Item Type: True-False   Objective: 9   Section 7.3
    Level: A: Standard   Tag: None

_____ The moon Io has only one-tenth the volcanic activity of the earth.

F

33) Item Type: Multiple Choice   Objective: 1   Section 7.1
    Level: B: Enriched   Tag: None

Knowledge of temperature and pressure conditions in the earth's mantle is based on the analysis of

a) direct measurements.   b) seismic wave behavior.

c) crustal magnetic properties.   d) volcanic bulges.

b

34) Item Type: Multiple Choice   Objective: 1   Section 7.1
    Level: B: Enriched   Tag: None

Which of the following is most likely to occur in an area of the asthenosphere where surrounding rock exerts less-than-normal pressure?

a) violent volcanic eruptions   b) magma formation

c) plate subduction   d) caldera formation

b

35) Item Type: Multiple Choice   Objective: 1   Section 7.1
     Level: A: Standard   Tag: None

Most magma forms in

a) continental crust.

b) the lithosphere.

c) the asthenosphere.

d) oceanic crust.

c

36) Item Type: Multiple Choice   Objective: 1   Section 7.1
     Level: B: Enriched   Tag: None

Estimates of the earth's inner temperatures and pressures are primarily based on the

a) drift rate of lithospheric plates.

b) composition of felsic magmas.

c) location of volcanic islands.

d) behavior of seismic waves.

d

37) Item Type: Multiple Choice   Objective: 1   Section 7.1
     Level: A: Standard   Tag: None

Great pressure in the asthenosphere keeps the rock there mostly

a) felsic.

b) molten.

c) solid.

d) gaseous.

c

38) Item Type: Multiple Choice   Objective: 2   Section 7.1
     Level: A: Standard   Tag: None

An opening in the earth's surface through which molten rock flows is called a

a) vent.

b) caldera.

c) mantle.

d) fault.

a

39) Item Type: Multiple Choice   Objective: 2   Section 7.1
     Level: B: Enriched   Tag: None

What usually happens to magma after it forms in the earth's mantle?

a) It spreads out sideways.

b) It sinks deeper down.

c) It stays where it forms.

d) It rises through cracks.

d

40) Item Type: Multiple Choice   Objective: 2   Section 7.1
     Level: A: Standard   Tag: None

An opening on the earth's surface through which molten rock flows and the material that builds up around the opening together form a

a) subduction zone.   b) trench.

c) convergent boundary.   d) volcano.

d

41) Item Type: Multiple Choice   Objective: 3   Section 7.1
     Level: B: Enriched   Tag: Diagram 0771

The area labeled **X** in the diagram is located over

a) a caldera.   b) a hot spot.

c) a mid-ocean ridge.   d) an island arc.

b

42) Item Type: Multiple Choice   Objective: 3   Section 7.1
     Level: A: Standard   Tag: None

What happens when a plate with oceanic crust meets a plate with continental crust?

a) The continental plate is subducted.   b) New crust forms over a hot spot.

c) The oceanic plate is subducted.   d) New crust forms under a hot spot.

c

43) Item Type: Multiple Choice   Objective: 3   Section 7.1
     Level: B: Enriched   Tag: None

The Aleutian Islands have formed along a

a) subduction zone.   b) hot-spot path.

c) mid-ocean ridge.   d) continental margin.

a

44) Item Type: Multiple Choice   Objective: 3   Section 7.1
     Level: A: Standard   Tag: None

As a result of the subduction of oceanic crust under a continent, magma is most likely to erupt from

a) an oceanic ridge.   b) an oceanic trench.

c) an island arc.   d) a volcanic cone.

d

45) Item Type: Multiple Choice Objective: 3 Section 7.1
Level: A: Standard Tag: None

A string of volcanoes that forms along a trench is called

a) an island arc.

b) a fissure.

c) a mid-ocean ridge.

d) a subducted plate.

a

46) Item Type: Multiple Choice Objective: 3 Section 7.1
Level: B: Enriched Tag: None

What type of volcanic feature is represented by Iceland?

a) a mid-ocean ridge

b) a subduction zone

c) a tsunami

d) an island arc

a

47) Item Type: Multiple Choice Objective: 3 Section 7.1
Level: A: Standard Tag: None

Iceland was formed over a

a) mid-ocean ridge.

b) subduction zone.

c) convergent boundary.

d) caldera.

a

48) Item Type: Multiple Choice Objective: 4 Section 7.2
Level: A: Standard Tag: None

Magma that erupts under water often forms rounded formations called

a) aa lava.

b) pillow lava.

c) volcanic bombs.

d) pahoehoe lava.

b

49) Item Type: Multiple Choice Objective: 4 Section 7.2
Level: A: Standard Tag: None

What type of lava usually flows out of oceanic volcanoes?

a) felsic   b) aa   c) mafic   d) pahoehoe

c

50) Item Type: Multiple Choice   Objective: 4   Section 7.2
Level: A: Standard   Tag: None

Felsic lava is usually identified by its color and its

a) silica content.

b) rounded shapes.

c) ropey texture.

d) sulfur content.

a

51) Item Type: Multiple Choice   Objective: 4   Section 7.2
Level: A: Standard   Tag: None

Which volcanic material would most likely be produced by a violent eruption?

a) tephra   b) pillow lava   c) aa   d) pahoehoe

a

52) Item Type: Multiple Choice   Objective: 4   Section 7.2
Level: A: Standard   Tag: None

The type of lava that is rich in silica, iron, and magnesium is

a) pillow.   b) felsic.   c) mafic.   d) aa.

b

53) Item Type: Multiple Choice   Objective: 4   Section 7.2
Level: B: Enriched   Tag: None

Pillow lava is most commonly found

a) in calderas.

b) along deep trenches.

c) in stratovolcanoes.

d) along mid-ocean ridges.

d

54) Item Type: Multiple Choice   Objective: 5   Section 7.2
Level: B: Enriched   Tag: None

The easiest way to distinguish between volcanic ash and volcanic dust particles is to compare their

a) color.   b) weight.   c) diameter.   d) density.

c

55) Item Type: Multiple Choice   Objective: 5   Section 7.2
    Level: A: Standard   Tag: None

Large tephra that is round or spindle-shaped is called volcanic

a) ash.   b) bombs.   c) bulges.   d) blocks.

b

56) Item Type: Multiple Choice   Objective: 6   Section 7.2
    Level: A: Standard   Tag: None

The broad volcanic feature formed by quiet eruptions of thin lava flows is called a

a) shield cone.   b) cinder cone.

c) rift.   d) stratovolcano.

a

57) Item Type: Multiple Choice   Objective: 6   Section 7.2
    Level: B: Enriched   Tag: None

Cinder cones are generally formed when volcanoes eject

a) thin lava.   b) solid fragments.   c) pillow lava.   d) gases.

b

58) Item Type: Multiple Choice   Objective: 6   Section 7.2
    Level: A: Standard   Tag: None

Which of the following generally results from a quiet lava eruption?

a) cinder cone   b) surface bulge

c) shield cones   d) stratovolcano

c

59) Item Type: Multiple Choice   Objective: 6   Section 7.2
    Level: A: Standard   Tag: None

The material that collects around the vent of a volcano as a result of eruptions forms three
major types of volcanic

a) fissures.   b) calderas.   c) craters.   d) cones.

d

60) Item Type: Multiple Choice   Objective: 6   Section 7.2
Level: B: Enriched   Tag: Diagram 0773

What type of volcanic formation is represented by the diagram above?

a) stratovolcano   b) shield cone   c) caldera   d) cinder cone

b

61) Item Type: Multiple Choice   Objective: 6   Section 7.2
Level: B: Enriched   Tag: Diagram 0773

The feature labeled **X** in the diagram above is a

a) volcanic bomb.   b) hot spot.   c) volcanic bulge.   d) crater.

d

62) Item Type: Multiple Choice   Objective: 6   Section 7.2
Level: A: Standard   Tag: None

Which of the following formations would most likely result from a single violent volcanic eruption?

a) shield cone   b) vent   c) cinder cone   d) caldera

c

63) Item Type: Multiple Choice   Objective: 6   Section 7.2
Level: A: Standard   Tag: None

What is formed when a magma chamber empties and collapses?

a) a crater   b) a fissure   c) a caldera   d) a vent

c

64) Item Type: Multiple Choice   Objective: 7   Section 7.2
Level: A: Standard   Tag: None

Seismographs can be useful in predicting volcanic eruptions because they measure

a) changes in surface bulging.   b) changes in gas composition.

c) temperature increases.   d) earthquake activity.

d

65) Item Type: Multiple Choice   Objective: 7   Section 7.2
Level: A: Standard   Tag: None

The catastrophic volcanic eruption that caused a series of tsunamis and a drop in global temperatures happened in

a) Japan.   b) Hawaii.   c) Krakatau.   d) Iceland.

c

66) Item Type: Multiple Choice   Objective: 7   Section 7.2
    Level: A: Standard   Tag: None

Before a volcanic eruption, seismic activity seems to

a) increase in frequency and decrease in intensity.

b) decrease in both frequency and intensity.

c) decrease in frequency and increase in intensity.

d) increase in both frequency and intensity.

d

67) Item Type: Multiple Choice   Objective: 7   Section 7.2
    Level: A: Standard   Tag: None

What causes bulging in the surface of volcanoes?

a) build up of volcanic blocks

b) upward movement of magma

c) changes in volcanic gases

d) tephra eruptions

b

68) Item Type: Multiple Choice   Objective: 8   Section 7.3
    Level: B: Enriched   Tag: None

The theory that Mars may still be volcanically active is based on evidence of

a) crust drift   b) seismic activity.

c) hot spots.   d) tephra formations.

b

69) Item Type: Multiple Choice   Objective: 8   Section 7.3
    Level: B: Enriched   Tag: None

What type of lava flow has been found on the earth's moon?

a) pahoehoe   b) pillow   c) felsic   d) basaltic

d

70)   Item Type: Multiple Choice   Objective: 8   Section 7.3
      Level: A: Standard   Tag: None

What type of volcanic formation is Olympus Mons?

a) shield cone   b) cinder cone

c) stratovolcano   d) oceanic volcano

a

71)   Item Type: Multiple Choice   Objective: 8   Section 7.3
      Level: B: Enriched   Tag: None

One of the features supporting the theory of volcanism on the moon is the presence of

a) smooth crater interiors.   b) continued eruption today.

c) volcanic cones.   d) abundant tephra.

a

72)   Item Type: Multiple Choice   Objective: 8   Section 7.3
      Level: A: Standard   Tag: None

Where is the volcano Olympus Mons located?

a) Io   b) the moon   c) Saturn   d) Mars

d

73)   Item Type: Multiple Choice   Objective: 9   Section 7.3
      Level: A: Standard   Tag: None

Which of the following planetary bodies has far more volcanic activity than the earth?

a) Mars   b) the moon   c) Io   d) Jupiter

c

74)   Item Type: Multiple Choice   Objective: 9   Section 7.3
      Level: A: Standard   Tag: None

Scientists believe that the material erupting from Io's volcanoes is primarily

a) felsic material.   b) sulfur and sulfur dioxide.

c) mafic material.   d) iron and carbon dioxide.

b

75) Item Type: Multiple Choice   Objective: 9   Section 7.3
     Level: A: Standard   Tag: None

As a result of volcanic activity, Io's surface is colored

a) jet black.

b) ashy gray.

c) bright yellow-red.

d) dull yellow-brown.

c

76) Item Type: Multiple Choice   Objective: 9   Section 7.3
     Level: A: Standard   Tag: None

The heat that causes the melting of Io's interior is probably generated by

a) friction from surface movement.

b) solar radiation.

c) constant meteorite bombardment.

d) tectonic activity.

a

77) Item Type: Multiple Choice   Objective: 9   Section 7.3
     Level: A: Standard   Tag: None

Which of the following is probably the most volcanically active body in the solar system?

a) the earth   b) Io   c) the moon   d) Mars

b

78) Item Type: Multiple Choice   Objective: 9   Section 7.3
     Level: A: Standard   Tag: None

The volcanic material that erupts on Io is generally characterized by

a) black, tarlike surfaces.

b) small, smooth craters.

c) giant shield volcanoes.

d) large, umbrella-shaped plumes.

d

79) Item Type: Fill in the Blank   Objective: 1   Section 7.1
     Level: A: Standard   Tag: None

When solid rock in the earth's mantle melts, it forms a liquid rock known as

_____.

magma

80) Item Type: Fill in the Blank   Objective: 1   Section 7.1
Level: A: Standard   Tag: None

Geologists believe that magma forms in areas where the surrounding rock exerts

less-than-normal _____.

| pressure ■ |

81) Item Type: Fill in the Blank   Objective: 2   Section 7.1
Level: A: Standard   Tag: None

The general name for magma that erupts onto the earth's surface is

_____.

| lava ■ |

82) Item Type: Fill in the Blank   Objective: 2   Section 7.1
Level: A: Standard   Tag: None

Any activity that includes the movement of magma toward the surface of the earth is called

_____.

| volcanism ■ |

83) Item Type: Fill in the Blank   Objective: 3   Section 7.1
Level: A: Standard   Tag: None

Areas of volcanism within lithospheric plates are known as _____.

| hot spots ■ |

84) Item Type: Fill in the Blank   Objective: 3   Section 7.1
Level: A: Standard   Tag: None

The major zone of active volcanoes that encircles the Pacific Ocean is called the

_____.

| Pacific Ring of Fire ■ |

85) Item Type: Fill in the Blank   Objective: 4   Section 7.2
Level: A: Standard   Tag: None

Lava with a wrinkled surface that forms when mafic lava hardens is known as

_____.

| pahoehoe ■ |

86) Item Type: Fill in the Blank   Objective: 4   Section 7.2
Level: A: Standard   Tag: None

The global effect of dust and ash from volcanic eruptions is likely to be a drop in

_____.

temperature

87) Item Type: Fill in the Blank   Objective: 4   Section 7.2
Level: B: Enriched   Tag: None

The composition of felsic lava differs from that of mafic lava because felsic lava contains

more _____.

silica

88) Item Type: Fill in the Blank   Objective: 4   Section 7.2
Level: A: Standard   Tag: None

The thin lava that generally erupts from oceanic volcanoes is called

_____.

mafic lava

89) Item Type: Fill in the Blank   Objective: 4   Section 7.2
Level: A: Standard   Tag: None

Mafic lava that breaks into jagged chunks is called _____.

aa

90) Item Type: Fill in the Blank   Objective: 5   Section 7.2
Level: B: Enriched   Tag: Diagram 0772

The fragment type indicated by the asterisk in the table is _____.

lapilli

91) Item Type: Fill in the Blank   Objective: 5   Section 7.2
Level: A: Standard   Tag: None

The largest tephra formed from solid rock is known as a(n) _____.

volcanic block

92) Item Type: Fill in the Blank   Objective: 5   Section 7.2
    Level: A: Standard   Tag: None

Volcanic rock particles small enough to be carried by wind are dust and

_____.

| ash |

93) Item Type: Fill in the Blank   Objective: 5   Section 7.2
    Level: A: Standard   Tag: None

Round and spindle-shaped tephra are called _____.

| volcanic bombs |

94) Item Type: Fill in the Blank   Objective: 5   Section 7.2
    Level: A: Standard   Tag: None

The largest pieces of volcanic material formed from solid rock blasted from a volcano are

called _____.

| volcanic blocks |

95) Item Type: Fill in the Blank   Objective: 6   Section 7.2
    Level: A: Standard   Tag: None

Another name for a composite cone is a(n) _____.

| stratovolcano |

96) Item Type: Fill in the Blank   Objective: 8   Section 7.3
    Level: B: Enriched   Tag: None

The formation of most of the craters on the moon's surface probably resulted from

_____.

| meteor bombardment |

97) Item Type: Fill in the Blank   Objective: 8   Section 7.3
    Level: B: Enriched   Tag: None

The volcano Olympus Mons is an unusual example of the cone type called a(n)

_____.

| shield cone |

98) Item Type: Essay   Objective: 3   Section 7.1
    Level: B: Enriched   Tag: None

How are mountains formed when a plate with an oceanic crust meets one with a continental crust?

_____

_____

_____

_____

_____

When a plate with oceanic crust meets a plate with continental crust, the oceanic crust, which is the denser of the two, is forces beneath the continental crust. The plate with the continental crust buckles and folds, forming a line of mountains along the edge of the continent. Also, magma from the subducted plate rises through the continental crust, forming volcanic mountains.

99) Item Type: Essay   Objective: 3   Section 7.1
    Level: B: Enriched   Tag: None

Explain how new oceanic floor is formed at mid-ocean ridges.

_____

_____

_____

_____

_____

At mid-ocean ridges, fractures between the spreading plates reach down to the asthenosphere. Magma rises through these fractures and comes to the surface through rifts. The emerging lava then cools and forms new ocean floor at the mid-ocean ridges.

100) Item Type: Essay   Objective: 3   Section 7.1
Level: B: Enriched   Tag: None

Explain how new islands in the Hawaiian chain may form in the future as a result of hot-spot volcanism.

_____

_____

_____

_____

_____

The areas of volcanism within plates called hot spots remain stationary as the lithospheric plates above them continue to move. As the volcanoes that have formed are carried away from the hot spot, they cease activity because there is no longer any magma underneath to feed them. New volcanoes, however, may form where new crust has moved over the hot spot. These new volcanoes may form additional islands in the Hawaiian chain.

101) Item Type: Essay   Objective: 3   Section 7.1
Level: B: Enriched   Tag: None

Why are eruptions of oceanic volcanoes usually quieter than eruptions of continental volcanoes?

_____

_____

_____

_____

The composition of the lava that reaches the surface generally determines the force with which a particular volcano will erupt. Oceanic volcanoes are usually produced by mafic lava. Because gases can easily escape from mafic lava, eruptions of oceanic volcanoes are usually quieter than eruptions of continental volcanoes, which produce felsic lava. Felsic lavas contain a large quantity of gases that boil out of a volcano explosively when a vent of fissure opens up.

102) Item Type: Essay   Objective: 4   Section 7.2
     Level: B: Enriched   Tag: None

Describe the major effects of the eruption of Krakatau on the island group itself and on the rest of the globe.

_____

_____

_____

_____

> The tremendous volcanic explosion destroyed the islands, propelling ash and rock into the sky. A cloud of ash kept the region in darkness for several days. The eruptions also triggered large tsunamis, or tidal waves, which caused heavy damage and the loss of many lives on the islands of Java and Sumatra. Global temperatures dropped as a result of a dust and gas cloud that circled the earth for several years.

103) Item Type: Essay   Objective: 7   Section 7.2
     Level: B: Enriched   Tag: None

Why are small earthquakes important warning signals of volcanic eruptions?

_____

_____

_____

_____

_____

> Small earthquakes result from the growing pressure on the surrounding rocks as magma in the earth's mantle works its way upward. Temperature changes within the rock and the actual fracturing of the rock surrounding a volcano also contribute to small earthquakes. The number of earthquakes often increases until they occur almost continuously before a volcanic eruption.

      TEST ITEM LISTING **137**

104) Item Type: Essay   Objective: 8   Section 7.3
     Level: B: Enriched   Tag: None

What evidence is there that volcanoes on Mars may still be active?

_____

_____

_____

_____

_____

> Whether Martian volcanoes are active is a question scientists have yet to answer. However, Mars does seem to be seismically active. A *Viking* landing craft searching Mars for signs of life detected two geological events that produced waves similar to those of an earthquake. These "marsquakes" may also mean that the Red Planet is still volcanically active.

105) Item Type: Essay   Objective: 9   Section 7.3
     Level: B: Enriched   Tag: None

How do scientists explain the volcanic activity on Jupiter's moon Io?

_____

_____

_____

_____

_____

> Scientists believe that Jupiter's gravitational pull moves Io inward and outward in its orbit. This pulling back and forth causes the surface of Io to move in and out. Friction from this motion probably makes the inside of Io heat up, leading to melting, magma formation, and then volcanism.

Chapter 7

1) Tag Name: Diagram 0771

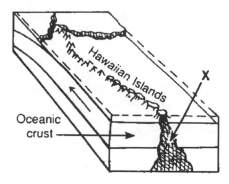

2) Tag Name: Diagram 0772

VOLCANIC ROCK
FRAGMENTS

Size	Type
less than 0.25 mm	dust
0.25 – 2 mm	
2 – 64 mm	*
more than 64 mm	bombs

3) Tag Name: Diagram 0773

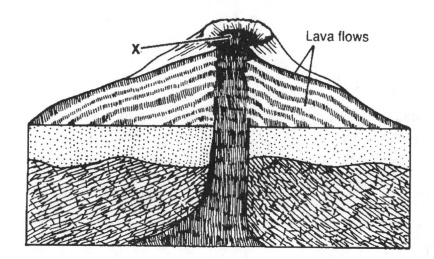

1) Item Type: True-False   Objective: 1   Section 8.1
    Level: A: Standard   Tag: None

_____ An element may be made of more than one type of atom.

F

2) Item Type: True-False   Objective: 1   Section 8.1
    Level: A: Standard   Tag: None

_____ Atoms of any one element are different from atoms of all other elements.

T

3) Item Type: True-False   Objective: 1   Section 8.1
    Level: A: Standard   Tag: None

_____ Physical properties describe how a substance interacts with other substances to produce different kinds of matter.

F

4) Item Type: True-False   Objective: 1   Section 8.1
    Level: A: Standard   Tag: None

_____ Elements cannot be broken down into simpler forms by ordinary chemical means.

T

5) Item Type: True-False   Objective: 1   Section 8.1
    Level: B: Enriched   Tag: None

_____ A chemical property of nitrogen is that it interacts with oxygen to form rust.

F

6) Item Type: True-False   Objective: 1   Section 8.1
    Level: A: Standard   Tag: None

_____ Many elements occur in pure form in the earth's crust.

F

7) Item Type: True-False   Objective: 2   Section 8.1
    Level: A: Standard   Tag: None

_____ Protons and neutrons form the nucleus of an atom.

T

8) Item Type: True-False   Objective: 2   Section 8.1
    Level: A: Standard   Tag: None

_____ Electrons and neutrons are kinds of subatomic particles.

T

9) Item Type: True-False   Objective: 3   Section 8.1
    Level: A: Standard   Tag: None

_____ The mass number of an atom is the sum of the number of protons and elections in that atom.

F

10) Item Type: True-False   Objective: 3   Section 8.1
    Level: A: Standard   Tag: None

_____ An isotope is a type of subatomic particle.

F

11) Item Type: True-False   Objective: 3   Section 8.1
    Level: A: Standard   Tag: None

_____ For a given atom in an element, the mass number minus the atomic number equals the number of neutrons in the atom.

T

12) Item Type: True-False   Objective: 4   Section 8.1
    Level: A: Standard   Tag: None

_____ Isotopes are atoms of the same element that differ in atomic number.

F

13) Item Type: True-False   Objective: 4   Section 8.1
    Level: A: Standard   Tag: None

_____ All atoms of the same element contain the same number of neutrons.

F

14) Item Type: True-False   Objective: 5   Section 8.1
    Level: A: Standard   Tag: None

_____ In general, the particles that make up a gas move faster than those that make up a solid.

T

Chapter 8

15) Item Type: True-False   Objective: 5   Section 8.1
    Level: A: Standard   Tag: None

    ____   The particles of a solid are more tightly packed than those of a gas.

    T

16) Item Type: True-False   Objective: 5   Section 8.1
    Level: A: Standard   Tag: None

    ____   The particles in a liquid move faster when heated.

    T

17) Item Type: True-False   Objective: 6   Section 8.2
    Level: A: Standard   Tag: None

    ____   Water is an example of a diatomic molecule.

    F

18) Item Type: True-False   Objective: 6   Section 8.2
    Level: B: Enriched   Tag: None

    ____   Atoms in which the outermost energy level is filled form compounds easily with
           other elements.

    F

19) Item Type: True-False   Objective: 6   Section 8.2
    Level: A: Standard   Tag: None

    ____   Metals are generally good electrical conductors.

    T

20) Item Type: True-False   Objective: 6   Section 8.2
    Level: A: Standard   Tag: None

    ____   A compound is a new substance with properties different from those of the
           elements that make it up.

    T

21) Item Type: True-False   Objective: 7   Section 8.2
    Level: B: Enriched   Tag: None

    ____   A covalent bond is formed by the transfer of electrons.

    F

22) Item Type: True-False   Objective: 7   Section 8.2
    Level: A: Standard   Tag: None

_____ Electrons are shared in covalent compounds.

T

23) Item Type: True-False   Objective: 7   Section 8.2
    Level: A: Standard   Tag: None

_____ Ionic and covalent are types of bonds.

T

24) Item Type: True-False   Objective: 7   Section 8.2
    Level: A: Standard   Tag: None

_____ Objects with like charges attract each other.

F

25) Item Type: True-False   Objective: 8   Section 8.2
    Level: A: Standard   Tag: None

_____ In a chemical formula, a subscript number after the symbol for an element is used to indicate how many isotopes of that element are in a single molecule.

F

26) Item Type: True-False   Objective: 9   Section 8.2
    Level: B: Enriched   Tag: None

_____ A material that contains two or more substances that are not chemically combined is called a mixture.

T

27) Item Type: Multiple Choice   Objective: 1   Section 8.1
    Level: A: Standard   Tag: None

Approximately how many elements occur naturally in the earth?

a) 12   b) 75   c) 90   d) 120

c

28) Item Type: Multiple Choice   Objective: 1   Section 8.1
Level: A: Standard   Tag: None

A substance that cannot be broken down into a simpler form by ordinary chemical means is called

a) an element.   b) a compound.   c) a mixture.   d) an isotope.

a

29) Item Type: Multiple Choice   Objective: 1   Section 8.1
Level: A: Standard   Tag: None

Each element consists of one basic kind of

a) proton.   b) ion.   c) atom.   d) alloy.

c

30) Item Type: Multiple Choice   Objective: 1   Section 8.1
Level: A: Standard   Tag: None

The smallest unit of an element that has all the basic properties of the element is called

a) an atom.   b) an electron.   c) a nucleus.   d) an isotope.

a

31) Item Type: Multiple Choice   Objective: 1   Section 8.1
Level: B: Enriched   Tag: None

Different elements are made up of different kinds of

a) electrons.   b) protons.   c) molecules.   d) atoms.

d

32) Item Type: Multiple Choice   Objective: 2   Section 8.1
Level: A: Standard   Tag: None

An electron is a type of

a) isotope.   b) element.   c) subatomic particle.   d) chemical bond.

c

33) Item Type: Multiple Choice   Objective: 2   Section 8.1
Level: B: Enriched   Tag: None

Which of the following would a scientist most likely study by using a particle accelerator?

a) covalent bonds   b) metallic elements   c) leptons   d) ions

c

34) Item Type: Multiple Choice   Objective: 2   Section 8.1
Level: B: Enriched   Tag: None

Which of the following particles is the smallest?

a) neutron   b) electron   c) atom   d) quark

d

35) Item Type: Multiple Choice   Objective: 2   Section 8.1
Level: B: Enriched   Tag: None

Electrons are kept from being pulled into the nucleus of the atom because of

a) electrical charges.   b) constant motion.

c) the electron cloud.   d) the subatomic particles.

b

36) Item Type: Multiple Choice   Objective: 2   Section 8.1
Level: A: Standard   Tag: None

Which of the following has no positive or negative charge?

a) electron   b) proton   c) ion   d) neutron

d

37) Item Type: Multiple Choice   Objective: 2   Section 8.1
Level: A: Standard   Tag: None

The region at the center of the atom is called the

a) nucleus.   b) neutron.   c) energy level.   d) electron cloud.

a

38) Item Type: Multiple Choice   Objective: 2   Section 8.1
Level: A: Standard   Tag: None

Protons and neutrons of an atom are located in the

a) energy levels.   b) atomic number.   c) neutron.   d) nucleus.

d

39) Item Type: Multiple Choice   Objective: 2   Section 8.1
Level: A: Standard   Tag: None

Which of the following always has a negative charge?

a) a nucleus   b) a neutron   c) an ion   d) an electron

d

Chapter 8

40) Item Type: Multiple Choice   Objective: 3   Section 8.1
    Level: A: Standard   Tag: Diagram 0871

According to the table above, what is the mass number of the element fluorine (F)?

a) 9   b) 19   c) 27   d) 28

b

41) Item Type: Multiple Choice   Objective: 3   Section 8.1
    Level: B: Enriched   Tag: Diagram 0871

According to the table above, how many neutrons are in one atom of the element fluorine (F)?

a) 9   b) 10   c) 19   d) 27

b

42) Item Type: Multiple Choice   Objective: 3   Section 8.1
    Level: B: Enriched   Tag: Diagram 0872

According to the table above, how many electrons does the element silver (Ag) have?

a) 47   b) 61   c) 108   d) 148

a

43) Item Type: Multiple Choice   Objective: 3   Section 8.1
    Level: B: Enriched   Tag: Diagram 0872

According to the table above, how many neutrons could an atom of the element zinc (Zn) have?

a) 30   b) 36   c) 65   d) 95

b

44) Item Type: Multiple Choice   Objective: 3   Section 8.1
    Level: A: Standard   Tag: None

Atoms with an atomic number of 8 must have

a) 8 neutrons.   b) 8 protons.   c) 16 neutrons.   d) 16 protons.

b

45) Item Type: Multiple Choice   Objective: 3   Section 8.1
    Level: A: Standard   Tag: None

An atom that contains 3 protons, 4 neutrons, and 3 electrons has an atomic number of

a) 3.   b) 4.   c) 7.   d) 10.

a

46) Item Type: Multiple Choice   Objective: 3   Section 8.1
Level: B: Enriched   Tag: None

How many electrons must an atom with 3 protons, 4 neutrons, and an atomic number of 3 have?

a) 3   b) 4   c) 6   d) 7

a

47) Item Type: Multiple Choice   Objective: 3   Section 8.1
Level: B: Enriched   Tag: None

What is the mass number of an atom with 4 protons, 5 neutrons, and 4 electrons?

a) 4   b) 5   c) 9   d) 13

c

48) Item Type: Multiple Choice   Objective: 3   Section 8.1
Level: B: Enriched   Tag: None

What is the atomic number of an atom with 11 protons, 12 neutrons, and 11 electrons?

a) 11   b) 12   c) 23   d) 34

a

49) Item Type: Multiple Choice   Objective: 3   Section 8.1
Level: A: Standard   Tag: None

What is the atomic number of an atom that has 10 neutrons, 8 protons, and 8 electrons?

a) 8   b) 10   c) 16   d) 18

a

50) Item Type: Multiple Choice   Objective: 4   Section 8.1
Level: A: Standard   Tag: None

Atoms of the same element that have different numbers of neutrons are called

a) alloys.   b) compounds.   c) isotopes.   d) ions.

c

51) Item Type: Multiple Choice   Objective: 4   Section 8.1
Level: A: Standard   Tag: None

Which of the following differs for each isotope of a given element?

a) electrical charge   b) energy levels

c) atomic number   d) mass number

d

52) Item Type: Multiple Choice   Objective: 4   Section 8.1
    Level: A: Standard   Tag: None

The isotopes of an element have a different number of

a) protons.   b) neutrons.   c) ions.   d) electrons.

b

53) Item Type: Multiple Choice   Objective: 4   Section 8.1
    Level: A: Standard   Tag: None

The addition of neutrons will change an atom's

a) mass number.   b) atomic number.

c) electrical charge.   d) energy levels.

a

54) Item Type: Multiple Choice   Objective: 5   Section 8.1
    Level: B: Enriched   Tag: None

Which of the following has the most loosely packed atoms?

a) steam   b) ice   c) soil   d) water

a

55) Item Type: Multiple Choice   Objective: 5   Section 8.1
    Level: A: Standard   Tag: None

Which of the following substances would have the most tightly packed atoms at room temperature?

a) a book   b) a moving river   c) hydrogen gas   d) motor oil

a

56) Item Type: Multiple Choice   Objective: 6   Section 8.2
    Level: B: Enriched   Tag: None

What is the maximum number of electrons an element with two energy levels can have?

a) 2   b) 4   c) 8   d) 10

d

57) Item Type: Multiple Choice   Objective: 6   Section 8.2
    Level: B: Enriched   Tag: None

Which of the following is a diatomic molecule?

a) oxygen   b) carbon dioxide   c) water   d) copper

a

58) Item Type: Multiple Choice   Objective: 6   Section 8.2
Level: A: Standard   Tag: None

Which of the following are arranged in energy levels within an atom?

a) elements   b) electrons   c) neutrons   d) nuclei

b

59) Item Type: Multiple Choice   Objective: 6   Section 8.2
Level: A: Standard   Tag: None

A molecule is the smallest complete unit of

a) an atom.   b) an element.   c) a compound.   d) a nucleus.

c

60) Item Type: Multiple Choice   Objective: 7   Section 8.2
Level: A: Standard   Tag: None

An atom that carries an electrical charge is called an

a) isotope.   b) ion.   c) element.   d) alloy.

b

61) Item Type: Multiple Choice   Objective: 7   Section 8.2
Level: B: Enriched   Tag: None

Ionic compounds are usually formed by the transfer of electrons from a metal to

a) an isotope.   b) a nonmetal.   c) an alloy.   d) an ion.

b

62) Item Type: Multiple Choice   Objective: 7   Section 8.2
Level: B: Enriched   Tag: None

An atom that is chemically unreactive is most likely to have how many electrons?

a) 1   b) 5   c) 10   d) 12

c

63) Item Type: Multiple Choice   Objective: 7   Section 8.2
Level: B: Enriched   Tag: Diagram 0873

According to this diagram, which of the following most accurately describes the formation of sodium chloride (NaCl)?

a) Chlorine gains an electron.   b) Sodium gains a proton.

c) Chlorine loses an electron.   d) Sodium loses a proton.

a

64) Item Type: Multiple Choice   Objective: 7   Section 8.2
     Level: A: Standard   Tag: None

In a covalent bond, the nucleus of each atom is attracted to the

a) other atom's neutrons.   b) other atom's nucleus.

c) electrons that are shared.   d) protons that are shared.

c

65) Item Type: Multiple Choice   Objective: 7   Section 8.2
     Level: A: Standard   Tag: None

Which of the following is shared between atoms in a covalent bond?

a) ions   b) neutrons   c) molecules   d) electrons

d

66) Item Type: Multiple Choice   Objective: 8   Section 8.2
     Level: A: Standard   Tag: None

How many atoms of potassium (K) are represented by the formula K_2SO_4?

a) 1   b) 2   c) 6   d) 7

b

67) Item Type: Multiple Choice   Objective: 8   Section 8.2
     Level: A: Standard   Tag: None

How many atoms of oxygen (O) are represented by the formula $Na_2C_2O_4$?

a) 2   b) 4   c) 6   d) 8

b

68) Item Type: Multiple Choice   Objective: 8   Section 8.2
     Level: A: Standard   Tag: None

What is the total number of atoms in a molecule of $C_3H_4O_3$?

a) 4   b) 7   c) 10   d) 36

c

69) Item Type: Multiple Choice   Objective: 8   Section 8.2
     Level: A: Standard   Tag: None

How many atoms of sulfur (S) are represented by the formula H_2SO_4?

a) 1   b) 2   c) 4   d) 6

a

70) Item Type: Multiple Choice   Objective: 8   Section 8.2
Level: A: Standard   Tag: None

The number of which of the following is shown in a compound's chemical formula?

a) electrons   b) protons   c) atoms   d) isotopes

c

71) Item Type: Multiple Choice   Objective: 9   Section 8.2
Level: B: Enriched   Tag: None

A solution in which one solid substance is uniformly dispersed in another solid substance is

a) a chemical formula.   b) a diatomic molecule.

c) a metal.   d) an alloy.

d

72) Item Type: Multiple Choice   Objective: 9   Section 8.2
Level: A: Standard   Tag: None

Sea water, smog, and brass are all examples of

a) alloys.   b) mixtures.

c) ionic compounds.   d) covalent compounds.

b

73) Item Type: Multiple Choice   Objective: 9   Section 8.2
Level: B: Enriched   Tag: None

Which of the following is an alloy?

a) rubber   b) bronze   c) wood   d) iron

b

74) Item Type: Multiple Choice   Objective: 9   Section 8.2
Level: B: Enriched   Tag: None

Which of the following is an example of an alloy?

a) sea water   b) smoke   c) copper   d) brass

d

75) Item Type: Multiple Choice   Objective: 9   Section 8.2
    Level: A: Standard   Tag: None

Rocks made of two or more compounds are called

a) metals.   b) mixtures.   c) alloys.   d) solutions.

b

76) Item Type: Fill in the Blank   Objective: 2   Section 8.1
    Level: A: Standard   Tag: None

In an atom, the region of space in which electrons move is called the

_____.

electron cloud

77) Item Type: Fill in the Blank   Objective: 2   Section 8.1
    Level: B: Enriched   Tag: None

The most basic subatomic particles known to scientists are leptons and

_____.

quarks

78) Item Type: Fill in the Blank   Objective: 2   Section 8.1
    Level: A: Standard   Tag: None

A subatomic particle that has no electric charge is called a(n) _____.

neutron

79) Item Type: Fill in the Blank   Objective: 3   Section 8.1
    Level: B: Enriched   Tag: None

An atom that has 11 neutrons and an atomic number of 10 has a mass number of

approximately _____.

21

80) Item Type: Fill in the Blank   Objective: 3   Section 8.1
    Level: A: Standard   Tag: None

The chart used to classify elements is called the _____.

periodic table

81) Item Type: Fill in the Blank   Objective: 5   Section 8.1
    Level: A: Standard   Tag: None

The state of matter that has no definite shape or volume is a(n) _____.

gas

82) Item Type: Fill in the Blank   Objective: 5   Section 8.1
    Level: A: Standard   Tag: None

A substance with a definite volume but no definite shape is a(n) _____.

liquid

83) Item Type: Fill in the Blank   Objective: 5   Section 8.1
    Level: A: Standard   Tag: None

The physical state of matter in which particles are packed tightly together in fixed positions

is a(n) _____.

solid

84) Item Type: Fill in the Blank   Objective: 5   Section 8.1
    Level: A: Standard   Tag: None

Matter that has a definite shape and is made up of tightly packed particles is classified as

a(n) _____.

solid

85) Item Type: Fill in the Blank   Objective: 6   Section 8.2
    Level: A: Standard   Tag: None

The most stable atoms have filled _____.

outermost energy levels

86) Item Type: Fill in the Blank   Objective: 6   Section 8.2
    Level: A: Standard   Tag: None

The smallest complete unit of a compound is a(n) _____.

molecule

87) Item Type: Fill in the Blank   Objective: 6   Section 8.2
    Level: A: Standard   Tag: None

Molecules made up of two atoms are called _____.

diatomic molecules

88) Item Type: Fill in the Blank   Objective: 6   Section 8.2
    Level: B: Enriched   Tag: None

Whether an element is a metal or a nonmetal is determined by the number and arrangement

of an atom's _____.

electrons

89) Item Type: Fill in the Blank   Objective: 6   Section 8.2
    Level: A: Standard   Tag: None

Atoms of two or more elements that are chemically combined form a(n)

_____.

compound

90) Item Type: Fill in the Blank   Objective: 7   Section 8.2
    Level: A: Standard   Tag: None

A chlorine atom that carries a charge is called a chloride _____.

ion

91) Item Type: Fill in the Blank   Objective: 7   Section 8.2
    Level: A: Standard   Tag: None

Atoms in all compounds are held together by forces called _____.

chemical bonds

92) Item Type: Fill in the Blank   Objective: 7   Section 8.2
    Level: A: Standard   Tag: None

Atoms with only one, two, or three electrons in the outermost level are generally classified

as _____.

metals

93) Item Type: Fill in the Blank   Objective: 7   Section 8.2
    Level: A: Standard   Tag: None

The type of chemical bond that is responsible for the formation of the compound sodium

chloride is called a(n) _____.

ionic bond

94)   Item Type: Fill in the Blank   Objective: 7   Section 8.2
      Level: A: Standard   Tag: None

The type of bond that joins the hydrogen and oxygen atoms in a water molecule is called

a(n) _____.

covalent bond

95)   Item Type: Fill in the Blank   Objective: 8   Section 8.2
      Level: A: Standard   Tag: None

The written notation used to represent the relative number of atoms of each element in a

compound is called a(n) _____.

chemical formula

96)   Item Type: Fill in the Blank   Objective: 9   Section 8.2
      Level: B: Enriched   Tag: None

Solid solutions of two or more metals are called _____.

alloys

97)   Item Type: Fill in the Blank   Objective: 9   Section 8.2
      Level: A: Standard   Tag: None

A mixture in which one substance is uniformly distributed in another substance is called

a(n) _____.

solution

98) Item Type: Essay   Objective: 4   Section 8.1
    Level: B: Enriched   Tag: None

Explain how isotopes of the same element differ from each other.

_____

_____

_____

_____

_____

Atoms of the same element that differ from each other in mass number are called isotopes. Isotopes of any given atom have the same number of protons but a different number of neutrons.

99) Item Type: Essay   Objective: 5   Section 8.1
    Level: A: Standard   Tag: None

Describe what happens to a solid when it is melted by heat.

_____

_____

_____

_____

_____

Adding heat causes the particles in a solid to move faster. When enough energy has been added, the individual particles in the solid move so rapidly that they break out of their fixed positions. This process results in the material becoming a liquid.

100) Item Type: Essay   Objective: 9   Section 8.2
Level: A: Standard   Tag: None

Explain the meanings of the terms *mixture* and *solution*.

_____

_____

_____

_____

_____

A mixture is any material that contains two or more substances that are not chemically combined. The substances in a mixture keep their individual properties. A solution is a mixture in which one substance is uniformly dispersed in another substance.

101) Item Type: Essay   Objective: 9   Section 8.2
Level: A: Standard   Tag: None

Explain what is meant by the term *mixture*.

_____

_____

_____

_____

_____

A mixture is a material that contains two or more substances that are not chemically combined. The substances in a mixture keep their individual properties. Therefore, unlike a compound, a mixture can be separated into its parts by physical means.

1) Tag Name: Diagram 0871

Noble Gases

4.00 ^{2} **He** Helium 2

14.01 $^2_5$ **N** Nitrogen 7	16.00 $^2_6$ **O** Oxygen 8	19.00 $^2_7$ **F** Fluorine 9	20.18 $^2_8$ **Ne** Neon 10
30.97 $^{2}_{8,5}$ **P** Phosphorus 15	32.07 $^2_{8,6}$ **S** Sulfur 16	35.45 $^2_{8,7}$ **Cl** Chlorine 17	39.95 $^2_{8,8}$ **Ar** Argon 18

2) Tag Name: Diagram 0872

63.55 ^{2,8,18,1} **Cu** Copper 29	65.39 ^{2,8,18,2} **Zn** Zinc 30	69.72 ^{2,8,18,3} **Ga** Gallium 31
107.9 ^{2,8,18,18,1} **Ag** Silver 47	112.4 ^{2,8,16,18,2} **Cd** Cadmium 48	114.8 ^{2,8,18,18,3} **In** Indium 49
197.0 ^{2,8,18,32,18,1} **Au** Gold 79	200.6 ^{2,8,18,32,18,2} **Hg** Mercury 80	204.4 ^{2,8,18,32,18,3} **Tl** Thallium 81

3) Tag Name: Diagram 0873

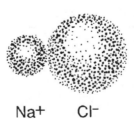

Na+   Cl−

Chapter 9

1) Item Type: True-False   Objective: 1   Section 9.1
      Level: A: Standard   Tag: None

     ____ Minerals are organic solids that are formed in the earth.

F

2) Item Type: True-False   Objective: 1   Section 9.1
      Level: A: Standard   Tag: None

     ____ A mineral is any solid, crystalline substance that occurs naturally.

F

3) Item Type: True-False   Objective: 1   Section 9.1
      Level: A: Standard   Tag: None

     ____ Oil and coal are classified as silicate minerals.

F

4) Item Type: True-False   Objective: 1   Section 9.1
      Level: A: Standard   Tag: None

     ____ All minerals occur naturally in the earth.

T

5) Item Type: True-False   Objective: 1   Section 9.1
      Level: A: Standard   Tag: None

     ____ Mineral mines are generally located deep underground.

T

6) Item Type: True-False   Objective: 1   Section 9.1
      Level: A: Standard   Tag: None

     ____ Mine inspectors investigate and report on mine accidents.

T

7) Item Type: True-False   Objective: 1   Section 9.1
      Level: A: Standard   Tag: None

     ____ Many minerals are liquids at room temperature.

F

8) Item Type: True-False   Objective: 2   Section 9.1
    Level: A: Standard   Tag: None

    ____   Silicate minerals make up more than 90 percent of the earth's crust.

    | T |

9) Item Type: True-False   Objective: 2   Section 9.1
    Level: A: Standard   Tag: None

    ____   All rock-forming minerals contain atoms of silicon (Si) and oxygen (O).

    | F |

10) Item Type: True-False   Objective: 2   Section 9.1
    Level: B: Enriched   Tag: None

    ____   Feldspars are minerals composed of atoms of silicon (Si), oxygen (O), and a metal.

    | T |

11) Item Type: True-False   Objective: 2   Section 9.1
    Level: B: Enriched   Tag: None

    ____   Plagioclase feldspar may contain sodium, calcium, or both.

    | T |

12) Item Type: True-False   Objective: 2   Section 9.1
    Level: A: Standard   Tag: None

    ____   Feldspars and quartz make up more than 50 percent of the earth's crust.

    | T |

13) Item Type: True-False   Objective: 3   Section 9.1
    Level: A: Standard   Tag: None

    ____   Sulfates and sulfides are types of nonsilicate minerals.

    | T |

14) Item Type: True-False   Objective: 3   Section 9.1
    Level: B: Enriched   Tag: None

    ____   Nonsilicate minerals can contain elements such as carbon, gold, and iron.

    | T |

Chapter 9

15) Item Type: True-False   Objective: 3   Section 9.1
    Level: A: Standard   Tag: None

    _____   All carbonates contain carbon and silicon atoms.

    F

16) Item Type: True-False   Objective: 3   Section 9.1
    Level: A: Standard   Tag: None

    _____   A compound that consists of chlorine combined with sodium is a halide.

    T

17) Item Type: True-False   Objective: 3   Section 9.1
    Level: B: Enriched   Tag: None

    _____   Sphalerite (ZnS) is a sulfate mineral.

    F

18) Item Type: True-False   Objective: 3   Section 9.1
    Level: A: Standard   Tag: None

    _____   One major group of nonsilicate minerals is the native elements.

    T

19) Item Type: True-False   Objective: 4   Section 9.1
    Level: B: Enriched   Tag: None

    _____   In silicate minerals, bonds may form between the oxygen atoms in the tetrahedron
            and other elements outside the tetrahedron.

    T

20) Item Type: True-False   Objective: 4   Section 9.1
    Level: A: Standard   Tag: None

    _____   The hardness of quartz is due primarily to bonds between oxygen and aluminum
            atoms.

    F

21) Item Type: True-False   Objective: 4   Section 9.1
    Level: A: Standard   Tag: None

    _____   The conditions under which minerals are produced usually allow single crystals to
            grow.

    F

22) Item Type: True-False   Objective: 4   Section 9.1
     Level: B: Enriched   Tag: None

____ In network silicates, each tetrahedron in the mineral is bonded to four other tetrahedra.

T

23) Item Type: True-False   Objective: 4   Section 9.1
     Level: A: Standard   Tag: None

____ A silicon-oxygen tetrahedron contains only one silicon atom.

T

24) Item Type: True-False   Objective: 5   Section 9.2
     Level: B: Enriched   Tag: None

____ Fluorite (CaF_2) is an example of a nonsilicate mineral.

T

25) Item Type: True-False   Objective: 5   Section 9.2
     Level: A: Standard   Tag: None

____ The streak test is a test of mineral density.

F

26) Item Type: True-False   Objective: 5   Section 9.2
     Level: A: Standard   Tag: None

____ Cleavage is important in identifying some minerals.

T

27) Item Type: True-False   Objective: 5   Section 9.2
     Level: A: Standard   Tag: None

____ Color is the most reliable indicator of the identity of a mineral.

F

28) Item Type: True-False   Objective: 5   Section 9.2
     Level: A: Standard   Tag: None

____ Mohs Scale is used to describe the density of minerals.

F

29) Item Type: True-False   Objective: 5   Section 9.2
     Level: A: Standard   Tag: None

_____ Cabochon-style gems are cut or ground to have faceted surfaces.

F

30) Item Type: True-False   Objective: 5   Section 9.2
     Level: A: Standard   Tag: None

_____ One of the six basic crystal shapes is called orthorhombic.

T

31) Item Type: True-False   Objective: 5   Section 9.2
     Level: A: Standard   Tag: None

_____ Large rocks have a higher density than small rocks.

F

32) Item Type: True-False   Objective: 5   Section 9.2
     Level: B: Enriched   Tag: None

_____ A hexagonal crystal system has three axes of the same length.

T

33) Item Type: True-False   Objective: 5   Section 9.2
     Level: A: Standard   Tag: None

_____ A mineral with a hardness of 5 on Mohs Scale will scratch a mineral with a hardness of 8.

F

34) Item Type: True-False   Objective: 6   Section 9.2
     Level: A: Standard   Tag: None

_____ Light rays are refracted or bent as they pass from one substance to a different substance.

T

35) Item Type: True-False   Objective: 6   Section 9.2
     Level: A: Standard   Tag: None

_____ Fluorescent minerals glow while being exposed to ultraviolet light.

T

36) Item Type: True-False   Objective: 6   Section 9.2
     Level: A: Standard   Tag: None

_____ All minerals have the same basic crystal shape.

F

37) Item Type: Multiple Choice   Objective: 1   Section 9.1
     Level: A: Standard   Tag: None

Which of the following is an example of a mineral?

a) coal   b) concrete   c) steel   d) quartz

d

38) Item Type: Multiple Choice   Objective: 1   Section 9.1
     Level: A: Standard   Tag: None

The main objective of a mining engineer is to

a) operate heavy machinery at a mine.   b) make mining safe and profitable.

c) locate valuable mineral deposits.   d) identify minerals found in mines.

b

39) Item Type: Multiple Choice   Objective: 2   Section 9.1
     Level: A: Standard   Tag: None

The ten most common minerals make up what percentage of the earth's crust?

a) 90%   b) 50%   c) 20%   d) 10%

a

40) Item Type: Multiple Choice   Objective: 2   Section 9.1
     Level: A: Standard   Tag: None

The two most abundant elements in common minerals are

a) calcium and manganese.   b) silicon and oxygen.

c) iron and magnesium.   d) carbon and potassium.

b

41) Item Type: Multiple Choice   Objective: 3   Section 9.1
     Level: A: Standard   Tag: None

The minerals gold and copper are examples of

a) silicate minerals.   b) organic compounds.

c) native elements.   d) radioactive substances.

c

42) Item Type: Multiple Choice   Objective: 3   Section 9.1
Level: B: Enriched   Tag: None

Which of the following is a nonsilicate mineral?

a) halite   b) quartz   c) feldspar   d) talc

a

43) Item Type: Multiple Choice   Objective: 3   Section 9.1
Level: A: Standard   Tag: None

Which of the following is one of the six major groups of nonsilicate minerals?

a) olivines   b) feldspars   c) pyroxenes   d) carbonates

d

44) Item Type: Multiple Choice   Objective: 4   Section 9.1
Level: A: Standard   Tag: None

How many oxygen atoms are in a silicon-oxygen tetrahedron?

a) 2   b) 3   c) 4   d) 5

c

45) Item Type: Multiple Choice   Objective: 4   Section 9.1
Level: A: Standard   Tag: None

Silicon-oxygen tetrahedra linked only by atoms of elements other than silicon and oxygen make up

a) sheet silicates.   b) chain silicates.

c) network silicates.   d) ionic silicates.

d

46) Item Type: Multiple Choice   Objective: 4   Section 9.1
Level: A: Standard   Tag: None

How many silicon atoms are in a basic silicon-oxygen tetrahedron?

a) 1   b) 2   c) 3   d) 4

a

47) Item Type: Multiple Choice   Objective: 4   Section 9.1
Level: B: Enriched   Tag: None

Minerals made up of single chains of Si-O tetrahedra are called

a) pyroxenes.   b) amphiboles.   c) crystals.   d) sheets.

a

48) Item Type: Multiple Choice   Objective: 4   Section 9.1
Level: B: Enriched   Tag: Diagram 0971

The structure presented in the diagram is an example of a

a) tetragonal crystal.   b) hexagonal crystal.

c) triclinic crystal.   d) cubic crystal.

d

49) Item Type: Multiple Choice   Objective: 4   Section 9.1
Level: B: Enriched   Tag: None

In network silicates each tetrahedron is bonded to how many other tetrahedra?

a) 4   b) 3   c) 2   d) 1

a

50) Item Type: Multiple Choice   Objective: 4   Section 9.1
Level: B: Enriched   Tag: None

Which of the following is made of single-chain tetrahedra?

a) halite   b) amphibole   c) pyroxene   d) carbonate

c

51) Item Type: Multiple Choice   Objective: 4   Section 9.1
Level: B: Enriched   Tag: None

Which of the following minerals is a network silicate?

a) olivine   b) halite   c) mica   d) feldspar

d

52) Item Type: Multiple Choice   Objective: 4   Section 9.1
Level: B: Enriched   Tag: None

Which of the following contain double chains of silicon-oxygen tetrahedra?

a) oxides   b) amphiboles   c) sulfides   d) pyroxenes

b

53) Item Type: Multiple Choice   Objective: 5   Section 9.2
Level: B: Enriched   Tag: Diagram 0973

According to the table above, which of the following minerals is white in powdered form?

a) garnet   b) pyrite   c) augite   d) sphalerite

a

54) Item Type: Multiple Choice   Objective: 5   Section 9.2
Level: A: Standard   Tag: None

The heft or relative weight of a mineral sample is directly related to the mineral's

a) luster.   b) cleavage.   c) density.   d) hardness.

c

55) Item Type: Multiple Choice   Objective: 5   Section 9.2
Level: B: Enriched   Tag: None

Diamond has which of the following types of luster?

a) metallic   b) brilliant   c) glassy   d) pearly

b

56) Item Type: Multiple Choice   Objective: 5   Section 9.2
Level: B: Enriched   Tag: Diagram 0975

According to the table, what is the approximate hardness of a mineral that scratches quartz and can be scratched by topaz?

a) 6.5   b) 7.5   c) 8.5   d) 9.5

b

57) Item Type: Multiple Choice   Objective: 5   Section 9.2
Level: B: Enriched   Tag: None

The properties of a mineral are primarily a result of its

a) hardness.   b) density.

c) chemical composition.   d) cleavage pattern.

c

58) Item Type: Multiple Choice   Objective: 5   Section 9.2
Level: B: Enriched   Tag: None

A mineral that splits into even sheets shows which of the following properties?

a) low density   b) consistent streak

c) good cleavage   d) triclinic crystal shape

c

59) Item Type: Multiple Choice   Objective: 5   Section 9.2
Level: B: Enriched   Tag: None

A mineral in which all atoms are bonded tightly together will most likely be

a) metallic.   b) hard.   c) dense.   d) magnetic.

b

60) Item Type: Multiple Choice   Objective: 5   Section 9.2
Level: B: Enriched   Tag: None

Ruby is a chromium-bearing variety of

a) corundum.   b) diamond.   c) feldspar.   d) quartz.

a

61) Item Type: Multiple Choice   Objective: 6   Section 9.2
Level: A: Standard   Tag: None

Needles of early compasses were often constructed of

a) lodestone.   b) calcite.   c) talc.   d) uranium.

a

62) Item Type: Multiple Choice   Objective: 6   Section 9.2
Level: A: Standard   Tag: None

The most common magnetic mineral is

a) hematite.   b) magnetite.   c) halite.   d) uranium.

b

63) Item Type: Multiple Choice   Objective: 6   Section 9.2
Level: A: Standard   Tag: None

The ability of a mineral to glow during and after exposure to ultraviolet light is called

a) phosphorescence. b) luminescence. c) fluorescence. d) radioactivity.

c

64) Item Type: Multiple Choice   Objective: 6   Section 9.2
Level: A: Standard   Tag: None

The release of subatomic particles and energy from an atom is called

a) magnetism.   b) fluorescence. c) radioactivity. d) phosphorescence.

c

65) Item Type: Multiple Choice   Objective: 6   Section 9.2
Level: B: Enriched   Tag: None

Which of the following is an ore containing unstable nuclei?

a) gypsum   b) feldspar   c) magnetite   d) pitchblende

d

66) Item Type: Multiple Choice   Objective: 6   Section 9.2
Level: A: Standard   Tag: None

Minerals that continue to glow after exposure to ultraviolet light are called

a) brilliant.   b) fluorescent.   c) radioactive.   d) phosphorescent.

d

67) Item Type: Fill in the Blank   Objective: 1   Section 9.1
Level: A: Standard   Tag: None

The two main types of minerals are _____.

silicate and nonsilicate

68) Item Type: Fill in the Blank   Objective: 2   Section 9.1
Level: B: Enriched   Tag: None

Plagioclase feldspar may contain both sodium and _____.

calcium

69) Item Type: Fill in the Blank   Objective: 2   Section 9.1
Level: A: Standard   Tag: None

The most common silicate mineral is _____.

feldspar

70) Item Type: Fill in the Blank   Objective: 2   Section 9.1
Level: A: Standard   Tag: None

Common minerals that make up the earth's crust are called _____.

rock-forming minerals

71) Item Type: Fill in the Blank   Objective: 2   Section 9.1
Level: A: Standard   Tag: None

All silicate minerals contain atoms of silicon and _____.

oxygen

72) Item Type: Fill in the Blank   Objective: 2   Section 9.1
     Level: A: Standard   Tag: None

The two main groups of minerals are silicate minerals and _____.

nonsilicate minerals

73) Item Type: Fill in the Blank   Objective: 2   Section 9.1
     Level: A: Standard   Tag: None

Orthoclase feldspar is formed when silicon and oxygen atoms combine with the metal

_____.

potassium

74) Item Type: Fill in the Blank   Objective: 3   Section 9.1
     Level: A: Standard   Tag: None

Elements that are commonly found uncombined with other elements are called

_____.

native elements

75) Item Type: Fill in the Blank   Objective: 4   Section 9.1
     Level: B: Enriched   Tag: None

The crystalline structure of pyroxenes consists of _____.

tetrahedral chains

76) Item Type: Fill in the Blank   Objective: 4   Section 9.1
     Level: A: Standard   Tag: None

In network silicates, how many of the oxygen atoms in the silicon-oxygen tetrahedron are

shared with other tetrahedra? _____

four

77) Item Type: Fill in the Blank   Objective: 5   Section 9.2
     Level: B: Enriched   Tag: None

The formula for calculating the density of a mineral is _____.

density = mass/volume

78) Item Type: Fill in the Blank   Objective: 6   Section 9.2
    Level: A: Standard   Tag: None

The bending of light rays as they pass through a mineral is referred to as

_____.

refraction

79) Item Type: Fill in the Blank   Objective: 6   Section 9.2
    Level: B: Enriched   Tag: None

What special property does the mineral pitchblende have? _____

radioactivity

80) Item Type: Essay   Objective: 4   Section 9.1
    Level: B: Enriched   Tag: None

Explain why quartz is harder than feldspar.

_____

_____

_____

_____

_____

Quartz is a network silicate containing only silicon-oxygen tetrahedra; the bonds between the tetrahedra are very strong, making quartz extremely hard. Feldspar is also a network silicate. Unlike quartz, however, some tetrahedra in feldspar have atoms of metal instead of silicon. The bonds between these atoms are weaker than those between silicon and oxygen. Therefore, feldspar is not as hard as quartz.

Chapter 9

1)   Tag Name: Diagram 0971

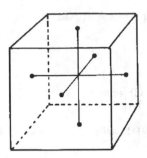

2)   Tag Name: Diagram 0973

PROPERTIES OF MINERALS

Mineral	Hardness	Streak	Color	Luster
augite	5–6	greenish-gray	green to black	nonmetallic
garnet	6.5–7.5	white	dark red	nonmetallic
magnetite	5–6	black	iron black	metallic
pyrite	6–6.5	green to black	brass yellow	metallic
sphalerite	3.5	reddish-brown	brown to black	nonmetallic

3)   Tag Name: Diagram 0975

Mineral	Hardness
feldspar	6
quartz	7
topaz	8
corundum	9

1) Item Type: True-False   Objective: 1   Section 10.1
    Level: A: Standard   Tag: None

        _____   The three types of rock are igneous, sedimentary, and extrusive.

    F

2) Item Type: True-False   Objective: 1   Section 10.1
    Level: A: Standard   Tag: None

        _____   Any one of the three types of rock can be changed into any other class.

    T

3) Item Type: True-False   Objective: 1   Section 10.1
    Level: A: Standard   Tag: None

        _____   Rocks are classified according to how they were formed.

    T

4) Item Type: True-False   Objective: 1   Section 10.1
    Level: A: Standard   Tag: None

        _____   Sedimentary rock is the parent material for all rocks.

    F

5) Item Type: True-False   Objective: 2   Section 10.1
    Level: A: Standard   Tag: None

        _____   Volcanism is the changing of one type of rock to another by heat, pressure, and
                chemical processes.

    F

6) Item Type: True-False   Objective: 2   Section 10.1
    Level: A: Standard   Tag: None

        _____   Igneous rock is produced when magma cools.

    T

7) Item Type: True-False   Objective: 2   Section 10.1
    Level: A: Standard   Tag: None

        _____   High temperatures can change sedimentary rock directly into magma.

    T

8)   Item Type: True-False   Objective: 2   Section 10.1
     Level: B: Enriched   Tag: None

____   Metamorphic rock may form from preexisting metamorphic rock.

T

9)   Item Type: True-False   Objective: 2   Section 10.1
     Level: A: Standard   Tag: None

____   According to the rock cycle, metamorphic rock forms only from sedimentary rock.

F

10) Item Type: True-False   Objective: 3   Section 10.2
     Level: A: Standard   Tag: None

____   A porphyry is an igneous rock with a mixture of small and large grains.

T

11) Item Type: True-False   Objective: 3   Section 10.2
     Level: A: Standard   Tag: None

____   Volcanic glass forms when magma cools very quickly.

T

12) Item Type: True-False   Objective: 3   Section 10.2
     Level: A: Standard   Tag: None

____   Magma that cools quickly produces rocks that have a fine-grained texture.

T

13) Item Type: True-False   Objective: 3   Section 10.2
     Level: A: Standard   Tag: None

____   Obsidian is an extrusive rock.

T

14) Item Type: True-False   Objective: 3   Section 10.2
     Level: A: Standard   Tag: None

____   In general, the faster magma cools, the smaller the resulting crystal size will be.

T

15) Item Type: True-False   Objective: 3   Section 10.2
    Level: A: Standard   Tag: None

_____ In general, the texture of intrusive igneous rocks is of a finer grain than that of extrusive igneous rocks.

F

16) Item Type: True-False   Objective: 4   Section 10.2
    Level: A: Standard   Tag: None

_____ Mafic rocks are light-colored rocks that are high in silica.

F

17) Item Type: True-False   Objective: 5   Section 10.2
    Level: A: Standard   Tag: None

_____ Laccoliths often form small, dome-shaped mountains.

T

18) Item Type: True-False   Objective: 5   Section 10.2
    Level: A: Standard   Tag: None

_____ Dikes are produced by volcanic activity.

T

19) Item Type: True-False   Objective: 5   Section 10.2
    Level: B: Enriched   Tag: None

_____ On the moon's surface, maria are formed by sedimentary rock.

F

20) Item Type: True-False   Objective: 5   Section 10.2
    Level: A: Standard   Tag: None

_____ A sill is an igneous structure that cuts across rock layers.

F

21) Item Type: True-False   Objective: 5   Section 10.2
    Level: A: Standard   Tag: None

_____ Batholiths are metamorphic rock structures.

F

22) Item Type: True-False   Objective: 5   Section 10.2
Level: A: Standard   Tag: None

_____ Dikes are often associated with volcanic necks.

T

23) Item Type: True-False   Objective: 5   Section 10.2
Level: B: Enriched   Tag: None

_____ The dark maria of the moon were created when lava flowed into meteor craters.

T

24) Item Type: True-False   Objective: 5   Section 10.2
Level: A: Standard   Tag: None

_____ Sills often form when magma flows between rock layers.

T

25) Item Type: True-False   Objective: 6   Section 10.3
Level: A: Standard   Tag: None

_____ Limestone can form as a chemical sedimentary rock or as an organic sedimentary rock.

T

26) Item Type: True-False   Objective: 6   Section 10.3
Level: A: Standard   Tag: None

_____ Conglomerates are formed from angular, gravel-sized fragments with sharp corners.

F

27) Item Type: True-False   Objective: 6   Section 10.3
Level: A: Standard   Tag: None

_____ Some sedimentary rocks are formed from the remains of organisms.

T

28) Item Type: True-False   Objective: 6   Section 10.3
Level: A: Standard   Tag: None

_____ Shale consists of clay-sized particles that are pressed into flat layers.

T

29) Item Type: True-False   Objective: 6   Section 10.3
     Level: A: Standard   Tag: None

_____ Stratification occurs when the same particle type is deposited over a long period of time.

F

30) Item Type: True-False   Objective: 6   Section 10.3
     Level: A: Standard   Tag: None

_____ Halite is a clastic sedimentary rock.

F

31) Item Type: True-False   Objective: 6   Section 10.3
     Level: A: Standard   Tag: None

_____ Organic sedimentary rock is composed of the remains of decaying organisms.

T

32) Item Type: True-False   Objective: 7   Section 10.3
     Level: B: Enriched   Tag: None

_____ Fossils are commonly found in gabbro.

F

33) Item Type: True-False   Objective: 7   Section 10.3
     Level: A: Standard   Tag: None

_____ Cavities in sandstone that contain quartz or calcite crystals are called geodes.

T

34) Item Type: True-False   Objective: 7   Section 10.3
     Level: A: Standard   Tag: None

_____ Fossils are most commonly preserved in sedimentary rock.

T

35) Item Type: True-False   Objective: 8   Section 10.4
     Level: A: Standard   Tag: None

_____ Metamorphism that occurs over an area of thousands of square kilometers during periods of tectonic activity is called contact metamorphism.

F

36) Item Type: True-False   Objective: 8   Section 10.4
     Level: B: Enriched   Tag: None

_____ Metamorphism can change the final composition of rock.

T

37) Item Type: True-False   Objective: 8   Section 10.4
     Level: A: Standard   Tag: None

_____ Most metamorphic rock forms near the earth's surface.

F

38) Item Type: True-False   Objective: 8   Section 10.4
     Level: B: Enriched   Tag: None

_____ During metamorphism, minerals may change in size.

T

39) Item Type: True-False   Objective: 8   Section 10.4
     Level: A: Standard   Tag: None

_____ Most metamorphic rock is formed by tectonic activity.

T

40) Item Type: True-False   Objective: 8   Section 10.4
     Level: A: Standard   Tag: None

_____ Metamorphism results from heat, pressure, and chemical processes.

T

41) Item Type: True-False   Objective: 8   Section 10.4
     Level: A: Standard   Tag: None

_____ Metamorphism can create a hard and durable rock from a softer sedimentary rock.

T

42) Item Type: True-False   Objective: 9   Section 10.4
     Level: A: Standard   Tag: None

_____ Unfoliated metamorphic rocks do not contain layers of crystals.

T

43) Item Type: Multiple Choice   Objective: 1   Section 10.1
Level: A: Standard   Tag: None

The parent material for all rocks is

a) silica.   b) quartz.   c) granite.   d) magma.

d

44) Item Type: Multiple Choice   Objective: 1   Section 10.1
Level: A: Standard   Tag: None

Metamorphism is best defined as the

a) compaction and cementation of rock fragments.

b) precipitation of minerals dissolved in water.

c) solidification of magma by cooling.

d) changing of a rock by heat and pressure.

d

45) Item Type: Multiple Choice   Objective: 1   Section 10.1
Level: A: Standard   Tag: None

The processes responsible for changing sediments into sedimentary rock are compaction and

a) foliation.   b) cementation.   c) intrusion.   d) stratification.

b

46) Item Type: Multiple Choice   Objective: 1   Section 10.1
Level: A: Standard   Tag: None

Which of the following types of rock is produced by magma that cools deep below the earth's crust?

a) extrusive igneous   b) intrusive igneous

c) foliated metamorphic   d) chemical sedimentary

b

47) Item Type: Multiple Choice   Objective: 1   Section 10.1
Level: A: Standard   Tag: None

Which of the following is an igneous rock?

a) limestone   b) gypsum   c) gneiss   d) basalt

d

48) Item Type: Multiple Choice   Objective: 2   Section 10.1
Level: A: Standard   Tag: None

Which of the following describes the process by which sedimentary rock becomes metamorphic rock?

a) weathering   b) erosion

c) intense heat and pressure   d) cooling and solidifying

c

49) Item Type: Multiple Choice   Objective: 3   Section 10.2
Level: A: Standard   Tag: None

The size and arrangement of crystalline grains in igneous rock is called

a) density.   b) texture.   c) hardness.   d) luster.

b

50) Item Type: Multiple Choice   Objective: 3   Section 10.2
Level: A: Standard   Tag: None

Which of the following is an indication of crystal size in a mineral?

a) hardness   b) texture   c) angularity   d) luster

b

51) Item Type: Multiple Choice   Objective: 4   Section 10.2
Level: A: Standard   Tag: None

An igneous rock with a mixture of large and small grains is

a) a porphyry.   b) an intrusion.   c) an extrusion.   d) a breccia.

a

52) Item Type: Multiple Choice   Objective: 4   Section 10.2
Level: B: Enriched   Tag: None

Which of the following is a fine-grained member of the granite family?

a) andesite   b) diorite   c) rhyolite   d) marble

c

53) Item Type: Multiple Choice   Objective: 4   Section 10.2
Level: B: Enriched   Tag: None

Gabbro is most similar to which of the following rocks?

a) basalt   b) obsidian   c) granite   d) diorite

a

54) Item Type: Multiple Choice   Objective: 4   Section 10.2
Level: B: Enriched   Tag: None

Which of the following would most likely form from a mafic magma that cooled slowly underground?

a) basalt   b) gabbro   c) obsidian   d) pumice

b

55) Item Type: Multiple Choice   Objective: 4   Section 10.2
Level: B: Enriched   Tag: None

Which of the following rocks is a member of the diorite family?

a) andesite   b) gabbro   c) obsidian   d) granite

a

56) Item Type: Multiple Choice   Objective: 4   Section 10.2
Level: B: Enriched   Tag: None

Gabbro is an example of a

a) felsic, fine-grained rock.   b) felsic, coarse-grained rock.

c) mafic, fine-grained rock.   d) mafic, coarse-grained rock.

d

57) Item Type: Multiple Choice   Objective: 4   Section 10.2
Level: A: Standard   Tag: None

Which of the following is one of the three families of igneous rocks?

a) clastic   b) foliated   c) gabbro   d) diorite

d

58) Item Type: Multiple Choice   Objective: 4   Section 10.2
Level: A: Standard   Tag: None

Felsic rocks are high in

a) quartz.   b) silica.   c) biotite.   d) calcite.

b

59) Item Type: Multiple Choice   Objective: 5   Section 10.2
Level: A: Standard   Tag: None

Which of the following form the core of many major mountain ranges?

a) conglomerates   b) concretions   c) batholiths   d) extrusions

c

60) Item Type: Multiple Choice   Objective: 5   Section 10.2
Level: A: Standard   Tag: None

Which of the following is most likely to form the core of a mountain range?

a) sill   b) dike   c) laccolith   d) batholith

d

61) Item Type: Multiple Choice   Objective: 5   Section 10.2
Level: B: Enriched   Tag: Diagram 1072

In the diagram above, the feature labeled **1** is an example of a

a) lava plateau.   b) laccolith.   c) geode.   d) batholith.

b

62) Item Type: Multiple Choice   Objective: 5   Section 10.2
Level: B: Enriched   Tag: Diagram 1072

In the diagram above, the feature labeled **2** is an example of a

a) sill.   b) dike.   c) plug.   d) neck.

a

63) Item Type: Multiple Choice   Objective: 5   Section 10.2
Level: B: Enriched   Tag: Diagram 1073

The structure labeled **X** in the diagram above is a

a) sill.   b) dike.   c) stock.   d) batholith.

a

64) Item Type: Multiple Choice   Objective: 5   Section 10.2
Level: B: Enriched   Tag: Diagram 1073

The structure labeled **Y** in the diagram above is a

a) sill.   b) dike.   c) laccolith.   d) batholith.

b

65) Item Type: Multiple Choice   Objective: 5   Section 10.2
Level: B: Enriched   Tag: None

Magma that cools deep below the earth's crust forms what type of rock?

a) clastic   b) intrusive   c) stratified   d) extrusive

b

66) Item Type: Multiple Choice   Objective: 5   Section 10.2
    Level: A: Standard   Tag: None

The structure formed when magma flows out through cracks in the earth's surface and spreads out over a large area is called a

a) volcanic neck.   b) sill.   c) laccolith.   d) lava plateau.

d

67) Item Type: Multiple Choice   Objective: 6   Section 10.3
    Level: B: Enriched   Tag: None

The white cliffs of Dover, England, are composed of

a) coal.   b) halite.   c) basalt.   d) chalk.

d

68) Item Type: Multiple Choice   Objective: 6   Section 10.3
    Level: B: Enriched   Tag: None

To which of the following groups does breccia belong?

a) foliated metamorphic   b) unfoliated metamorphic

c) clastic sedimentary   d) chemical sedimentary

c

69) Item Type: Multiple Choice   Objective: 6   Section 10.3
    Level: A: Standard   Tag: None

Which of the following is an organic sedimentary rock?

a) basalt   b) coal   c) conglomerate   d) sandstone

b

70) Item Type: Multiple Choice   Objective: 6   Section 10.3
    Level: A: Standard   Tag: None

Chalk is a type of

a) coal.   b) quartzite.   c) limestone.   d) shale.

c

71) Item Type: Multiple Choice   Objective: 6   Section 10.3
    Level: A: Standard   Tag: None

A rock composed of cemented, rounded pebbles is a

a) porphyry.   b) shale.   c) breccia.   d) conglomerate.

d

72) Item Type: Multiple Choice   Objective: 7   Section 10.3
    Level: A: Standard   Tag: None

The layering of sedimentary rock with coarse grains at the bottom and fine grains at the top is

a) foliation.   b) concretion.   c) graded bedding.   d) cross-bedding.

c

73) Item Type: Multiple Choice   Objective: 7   Section 10.3
    Level: A: Standard   Tag: None

Ripple marks in sandstone may form by the action of

a) wind.   b) magma.   c) heat.   d) intrusion.

a

74) Item Type: Multiple Choice   Objective: 7   Section 10.3
    Level: A: Standard   Tag: None

Ripple marks in rocks are formed by

a) moving wind or water.   b) drying and shrinking.

c) heating and cooling.   d) contact with magma.

a

75) Item Type: Multiple Choice   Objective: 8   Section 10.4
    Level: B: Enriched   Tag: None

Regional metamorphism occurs as a result of

a) tectonic activity.   b) volcanic eruptions.

c) earthquakes.   d) sedimentation.

a

76) Item Type: Multiple Choice   Objective: 8   Section 10.4
    Level: B: Enriched   Tag: None

The changing of rock by pressure from colliding tectonic plates is called

a) igneous intrusion.   b) igneous extrusion.

c) contact metamorphism.   d) regional metamorphism.

d

77) Item Type: Multiple Choice   Objective: 8   Section 10.4
    Level: A: Standard   Tag: None

Where does most metamorphic rock form?

a)   deep below the earth's surface   b)   within volcanoes

c)   on the earth's surface   d)   on lake beds

a

78) Item Type: Multiple Choice   Objective: 9   Section 10.4
    Level: A: Standard   Tag: None

Which of the following is classified as a metamorphic rock?

a) basalt   b) diorite   c) limestone   d) schist

d

79) Item Type: Fill in the Blank   Objective: 1   Section 10.1
    Level: A: Standard   Tag: None

Rocks are classified into three groups based on how the rocks are

_____.

formed

80) Item Type: Fill in the Blank   Objective: 2   Section 10.1
    Level: B: Enriched   Tag: Diagram 1071

The type of rock represented by rectangle **A** in the diagram is _____.

igneous

81) Item Type: Fill in the Blank   Objective: 2   Section 10.1
    Level: B: Enriched   Tag: Diagram 1071

In the diagram above, the arrow labeled **B** represents heat and _____.

pressure

82) Item Type: Fill in the Blank   Objective: 2   Section 10.1
    Level: A: Standard   Tag: None

Rocks are changed from one type to another in a series of changes called the rock

_____.

cycle

83) Item Type: Fill in the Blank   Objective: 3   Section 10.2
    Level: A: Standard   Tag: None

The texture of igneous rock is determined by the rate at which magma

_____.

cools

84) Item Type: Fill in the Blank   Objective: 3   Section 10.2
    Level: A: Standard   Tag: None

Obsidian forms through the extremely rapid cooling of magma, which prevents the

formation of _____.

crystals

85) Item Type: Fill in the Blank   Objective: 4   Section 10.2
    Level: A: Standard   Tag: None

Igneous rocks that are dark colored and contain little silica are called

_____.

mafic

86) Item Type: Fill in the Blank   Objective: 4   Section 10.2
    Level: A: Standard   Tag: None

Igneous rocks that are light colored and are high in silica are called

_____.

felsic

87) Item Type: Fill in the Blank   Objective: 5   Section 10.2
    Level: B: Enriched   Tag: None

A batholith-like structure that covers an area less than 100 km^2 is called a(n)

_____.

stock

88) Item Type: Fill in the Blank   Objective: 7   Section 10.3
     Level: A: Standard   Tag: None

Remains or traces of ancient plants and animals that are preserved in rock are called

_____.

| fossils |

89) Item Type: Fill in the Blank   Objective: 7   Section 10.3
     Level: A: Standard   Tag: None

When sediment deposition occurs in curved slopes, the inclined layers are said to be

_____.

| cross-bedded |

90) Item Type: Fill in the Blank   Objective: 7   Section 10.3
     Level: B: Enriched   Tag: None

Nodules that form when minerals precipitated from solutions build up around existing rock

particles are called _____.

| concretions |

91) Item Type: Fill in the Blank   Objective: 7   Section 10.3
     Level: A: Standard   Tag: None

A sedimentary rock that contains quartz or calcite crystals in a hollow core is a(n)

_____.

| geode |

92) Item Type: Fill in the Blank   Objective: 7   Section 10.3
     Level: B: Enriched   Tag: Diagram 1074

The crystal-filled rock in the diagram is a(n) _____.

| geode |

93) Item Type: Fill in the Blank   Objective: 8   Section 10.4
     Level: B: Enriched   Tag: None

Quartzite is produced by the metamorphism of _____.

| sandstone |

94) Item Type: Fill in the Blank   Objective: 8   Section 10.4
   Level: A: Standard   Tag: None

The type of metamorphism that results from the heat of magma is called

_____.

| contact metamorphism ■ |

95) Item Type: Fill in the Blank   Objective: 8   Section 10.4
   Level: A: Standard   Tag: None

Metamorphism that occurs over large areas is called _____.

| regional metamorphism ■ |

96) Item Type: Fill in the Blank   Objective: 9   Section 10.4
   Level: A: Standard   Tag: Diagram 1075

The appearance of the metamorphic rock in the diagram indicates that it should be

classified as _____.

| foliated ■ |

97) Item Type: Fill in the Blank   Objective: 9   Section 10.4
   Level: A: Standard   Tag: None

Slate is formed when great pressure acts on the sedimentary rock _____.

| shale ■ |

98) Item Type: Essay   Objective: 2   Section 10.1
   Level: B: Enriched   Tag: None

Explain how igneous rock can change into sedimentary rock.

_____

_____

_____

_____

_____

| As igneous rock is worn away and broken down by agents such as water and wind, rock fragments form. These fragments can then be compacted and cemented to form sedimentary rock. ■ |

99) Item Type: Essay   Objective: 3   Section 10.2
    Level: B: Enriched   Tag: None

Why do most extrusive igneous rocks have small mineral crystals?

_____

_____

_____

_____

_____

The size of mineral crystals in igneous rock depends largely on how quickly or slowly the magma cools. Extrusive rock forms when magma cools on the earth's surface. Since surface temperatures are relatively cool, the magma cools rapidly, which prevents the formation of large crystals.

100) Item Type: Essay   Objective: 9   Section 10.4
    Level: A: Standard   Tag: None

How may foliation occur in metamorphic rock?

_____

_____

_____

_____

Foliation may occur in one of two ways. Pressure may flatten the mineral crystals in the original rock, squeezing them into parallel bands. Foliation may also occur as minerals of different densities separate into bands.

Chapter 10

1) Tag Name: Diagram 1071

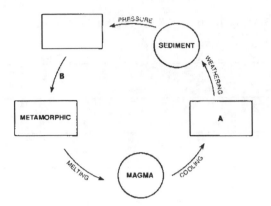

2) Tag Name: Diagram 1072

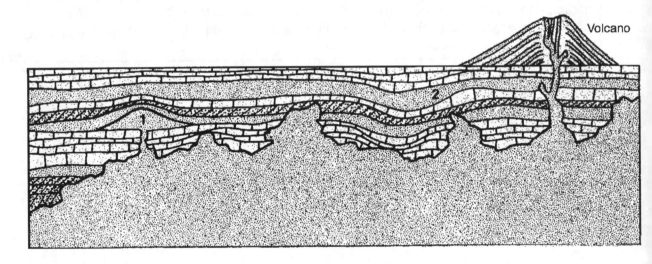

3) Tag Name: Diagram 1073

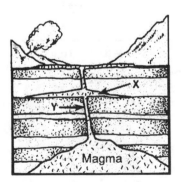

4)   Tag Name: Diagram 1074

5)   Tag Name: Diagram 1075

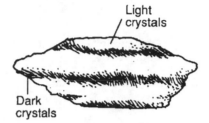

Chapter 11

1) Item Type: True-False   Objective: 1   Section 11.1
Level: A: Standard   Tag: None

_____ Placer deposits are formed by contact metamorphism.

F

2) Item Type: True-False   Objective: 1   Section 11.1
Level: A: Standard   Tag: None

_____ A lode is a layer of minerals that has been deposited in a stream bed.

F

3) Item Type: True-False   Objective: 1   Section 11.1
Level: A: Standard   Tag: None

_____ Nuclear fuel is a nonrenewable resource.

T

4) Item Type: True-False   Objective: 1   Section 11.1
Level: A: Standard   Tag: None

_____ Renewable resources are resources that are replaced faster than they are used.

T

5) Item Type: True-False   Objective: 1   Section 11.1
Level: A: Standard   Tag: None

_____ Placer deposits are most likely to form in turbulent river rapids.

F

6) Item Type: True-False   Objective: 2   Section 11.1
Level: A: Standard   Tag: None

_____ Halite, sulfur, and diamond are nonmetallic minerals used as building materials.

F

7) Item Type: True-False   Objective: 2   Section 11.1
Level: A: Standard   Tag: None

_____ The ocean floor is a largely untapped source of mineral resources.

T

8) Item Type: True-False   Objective: 2   Section 11.1
    Level: A: Standard   Tag: None

    _____ Minerals are renewable resources.

    | F | ■ |

9) Item Type: True-False   Objective: 2   Section 11.1
    Level: A: Standard   Tag: None

    _____ Air and water are nonrenewable resources.

    | F | ■ |

10) Item Type: True-False   Objective: 2   Section 11.1
    Level: A: Standard   Tag: None

    _____ Mercury and quartz are examples of gemstones.

    | F | ■ |

11) Item Type: True-False   Objective: 3   Section 11.2
    Level: A: Standard   Tag: None

    _____ Anthracite is a light-colored mineral commonly used as a gemstone.

    | F | ■ |

12) Item Type: True-False   Objective: 3   Section 11.2
    Level: A: Standard   Tag: None

    _____ Coal is produced by the process of carbonization.

    | T | ■ |

13) Item Type: True-False   Objective: 4   Section 11.2
    Level: A: Standard   Tag: None

    _____ Microorganisms that lived in ancient oceans and lakes are an important source of
    hydrocarbons.

    | T | ■ |

14) Item Type: True-False   Objective: 4   Section 11.2
    Level: A: Standard   Tag: None

    _____ Natural gas is formed from the remains of living things.

    | T | ■ |

15) Item Type: True-False   Objective: 4   Section 11.2
    Level: A: Standard   Tag: None

_____ Natural gas is often found above petroleum reservoirs.

T

16) Item Type: True-False   Objective: 4   Section 11.2
    Level: A: Standard   Tag: None

_____ Rock through which liquids cannot flow is called impermeable rock.

T

17) Item Type: True-False   Objective: 5   Section 11.2
    Level: A: Standard   Tag: None

_____ Petroleum is used to produce plastics, detergents, and some medicines and
insecticides.

T

18) Item Type: True-False   Objective: 5   Section 11.2
    Level: A: Standard   Tag: None

_____ Synthetic rubber is commonly made from bauxite.

F

19) Item Type: True-False   Objective: 6   Section 11.2
    Level: A: Standard   Tag: None

_____ Fossil fuels are available in limitless supplies.

F

20) Item Type: True-False   Objective: 7   Section 11.2
    Level: A: Standard   Tag: None

_____ Sulfur dioxide is a pollutant that is commonly produced when fossil fuels are
burned.

T

21) Item Type: True-False   Objective: 7   Section 11.2
    Level: A: Standard   Tag: None

_____ Sulfur dioxide is often released into the atmosphere when coal is burned.

T

Chapter 11

22) Item Type: True-False   Objective: 8   Section 11.3
    Level: B: Enriched   Tag: None

____ Fuel rods for nuclear reactors contain two isotopes of uranium.

T

23) Item Type: True-False   Objective: 8   Section 11.3
    Level: A: Standard   Tag: None

____ The process of splitting the nucleus of a large atom into two or more smaller nuclei
is called nuclear fission.

T

24) Item Type: True-False   Objective: 8   Section 11.3
    Level: B: Enriched   Tag: None

____ Uranium-238 is the radioactive isotope used for nuclear fission.

F

25) Item Type: True-False   Objective: 8   Section 11.3
    Level: A: Standard   Tag: None

____ Nuclear fuel produces radiation that is harmful to living cells.

T

26) Item Type: True-False   Objective: 9   Section 11.3
    Level: A: Standard   Tag: None

____ Nuclear fusion is the combining of atomic nuclei.

T

27) Item Type: True-False   Objective: 11   Section 11.4
    Level: A: Standard   Tag: None

____ Energy obtained directly from heat in the earth's crust is called geothermal energy.

T

28) Item Type: True-False   Objective: 11   Section 11.4
    Level: A: Standard   Tag: None

____ Technology for using geothermal energy is not yet available.

F

29) Item Type: True-False   Objective: 12   Section 11.4
    Level: A: Standard   Tag: None

    _____   All of the world's energy needs could easily be met by modern wind-driven generators.

    F

30) Item Type: Multiple Choice   Objective: 1   Section 11.1
    Level: A: Standard   Tag: None

    Which of the following is a native element?

    a) chalcopyrite   b) copper   c) bauxite   d) hydrocarbon

    b

31) Item Type: Multiple Choice   Objective: 1   Section 11.1
    Level: B: Enriched   Tag: None

    Minerals that have shiny surfaces and are good conductors of electricity are

    a) ores.   b) metals.   c) lodes.   d) gemstones.

    b

32) Item Type: Multiple Choice   Objective: 1   Section 11.1
    Level: A: Standard   Tag: Diagram 1171

    At which location in the diagram would a placer deposit most likely be found?

    a) *A*   b) *B*   c) *C*   d) *D*

    a

33) Item Type: Multiple Choice   Objective: 1   Section 11.1
    Level: A: Standard   Tag: None

    Hot mineral solutions flowing through cracks in rocks are most likely to produce

    a) veins.   b) placer deposits.   c) peat.   d) crude oil.

    a

34) Item Type: Multiple Choice   Objective: 1   Section 11.1
    Level: B: Enriched   Tag: None

    Magnetite and hematite are important sources of

    a) mercury.   b) lead.   c) iron.   d) copper.

    c

35) Item Type: Multiple Choice   Objective: 1   Section 11.1
Level: B: Enriched   Tag: None

Which of the following minerals is obtained from bauxite?

a) gold   b) aluminum   c) platinum   d) sulfur

b

36) Item Type: Multiple Choice   Objective: 1   Section 11.1
Level: B: Enriched   Tag: None

Contact metamorphism is caused primarily by

a) heat.   b) rain.   c) radioactivity.   d) wind.

a

37) Item Type: Multiple Choice   Objective: 1   Section 11.1
Level: A: Standard   Tag: None

Placer deposits form at the bottom of

a) stream beds.   b) oceans.   c) mountains.   d) magma bodies.

a

38) Item Type: Multiple Choice   Objective: 2   Section 11.1
Level: A: Standard   Tag: None

Which of the following minerals is often an ingredient in food?

a) gypsum   b) sphalerite   c) halite   d) hematite

c

39) Item Type: Multiple Choice   Objective: 2   Section 11.1
Level: B: Enriched   Tag: None

Which of the following is a source of glass?

a) bauxite   b) sphalerite   c) calcite   d) quartz

d

40) Item Type: Multiple Choice   Objective: 2   Section 11.1
Level: A: Standard   Tag: None

Nonmetallic minerals prized for their brilliance and color are called

a) metals.   b) placers.   c) ores.   d) gemstones.

d

41) Item Type: Multiple Choice   Objective: 2   Section 11.1
Level: A: Standard   Tag: None

In general, the most efficient conductors of heat and electricity are

a) metals.   b) petrochemicals.   c) gemstones.   d) ores.

a

42) Item Type: Multiple Choice   Objective: 2   Section 11.1
Level: A: Standard   Tag: None

Recycling is a means of

a) conserving minerals.   b) mining minerals.

c) locating minerals.   d) identifying minerals.

a

43) Item Type: Multiple Choice   Objective: 2   Section 11.1
Level: B: Enriched   Tag: None

Hematite is a source of

a) iron.   b) aluminum.   c) zinc.   d) diamonds.

a

44) Item Type: Multiple Choice   Objective: 2   Section 11.1
Level: A: Standard   Tag: None

Which of the following is a petrochemical product?

a) kaolinite   b) anthracite   c) galena   d) insecticide

d

45) Item Type: Multiple Choice   Objective: 3   Section 11.2
Level: B: Enriched   Tag: None

Which of the following is produced by the process of carbonization?

a) gold   b) coal   c) quartz   d) cinnabar

b

46) Item Type: Multiple Choice   Objective: 3   Section 11.2
Level: A: Standard   Tag: None

The process of coal formation began in ancient

a) magma bodies.   b) oceans.   c) swamps.   d) caves.

c

47) Item Type: Multiple Choice   Objective: 3   Section 11.2
    Level: B: Enriched   Tag: None

Bituminous coal is produced when extreme pressure is applied to

a) lignite.   b) anthracite.   c) natural gas.   d) crude oil.

a

48) Item Type: Multiple Choice   Objective: 3   Section 11.2
    Level: A: Standard   Tag: None

The hardest form of coal is

a) peat.   b) anthracite.   c) lignite.   d) bituminous coal.

b

49) Item Type: Multiple Choice   Objective: 3   Section 11.2
    Level: B: Enriched   Tag: None

Which of the following gases are produced during the formation of coal?

a) helium and nitrogen   b) methane and carbon dioxide

c) propane and oxygen   d) carbon monoxide and argon

b

50) Item Type: Multiple Choice   Objective: 3   Section 11.2
    Level: B: Enriched   Tag: None

Anthracite is a type of

a) geothermal energy source.   b) organic rock.

c) gemstone.   d) microorganism.

b

51) Item Type: Multiple Choice   Objective: 4   Section 11.2
    Level: B: Enriched   Tag: None

Petroleum deposits are most often associated with

a) synclines.   b) anticlines.   c) faults.   d) volcanoes.

b

52) Item Type: Multiple Choice   Objective: 4   Section 11.2
    Level: B: Enriched   Tag: Diagram 1172

In the diagram above, the layer labeled *X* is most likely composed of

a) lignite.   b) sandstone.   c) bauxite.   d) shale.

d

53) Item Type: Multiple Choice   Objective: 4   Section 11.2
    Level: B: Enriched   Tag: Diagram 1172

In the diagram above, which of the following is most likely to be found in the layer labeled **Y**?

a) gasoline   b) water   c) natural gas   d) peat

c

54) Item Type: Multiple Choice   Objective: 4   Section 11.2
    Level: B: Enriched   Tag: Diagram 1173

The structure labeled **X** in the diagram above is a

a) fault.   b) cap rock.   c) vein.   d) lode.

b

55) Item Type: Multiple Choice   Objective: 4   Section 11.2
    Level: A: Standard   Tag: None

Which of the following is produced by the accumulation of microorganisms on the ocean floor?

a) natural gas   b) gold deposits   c) uranium   d) coal

a

56) Item Type: Multiple Choice   Objective: 5   Section 11.2
    Level: B: Enriched   Tag: None

Most electricity in the United States is produced from

a) fossil fuels.   b) nuclear fuels.

c) solar energy.   d) hydroelectric energy.

a

57) Item Type: Multiple Choice   Objective: 5   Section 11.2
    Level: A: Standard   Tag: None

Crude oil is an important source of

a) geothermal energy.   b) petrochemicals.   c) cinnabar.   d) anthracite.

b

58) Item Type: Multiple Choice   Objective: 5   Section 11.2
    Level: A: Standard   Tag: None

Plastics, chemical fertilizers, shampoos, and synthetic fabrics are all produced from

a) metal ores.   b) natural gas.   c) coal.   d) petroleum.

d

59) Item Type: Multiple Choice   Objective: 6   Section 11.2
     Level: B: Enriched   Tag: None

According to scientists, what percentage of the petroleum in the United States has already been discovered?

a) 5%   b) 25%   c) 50%   d) 75%

d

60) Item Type: Multiple Choice   Objective: 6   Section 11.2
     Level: B: Enriched   Tag: None

Scientists estimate that at the present rate of use, the world's coal reserves will last approximately

a) 3,000 years.   b) 200 years.   c) 50 years.   d) 5 years.

b

61) Item Type: Multiple Choice   Objective: 7   Section 11.2
     Level: A: Standard   Tag: None

A serious environmental problem results from burning coal that contains a large amount of

a) hydrogen.   b) carbon.   c) oxygen.   d) sulfur.

d

62) Item Type: Multiple Choice   Objective: 7   Section 11.2
     Level: B: Enriched   Tag: None

Strip mining often causes harm to the environment by exposing

a) acidic rocks.   b) radioactive elements.

c) natural gas.   d) sulfur compounds.

a

63) Item Type: Multiple Choice   Objective: 8   Section 11.3
     Level: B: Enriched   Tag: None

Fission reactors are often cooled by circulating

a) air.   b) oxygen.   c) water.   d) oil.

c

64) Item Type: Multiple Choice   Objective: 8   Section 11.3
Level: B: Enriched   Tag: None

In nuclear fission reactors, nuclei are split by fast-moving

a) electrons.   b) neutrons.   c) uranium atoms.   d) hydrogen atoms.

b

65) Item Type: Multiple Choice   Objective: 8   Section 11.3
Level: A: Standard   Tag: None

Which of the following is an important product of fission reactions?

a) sound energy   b) electrical energy   c) chemical energy   d) heat energy

d

66) Item Type: Multiple Choice   Objective: 8   Section 11.3
Level: A: Standard   Tag: None

In nuclear reactors, electricity is produced when the turbines of generators are turned by

a) neutrons.   b) steam.   c) protons.   d) fuel pellets.

b

67) Item Type: Multiple Choice   Objective: 8   Section 11.3
Level: A: Standard   Tag: None

Fuel rods in nuclear reactors are made up of pellets of

a) uranium.   b) chromium.   c) platinum.   d) magnesium.

a

68) Item Type: Multiple Choice   Objective: 8   Section 11.3
Level: B: Enriched   Tag: None

A common method used to dispose of nuclear waste is to store it in

a) cement vaults.   b) mountain caves.
c) salt mines.   d) coal mines.

c

69) Item Type: Multiple Choice   Objective: 8   Section 11.3
Level: A: Standard   Tag: None

The fuel for nuclear fusion reactors is

a) plutonium.   b) chromium.   c) hydrogen.   d) fossil fuels.

c

70) Item Type: Multiple Choice   Objective: 8   Section 11.3
Level: B: Enriched   Tag: None

Which of the following energy sources converts one element into another element?

a) active solar   b) hydroelectric   c) nuclear fission   d) geothermal

c

71) Item Type: Multiple Choice   Objective: 8   Section 11.3
Level: B: Enriched   Tag: None

One advantage of nuclear fusion over nuclear fission as an energy source is that nuclear fission

a) is easily harnessed.   b) produces less harmful waste.

c) requires no fuel.   d) does not generate heat.

b

72) Item Type: Multiple Choice   Objective: 9   Section 11.3
Level: B: Enriched   Tag: None

Which of the following is a potential source of fuel for nuclear fusion reactions?

a) water   b) coal   c) uranium   d) petroleum

a

73) Item Type: Multiple Choice   Objective: 9   Section 11.3
Level: B: Enriched   Tag: Diagram 1174

Which of the following reactions is represented by this diagram?

a) decomposition   b) carbonation   c) nuclear fission   d) nuclear fusion

d

74) Item Type: Multiple Choice   Objective: 9   Section 11.3
Level: A: Standard   Tag: None

The sun produces its high surface temperature through which type of energy?

a) geothermal   b) nuclear fusion   c) hydroelectric   d) nuclear fission

b

75) Item Type: Multiple Choice   Objective: 10   Section 11.4
Level: A: Standard   Tag: None

Using a rooftop collector to capture the sun's energy is an example of

a) nuclear fusion.

b) nuclear fission.

c) active solar heating.

d) passive solar heating.

c

76) Item Type: Multiple Choice   Objective: 10   Section 11.4
Level: A: Standard   Tag: None

Sunlight can be focused onto liquid saltpeter in order to make use of

a) nuclear energy.

b) geothermal energy.

c) solar energy.

d) hydroelectric energy.

c

77) Item Type: Multiple Choice   Objective: 11   Section 11.4
Level: B: Enriched   Tag: None

Power plants using geothermal energy are used extensively in

a) England.   b) Iceland.   c) Mexico.   d) Canada.

b

78) Item Type: Multiple Choice   Objective: 11   Section 11.4
Level: A: Standard   Tag: None

Water that is heated as it flows through hot rocks is a source of

a) geothermal energy.

b) fossil fuels.

c) solar energy.

d) hydroelectric energy.

a

79) Item Type: Multiple Choice   Objective: 11   Section 11.4
Level: A: Standard   Tag: None

The energy source for geothermal energy is

a) magma.   b) the sun.   c) wind.   d) radioactive ores.

a

80) Item Type: Multiple Choice   Objective: 12   Section 11.4
    Level: A: Standard   Tag: None

Hydroelectric energy is energy produced by

a) hot geysers.

b) burning coal.

c) fission reactions.

d) running water.

d

81) Item Type: Multiple Choice   Objective: 12   Section 11.4
    Level: A: Standard   Tag: None

What percentage of the electricity used in the United States comes from hydroelectric plants?

a) 4%   b) 11%   c) 27%   d) 53%

b

82) Item Type: Multiple Choice   Objective: 12   Section 11.4
    Level: A: Standard   Tag: None

In power plants, the force of steam is used to spin turbines primarily in order to

a) generate electricity.

b) cool the reactor.

c) generate heat.

d) drive machinery.

a

83) Item Type: Multiple Choice   Objective: 12   Section 11.4
    Level: A: Standard   Tag: None

A disadvantage of wind-driven generators is that

a) fuel is expensive.

b) wind energy is inefficient.

c) the wind does not always blow.

d) generators pollute the atmosphere.

c

84) Item Type: Multiple Choice   Objective: 12   Section 11.4
    Level: A: Standard   Tag: None

The energy in ocean tides is used to produce electricity by using the sea water to

a) produce steam.

b) turn turbines.

c) fuel fusion reactors.

d) store solar energy.

b

85) Item Type: Fill in the Blank   Objective: 1   Section 11.1
Level: A: Standard   Tag: None

A narrow, fingerlike band of a mineral is called a(n) _____.

vein

86) Item Type: Fill in the Blank   Objective: 3   Section 11.2
Level: A: Standard   Tag: None

Fossil fuels are made up of compounds of carbon and hydrogen called

_____.

hydrocarbons

87) Item Type: Fill in the Blank   Objective: 4   Section 11.2
Level: A: Standard   Tag: None

The rock immediately above a deposit of petroleum is called a(n) _____.

cap rock

88) Item Type: Fill in the Blank   Objective: 4   Section 11.2
Level: B: Enriched   Tag: None

Natural gas is often found immediately above deposits of _____.

petroleum

89) Item Type: Fill in the Blank   Objective: 4   Section 11.2
Level: A: Standard   Tag: None

Petroleum is produced when organic matter is subjected to great heat and

_____.

pressure

90) Item Type: Fill in the Blank   Objective: 5   Section 11.2
Level: A: Standard   Tag: None

Fossil fuels are found in the earth in the form of crude oil or unrefined

_____.

petroleum

91) Item Type: Fill in the Blank   Objective: 8   Section 11.3
    Level: A: Standard   Tag: None

Fuel rods in a nuclear reactor are made from isotopes of the element

_____.

uranium

92) Item Type: Fill in the Blank   Objective: 8   Section 11.3
    Level: A: Standard   Tag: None

In nuclear fission reactors, atomic nuclei are split when they are struck by

_____.

neutrons

93) Item Type: Fill in the Blank   Objective: 9   Section 11.3
    Level: A: Standard   Tag: None

The sun's energy that reaches the earth is produced by the process of nuclear

_____.

fusion

94) Item Type: Fill in the Blank   Objective: 9   Section 11.3
    Level: B: Enriched   Tag: None

In the future, nuclear fusion reactors may be fueled by an almost limitless supply of

hydrogen atoms from the _____.

oceans

95) Item Type: Fill in the Blank   Objective: 10   Section 11.4
    Level: A: Standard   Tag: None

The two types of solar energy systems are called _____.

active and passive

96) Item Type: Fill in the Blank   Objective: 10   Section 11.4
    Level: A: Standard   Tag: Diagram 1175

What type of energy is the building in the diagram designed to take advantage of?

_____.

solar

97) Item Type: Fill in the Blank   Objective: 10   Section 11.4
Level: B: Enriched   Tag: None

A rooftop box designed to capture the energy in sunlight is called a(n)

_____.

| solar collector ■ |

98) Item Type: Fill in the Blank   Objective: 11   Section 11.4
Level: B: Enriched   Tag: None

Which type of energy makes use of the energy found in magma? _____

| geothermal energy ■ |

99) Item Type: Fill in the Blank   Objective: 11   Section 11.4
Level: A: Standard   Tag: None

In a geothermal generating plant, energy is produced when steam turns a(n)

_____.

| turbine ■ |

100) Item Type: Essay   Objective: 3   Section 11.2
Level: B: Enriched   Tag: None

Explain why coal is a more efficient heat source than peat.

_____

_____

_____

_____

_____

_____

| Coal is formed when peat is subjected to pressure from overlying sediments. This pressure forces water out of the peat, leaving a material that is more dense and that contains a higher carbon content than the original peat. Because of the coal's higher density and carbon content, it burns longer and hotter than peat. ■ |

Chapter 11

101) Item Type: Essay   Objective: 3   Section 11.2
Level: B: Enriched   Tag: None

How does the process of carbonization produce coal from the remains of plants?

_____

_____

_____

_____

_____

Carbonization occurs when partially decomposed plants are buried in swamp mud. Bacteria consume some of the plant material. These bacteria then release marsh gas, which includes methane (CH_4) and carbon dioxide (CO_2). As the gas escapes, the original compounds present in the plants gradually change, and only carbon remains.

102) Item Type: Essay   Objective: 6   Section 11.2
Level: A: Standard   Tag: None

Describe how substitution and recycling can be used to conserve mineral resources.

_____

_____

_____

_____

_____

One way to conserve minerals is to use more abundant materials, such as plastic, in place of minerals. Another way to conserve minerals is to recycle them, or use them over again. Metals such as iron, copper, and aluminum can be recycled.

103) Item Type: Essay   Objective: 7   Section 11.2
Level: B: Enriched   Tag: None

Explain how strip mining can pose a threat to life in nearby rivers and streams.

_____

_____

_____

_____

_____

During strip mining, topsoil and rocks that are displaced to expose the coal are left in steep slopes. When wet, rocks exposed during mining give off acids. Rain may carry the acids into nearby rivers and streams, causing harm to living things there.

104) Item Type: Essay   Objective: 8   Section 11.3
Level: B: Enriched   Tag: None

In nuclear reactors, the rate of the fission reaction is controlled by raising and lowering rods made from materials that absorb neutrons. How do these rods control the nuclear fission reaction?

_____

_____

_____

_____

A nuclear fission reaction begins when U-235 atoms are struck by neutrons. The U-235 atoms split, releasing energy and more neutrons. These newly released neutrons strike more U-235 atoms, which release more energy and more neutrons to continue the chain reaction. The reaction can be slowed by lowering a neutron-absorbing material into the reactor. When the material is raised, neutrons become available again for the chain reaction.

105) Item Type: Essay   Objective: 12   Section 11.4
      Level: A: Standard   Tag: None

Explain how movements of ocean water can be used to produce electricity.

_____

_____

_____

_____

_____

Twice each day, water in the oceans moves towards and away from the shore. These movements are called tides. High tide occurs when the water reaches its highest point along the shore, low tide when it reaches its lowest point. To make use of this tidal power, dams trap the water at high tide and release it at low tide. The water spins the turbines of electric generators to produce electricity.

1) Tag Name: Diagram 1171

2) Tag Name: Diagram 1172

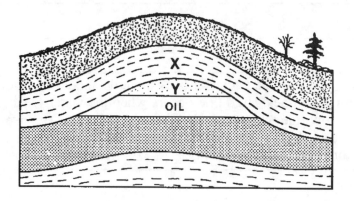

3) Tag Name: Diagram 1173

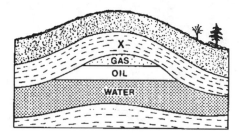

4) Tag Name: Diagram 1174

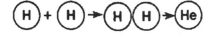

5) Tag Name: Diagram 1175

1) Item Type: True-False   Objective: 1   Section 12.1
   Level: A: Standard   Tag: None

_____ Joints are often formed when the pressure on rock decreases.

T

2) Item Type: True-False   Objective: 1   Section 12.1
   Level: A: Standard   Tag: None

_____ Plants can act as both mechanical and chemical weathering agents.

T

3) Item Type: True-False   Objective: 1   Section 12.1
   Level: A: Standard   Tag: None

_____ Joints usually develop in granite when the pressure on it increases.

F

4) Item Type: True-False   Objective: 1   Section 12.1
   Level: A: Standard   Tag: None

_____ Abrasion is a chemical process.

F

5) Item Type: True-False   Objective: 2   Section 12.1
   Level: A: Standard   Tag: None

_____ Oxidation produces a reddish soil.

T

6) Item Type: True-False   Objective: 2   Section 12.1
   Level: A: Standard   Tag: None

_____ Rainwater is naturally acidic.

T

7) Item Type: True-False   Objective: 2   Section 12.1
   Level: B: Enriched   Tag: None

_____ The red color of soils in the southeastern United States results from the presence of aluminum.

F

8) Item Type: True-False   Objective: 2   Section 12.1
    Level: B: Enriched   Tag: None

    _____ Galvanized steel resists rusting.

    T

9) Item Type: True-False   Objective: 2   Section 12.1
    Level: A: Standard   Tag: None

    _____ Chemical weathering breaks down a rock without changing the composition of the rock.

    F

10) Item Type: True-False   Objective: 3   Section 12.2
    Level: A: Standard   Tag: None

    _____ Calcite-containing rocks resist chemical weathering.

    F

11) Item Type: True-False   Objective: 5   Section 12.2
    Level: A: Standard   Tag: None

    _____ Rocks weather most rapidly in a hot, dry climate.

    F

12) Item Type: True-False   Objective: 5   Section 12.2
    Level: A: Standard   Tag: None

    _____ Weathering of rocks occurs slowly in hot, humid climates.

    F

13) Item Type: True-False   Objective: 5   Section 12.2
    Level: A: Standard   Tag: None

    _____ Variable weather conditions can cause fractures that expose new rock surfaces.

    T

14) Item Type: True-False   Objective: 6   Section 12.3
    Level: A: Standard   Tag: None

    _____ Transported soils usually have three distinct horizons.

    F

15) Item Type: True-False   Objective: 6   Section 12.3
      Level: B: Enriched   Tag: None

_____ Weathered granite forms sandy soils.

T

16) Item Type: True-False   Objective: 6   Section 12.3
      Level: A: Standard   Tag: None

_____ The C horizon in a residual soil consists of partially weathered bedrock.

T

17) Item Type: True-False   Objective: 6   Section 12.3
      Level: A: Standard   Tag: None

_____ Transported soils have well-defined horizons.

F

18) Item Type: True-False   Objective: 6   Section 12.3
      Level: B: Enriched   Tag: None

_____ Beach sands are often produced by hydrolysis and oxidation.

F

19) Item Type: True-False   Objective: 6   Section 12.3
      Level: A: Standard   Tag: None

_____ The B horizon usually contains the most organic matter.

F

20) Item Type: True-False   Objective: 7   Section 12.3
      Level: B: Enriched   Tag: None

_____ Pedocal soils are less acidic than pedalfer soils.

T

21) Item Type: True-False   Objective: 7   Section 12.3
      Level: A: Standard   Tag: None

_____ Steep slopes are often too dry to support plant growth.

T

22) Item Type: True-False   Objective: 7   Section 12.3
     Level: A: Standard   Tag: None

_____ Desert soil consists mostly of regolith.

T

23) Item Type: True-False   Objective: 7   Section 12.3
     Level: A: Standard   Tag: None

_____ Arctic soils are produced mainly by mechanical weathering.

T

24) Item Type: True-False   Objective: 7   Section 12.3
     Level: B: Enriched   Tag: None

_____ Flat land that has good drainage tends to have thick residual soils.

T

25) Item Type: True-False   Objective: 7   Section 12.3
     Level: B: Enriched   Tag: None

_____ Tropical soils generally make good farmland.

F

26) Item Type: True-False   Objective: 7   Section 12.3
     Level: A: Standard   Tag: None

_____ Arctic soils are composed mainly of regolith.

T

27) Item Type: True-False   Objective: 8   Section 12.4
     Level: A: Standard   Tag: None

_____ Chemical and mechanical are two types of erosion.

F

28) Item Type: True-False   Objective: 8   Section 12.4
     Level: A: Standard   Tag: None

_____ Gravity is an important agent of erosion.

T

29) Item Type: True-False   Objective: 9   Section 12.4
    Level: B: Enriched   Tag: None

_____ The most rapid and sudden type of downslope mass movement is known as creep.

F

30) Item Type: True-False   Objective: 9   Section 12.4
    Level: A: Standard   Tag: None

_____ Mudflows occur most often on plains and peneplains.

F

31) Item Type: True-False   Objective: 9   Section 12.4
    Level: A: Standard   Tag: None

_____ Rockfalls and landslides are examples of mass movements.

T

32) Item Type: True-False   Objective: 10   Section 12.4
    Level: A: Standard   Tag: None

_____ Gullying often occurs when a slope is plowed for farming.

T

33) Item Type: True-False   Objective: 10   Section 12.4
    Level: A: Standard   Tag: None

_____ Strip-cropping generally accelerates erosion.

F

34) Item Type: True-False   Objective: 10   Section 12.4
    Level: A: Standard   Tag: None

_____ Crop rotation is a soil conservation method that often allows small gullies to fill in with soil.

T

35) Item Type: True-False   Objective: 11   Section 12.4
    Level: A: Standard   Tag: None

_____ Youthful mountains are characterized by rounded peaks and gentle slopes.

F

36) Item Type: Multiple Choice   Objective: 1   Section 12.1
Level: B: Enriched   Tag: None

Glaciers weather surface rocks mainly through the process of

a) oxidation.   b) carbonation.   c) abrasion.   d) decomposition.

c

37) Item Type: Multiple Choice   Objective: 1   Section 12.1
Level: A: Standard   Tag: None

Which of the following is a type of mechanical weathering?

a) oxidation   b) hydrolysis   c) carbonation   d) ice wedging

d

38) Item Type: Multiple Choice   Objective: 1   Section 12.1
Level: B: Enriched   Tag: None

When water freezes, its volume increases by about

a) 1%.   b) 10%.   c) 30%.   d) 50%.

b

39) Item Type: Multiple Choice   Objective: 1   Section 12.1
Level: A: Standard   Tag: None

The rounding of a rock caused by collisions with other rocks is called

a) abrasion.   b) gullying.   c) exfoliation.   d) creep.

a

40) Item Type: Multiple Choice   Objective: 2   Section 12.1
Level: B: Enriched   Tag: Diagram 1271

What type of weathering most likely caused the formations in the diagram?

a) exfoliation   b) abrasion   c) carbonation   d) oxidation

c

41) Item Type: Multiple Choice   Objective: 2   Section 12.1
Level: B: Enriched   Tag: None

Lichens and mosses erode rocks by producing

a) limestone.   b) oxides.   c) bases.   d) acids.

d

42) Item Type: Multiple Choice   Objective: 2   Section 12.1
Level: B: Enriched   Tag: None

Feldspar can be changed by hydrolysis into

a) limestone.   b) iron oxide.   c) kaolin.   d) carbonic acid.

c

43) Item Type: Multiple Choice   Objective: 2   Section 12.1
Level: A: Standard   Tag: None

Weathering in which thin sheets of rock flake off the surface is called

a) carbonation.   b) decomposition. c) oxidation.   d) exfoliation.

d

44) Item Type: Multiple Choice   Objective: 2   Section 12.1
Level: A: Standard   Tag: None

Uncontaminated rainwater is naturally acidic because of the presence of

a) silica.   b) carbon dioxide. c) oxygen.   d) iron oxide.

b

45) Item Type: Multiple Choice   Objective: 2   Section 12.1
Level: A: Standard   Tag: None

Rusting is an example of which of the following processes?

a) hydrolysis   b) oxidation   c) exfoliation   d) carbonation

b

46) Item Type: Multiple Choice   Objective: 2   Section 12.1
Level: A: Standard   Tag: None

Underground limestone caverns are most often produced by

a) hydrolysis.   b) carbonation.   c) oxidation.   d) exfoliation.

b

47) Item Type: Multiple Choice   Objective: 3   Section 12.2
Level: A: Standard   Tag: None

Which of the following generally weathers most rapidly?

a) limestone   b) obsidian   c) granite   d) gneiss

a

48) Item Type: Multiple Choice   Objective: 3   Section 12.2
    Level: A: Standard   Tag: None

Which of the following minerals is most resistant to weathering?

a) calcite   b) feldspar   c) sandstone   d) quartz

d

49) Item Type: Multiple Choice   Objective: 3   Section 12.2
    Level: A: Standard   Tag: None

The rate at which conglomerates weather is most dependent upon the rock's

a) cementing material.   b) age.   c) grain size.   d) texture.

a

50) Item Type: Multiple Choice   Objective: 4   Section 12.2
    Level: B: Enriched   Tag: None

Which of the following rocks is most likely to weather quickly?

a) an exposed rock on a plain   b) an exposed rock on a slope

c) a buried rock on a plain   d) a buried rock on a slope

b

51) Item Type: Multiple Choice   Objective: 4   Section 12.2
    Level: A: Standard   Tag: Diagram 1272

At which point in the diagram will weathering occur most rapidly?

a) *A*   b) *B*   c) *C*   d) *D*

c

52) Item Type: Multiple Choice   Objective: 5   Section 12.2
    Level: B: Enriched   Tag: None

Which of the following best accounts for the damage to Cleopatra's Needle after being moved to New York City?

a) carbonation   b) dry climate   c) pollution   d) cold climate

c

53) Item Type: Multiple Choice   Objective: 5   Section 12.2
    Level: A: Standard   Tag: None

Weathering is generally slow in climates with extended periods of

a) acid rain.   b) cold.   c) high winds.   d) humidity.

b

54) Item Type: Multiple Choice   Objective: 5   Section 12.2
Level: A: Standard   Tag: None

Which of the following climatic conditions is most likely to cause accelerated weathering?

a) hot and dry   b) extreme cold   c) freezes and thaws   d) steady rainfall

c

55) Item Type: Multiple Choice   Objective: 6   Section 12.3
Level: A: Standard   Tag: Diagram 1273

Which horizon in the diagram accumulates leached minerals?

a) *A*   b) *B*   c) *C*   d) *D*

b

56) Item Type: Multiple Choice   Objective: 6   Section 12.3
Level: B: Enriched   Tag: Diagram 1273

Which layer in the diagram represents the soil's parent rock?

a) *A*   b) *B*   c) *C*   d) *D*

d

57) Item Type: Multiple Choice   Objective: 6   Section 12.3
Level: A: Standard   Tag: None

Which of the following has the smallest particle size?

a) clay   b) silt   c) sand   d) gravel

a

58) Item Type: Multiple Choice   Objective: 6   Section 12.3
Level: A: Standard   Tag: None

Which of the following contains particles large enough to be felt but too small to be seen easily?

a) clay   b) silt   c) sand   d) gravel

b

59) Item Type: Multiple Choice   Objective: 7   Section 12.3
Level: B: Enriched   Tag: None

Gulf states and states east of the Mississippi River generally have a soil type known as

a) pedalfer.   b) laterite.   c) pedocal.   d) regolith.

a

60) Item Type: Multiple Choice   Objective: 7   Section 12.3
    Level: A: Standard   Tag: None

Humus is made up mainly of

a) weathered rock fragments.   b) organic material.

c) unjointed bedrock.   d) iron particles.

b

61) Item Type: Multiple Choice   Objective: 8   Section 12.4
    Level: A: Standard   Tag: None

Which of the following is a type of accelerated erosion?

a) leaching   b) terracing   c) gullying   d) wedging

c

62) Item Type: Multiple Choice   Objective: 9   Section 12.4
    Level: A: Standard   Tag: None

A pile of rocks that accumulates at the base of a slope is called

a) humus.   b) slump.   c) regolith.   d) talus.

d

63) Item Type: Multiple Choice   Objective: 9   Section 12.4
    Level: A: Standard   Tag: None

A mass movement in which the upper soil layers thaw and flow is called

a) tropical.   b) solifluction.   c) creep.   d) mudflow.

b

64) Item Type: Multiple Choice   Objective: 9   Section 12.4
    Level: A: Standard   Tag: None

In which of the following climates is solifluction most likely to occur?

a) tropical   b) desert   c) temperate   d) arctic

d

65) Item Type: Multiple Choice   Objective: 9   Section 12.4
    Level: A: Standard   Tag: None

Which of the following is usually the slowest type of mass movement?

a) landslide   b) slump   c) rockfall   d) creep

d

66) Item Type: Multiple Choice   Objective: 9   Section 12.4
      Level: A: Standard   Tag: None

Unstable movement of rock and soil along a curved slope is called a

a) mudflow.   b) creep.   c) landslide.   d) slump.

d

67) Item Type: Multiple Choice   Objective: 9   Section 12.4
      Level: A: Standard   Tag: None

Boulders dropping from a steep cliff is an example of which type of mass movement?

a) rockfall   b) sheet erosion   c) landslide   d) soil erosion

a

68) Item Type: Multiple Choice   Objective: 10   Section 12.4
      Level: A: Standard   Tag: None

Which farming method involves the construction of steplike ridges that follow the contour of a sloped field?

a) strip-cropping   b) contour plowing

c) terracing   d) crop rotation

c

69) Item Type: Multiple Choice   Objective: 11   Section 12.4
      Level: A: Standard   Tag: Diagram 1275

In this diagram, which area represents a plateau?

a) **A**   b) **B**   c) **C**   d) **D**

a

70) Item Type: Multiple Choice   Objective: 11   Section 12.4
      Level: B: Enriched   Tag: Diagram 1275

In the diagram above, the feature labeled **C** represents a

a) butte.   b) monadnock.   c) mesa.   d) peneplain.

b

71) Item Type: Multiple Choice   Objective: 11   Section 12.4
      Level: A: Standard   Tag: None

Small, tablelike areas that once formed a broad, flat landform are called

a) mesas.   b) plateaus.   c) peneplains.   d) terraces.

a

72) Item Type: Multiple Choice   Objective: 11   Section 12.4
Level: A: Standard   Tag: None

A flat landform that lies on top of horizontal rock layers near sea level is called a

a) peneplain.   b) plateau.   c) butte.   d) plain.

d

73) Item Type: Fill in the Blank   Objective: 1   Section 12.1
Level: B: Enriched   Tag: None

The type of mechanical weathering that is commonly caused by wind-carried sand is called

_____.

abrasion

74) Item Type: Fill in the Blank   Objective: 1   Section 12.1
Level: A: Standard   Tag: None

The type of mechanical weathering that occurs when water seeps into cracks in a rock and

freezes is called _____.

ice wedging

75) Item Type: Fill in the Blank   Objective: 1   Section 12.1
Level: A: Standard   Tag: None

Cracks that form in rocks due to expansion are called _____.

joints

76) Item Type: Fill in the Blank   Objective: 2   Section 12.1
Level: B: Enriched   Tag: None

The decomposition reaction in which hydrogen ions from water displace elements in a

mineral is called _____.

hydrolysis

77) Item Type: Fill in the Blank   Objective: 2   Section 12.1
Level: A: Standard   Tag: None

Rain is acidified by atmospheric levels of the two elements _____.

sulfur and nitrogen

78) Item Type: Fill in the Blank   Objective: 3   Section 12.2
Level: B: Enriched   Tag: None

What common mineral in limestone is most easily weathered by carbonation?

_____

calcite

79) Item Type: Fill in the Blank   Objective: 3   Section 12.2
Level: B: Enriched   Tag: None

Rocks containing calcite are most easily weathered by the chemical process of

_____.

carbonation

80) Item Type: Fill in the Blank   Objective: 3   Section 12.2
Level: A: Standard   Tag: None

Sandstones that are resistant to weathering are most likely cemented by

_____.

silicates

81) Item Type: Fill in the Blank   Objective: 4   Section 12.2
Level: B: Enriched   Tag: None

When a rock is split into eight smaller blocks, its surface area increases by a factor of

_____.

two

82) Item Type: Fill in the Blank   Objective: 4   Section 12.2
Level: A: Standard   Tag: None

Fractures and joints allow weathering to occur more rapidly by increasing a rock's

_____.

surface area

83) Item Type: Fill in the Blank   Objective: 4   Section 12.2
Level: A: Standard   Tag: None

Water that penetrates rock and then freezes causes weathering by the process of

_____.

ice wedging

84) Item Type: Fill in the Blank   Objective: 5   Section 12.2
Level: A: Standard   Tag: None

The four main factors that determine the rate of weathering are climate, composition,

exposure, and _____.

topography

85) Item Type: Fill in the Blank   Objective: 6   Section 12.3
Level: B: Enriched   Tag: Diagram 1274

In the diagram above, bedrock is represented by the layer labeled _____.

D

86) Item Type: Fill in the Blank   Objective: 6   Section 12.3
Level: B: Enriched   Tag: Diagram 1274

In this diagram above, organic matter is most concentrated in the layer labeled

_____.

A

87) Item Type: Fill in the Blank   Objective: 6   Section 12.3
Level: A: Standard   Tag: None

A dark, organic material in the soil produced by the decaying remains of plants and

animals is called _____.

humus

88) Item Type: Fill in the Blank   Objective: 6   Section 12.3
Level: A: Standard   Tag: None

The layers of a soil profile are called _____.

horizons

89) Item Type: Fill in the Blank   Objective: 7   Section 12.3
     Level: B: Enriched   Tag: None

What is the name for a soil containing iron and aluminum minerals that do not dissolve

easily in water? _____

| laterite ▪ |

90) Item Type: Fill in the Blank   Objective: 7   Section 12.3
     Level: B: Enriched   Tag: None

Pedocal soils are soils that contain large amounts of the mineral _____.

| calcium carbonate ▪ |

91) Item Type: Fill in the Blank   Objective: 7   Section 12.3
     Level: B: Enriched   Tag: None

The removal of which soil horizon would probably lead to the greatest reduction of soil

fertility? _____

| A horizon ▪ |

92) Item Type: Fill in the Blank   Objective: 7   Section 12.3
     Level: A: Standard   Tag: None

Thick soil that forms in tropical climate is called _____.

| laterite ▪ |

93) Item Type: Fill in the Blank   Objective: 8   Section 12.4
     Level: A: Standard   Tag: None

The process by which the products of weathering are transported is called

_____.

| erosion ▪ |

94) Item Type: Fill in the Blank   Objective: 8   Section 12.4
     Level: A: Standard   Tag: None

The primary agents of erosion are gravity, wind, glaciers, and _____.

| water ▪ |

95) Item Type: Fill in the Blank   Objective: 9   Section 12.4
    Level: A: Standard   Tag: None

The type of mass movement that occurs when a large block of soil and rock slides

downhill along a curved slope because of gravity is _____.

slump

96) Item Type: Fill in the Blank   Objective: 9   Section 12.4
    Level: A: Standard   Tag: None

Rock fragments accumulating in piles at the base of a slope are called

_____.

talus

97) Item Type: Fill in the Blank   Objective: 10   Section 12.4
    Level: A: Standard   Tag: None

The type of accelerated erosion that occurs when high winds or the removal of cover

plants causes parallel layers of topsoil to be stripped away is _____.

sheet erosion

98) Item Type: Fill in the Blank   Objective: 11   Section 12.4
    Level: A: Standard   Tag: None

What is the correct term for a knob of rock that rises up from a peneplain?

_____

monadnock

99) Item Type: Fill in the Blank   Objective: 11   Section 12.4
    Level: A: Standard   Tag: None

The landform that is created when a plateau is dissected into smaller tablelike areas is

called a(n) _____.

mesa

100) Item Type: Fill in the Blank   Objective: 11   Section 12.4
Level: A: Standard   Tag: None

The surface of a mountain that has been reduced by erosion to a low, almost featureless

surface near sea level is called a(n) _____.

peneplain

101) Item Type: Essay   Objective: 1   Section 12.1
Level: A: Standard   Tag: None

Describe how plants and animals contribute to the weathering of rocks.

_____

_____

_____

_____

_____

Plants and animals are important agents of mechanical weathering. The roots of plants can
work their way into cracks in rocks, creating pressure that wedges the rock apart. Some
plants also produce acids that chemically weather rocks. The digging activities of
burrowing animals contribute to weathering by exposing new rock surfaces to weathering
agents.

102) Item Type: Essay   Objective: 6   Section 12.3
Level: B: Enriched   Tag: None

How does a residual soil differ from a transported soil?

_____

_____

_____

_____

_____

Residual soil is soil that rests on top of its parent rock. Transported soils are soils that
have been carried away from their parent rock by agents such as water, wind, and
glaciers. Residual soils have well-defined horizons, while transported soils do not.

103) Item Type: Essay   Objective: 6   Section 12.3
Level: B: Enriched   Tag: None

How would a soil formed from a feldspar-rich rock differ from a soil formed from a quartz-rich rock?

_____

_____

_____

_____

_____

Rocks that are rich in feldspar form soils containing large amounts of clay; these soils consist of fine grains of silicate material containing aluminum and water. Rocks containing large amounts of quartz weather chemically to form sandy soils.

104) Item Type: Essay   Objective: 7   Section 12.3
Level: A: Standard   Tag: None

How is a desert soil similar to an arctic soil?

_____

_____

_____

_____

Both soils form mainly as a result of mechanical weathering. This leads to thin soils consisting mainly of rock fragments called regolith.

105) Item Type: Essay   Objective: 7   Section 12.3
Level: B: Enriched   Tag: None

What are pedocal soils, and under what circumstances do they form?

_____

_____

_____

_____

_____

> A pedocal contains large amounts of calcium carbonate. They are less acidic than pedalfers and are very fertile. Pedocals form in areas that receive less than 65 cm of rain per year.

106) Item Type: Essay   Objective: 8   Section 12.4
Level: A: Standard   Tag: None

Define the term *erosion,* and name four agents that cause it to occur.

_____

_____

_____

_____

_____

> Erosion is the process that transports the products of weathering away from where the weathering occurred. The agents of erosion are gravity, wind, glaciers, and water.

107) Item Type: Essay   Objective: 10   Section 12.4
     Level: A: Standard   Tag: None

Describe three farming methods that are used to reduce the erosion of farmland.

_____

_____

_____

_____

_____

Contour plowing: the soil is plowed in circular bands that follow the contour of the land
to reduce water erosion. Strip-cropping: cover crops are planted next to crops that expose
soil to erosion. Terracing: step-like ridges that follow the contours of a sloped field are
constructed to reduce erosion. Crop rotation: crops are rotated to help prevent gullying and
sheet erosion.

108) Item Type: Essay   Objective: 11   Section 12.4
     Level: B: Enriched   Tag: None

How can youthful mountains be distinguished from mature mountains?

_____

_____

_____

_____

_____

Youthful mountains are rugged and have sharp peaks and deep, narrow valleys. Mature
mountains have rounded peaks and gentle slopes.

Chapter 12

1)   Tag Name: Diagram 1271

2)   Tag Name: Diagram 1272

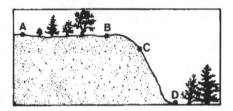

3)   Tag Name: Diagram 1273

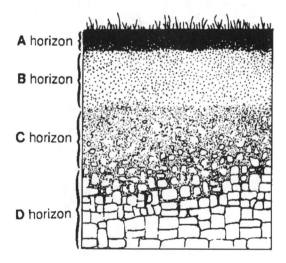

4)   Tag Name: Diagram 1274

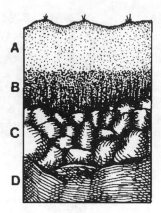

5)   Tag Name: Diagram 1275

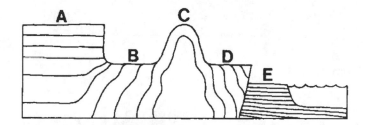

Chapter 13

1) Item Type: True-False   Objective: 1   Section 13.1
   Level: B: Enriched   Tag: None

_____ The continuous movement of water between the atmosphere and the earth's surface is called the hydrologic cycle.

T

2) Item Type: True-False   Objective: 1   Section 13.1
   Level: B: Enriched   Tag: None

_____ The continents lose approximately 70,000 km^3 of water each year through evapotranspiration.

T

3) Item Type: True-False   Objective: 1   Section 13.1
   Level: A: Standard   Tag: None

_____ The hydrologic cycle is also called the water cycle.

T

4) Item Type: True-False   Objective: 1   Section 13.1
   Level: A: Standard   Tag: None

_____ Most water evaporating from the earth's surface evaporates from rivers and lakes.

F

5) Item Type: True-False   Objective: 1   Section 13.1
   Level: A: Standard   Tag: None

_____ When water vapor rises in the atmosphere, it expands and cools.

T

6) Item Type: True-False   Objective: 1   Section 13.1
   Level: B: Enriched   Tag: None

_____ Galvanizing is a process used to inhibit rusting.

T

7) Item Type: True-False   Objective: 2   Section 13.1
   Level: A: Standard   Tag: None

_____ Deserts often form as a result of insufficient precipitation.

T

8) Item Type: True-False   Objective: 2   Section 13.1
   Level: A: Standard   Tag: None

   _____ Evapotranspiration increases with increasing temperature.

   | T |

9) Item Type: True-False   Objective: 2   Section 13.1
   Level: B: Enriched   Tag: None

   _____ A local water budget is always balanced.

   | F |

10) Item Type: True-False   Objective: 2   Section 13.1
    Level: B: Enriched   Tag: None

    _____ Evapotranspiration is higher at the poles than at the equator.

    | F |

11) Item Type: True-False   Objective: 2   Section 13.1
    Level: A: Standard   Tag: None

    _____ Irrigation is often necessary in areas that have high evapotranspiration.

    | T |

12) Item Type: True-False   Objective: 2   Section 13.1
    Level: A: Standard   Tag: None

    _____ Approximately 50 percent of global precipitation falls on oceans.

    | F |

13) Item Type: True-False   Objective: 3   Section 13.1
    Level: A: Standard   Tag: None

    _____ Most water used by industry is recycled.

    | F |

14) Item Type: True-False   Objective: 3   Section 13.1
    Level: A: Standard   Tag: None

    _____ Most water used by industry is returned to rivers and oceans as waste water.

    | T |

15) Item Type: True-False   Objective: 4   Section 13.2
      Level: A: Standard   Tag: None

_____ A divide is the area separating a stream's bed load from its suspended load.

F

16) Item Type: True-False   Objective: 4   Section 13.2
      Level: A: Standard   Tag: None

_____ Stream piracy decreases the size of a river system.

F

17) Item Type: True-False   Objective: 4   Section 13.2
      Level: A: Standard   Tag: None

_____ Feeder streams are also called tributaries.

T

18) Item Type: True-False   Objective: 4   Section 13.2
      Level: A: Standard   Tag: None

_____ A water gap occurs at an uplifted land area.

T

19) Item Type: True-False   Objective: 4   Section 13.2
      Level: A: Standard   Tag: None

_____ The path that a river follows is called a gully.

F

20) Item Type: True-False   Objective: 5   Section 13.2
      Level: A: Standard   Tag: None

_____ In a suspended stream load, sediment advances by sliding and rolling.

F

21) Item Type: True-False   Objective: 5   Section 13.2
      Level: A: Standard   Tag: None

_____ Bed loads commonly consist of gravel and stones.

T

22) Item Type: True-False   Objective: 5   Section 13.2
    Level: A: Standard   Tag: None

_____ Channel erosion is quicker in rivers with a small bed-load volume.

F

23) Item Type: True-False   Objective: 5   Section 13.2
    Level: A: Standard   Tag: None

_____ Minerals transported as ions and molecules make up a stream's dissolved load.

T

24) Item Type: True-False   Objective: 5   Section 13.2
    Level: A: Standard   Tag: None

_____ Erosion often increases the amount of land drained in a watershed.

T

25) Item Type: True-False   Objective: 6   Section 13.2
    Level: A: Standard   Tag: None

_____ Oxbow lakes are formed in potholes.

F

26) Item Type: True-False   Objective: 6   Section 13.2
    Level: A: Standard   Tag: None

_____ In general, mature rivers flow faster than youthful rivers.

F

27) Item Type: True-False   Objective: 7   Section 13.3
    Level: B: Enriched   Tag: None

_____ The decline in the prosperity of Bruges, Belgium, was due to the process of stream deposition.

T

28) Item Type: True-False   Objective: 7   Section 13.3
    Level: A: Standard   Tag: None

_____ As the velocity of a stream decreases, its load of sediment also decreases.

T

29) Item Type: True-False   Objective: 7   Section 13.3
    Level: A: Standard   Tag: None

_____ The tip of a fan-shaped deposit at the mouth of a stream usually points upstream.

T

30) Item Type: True-False   Objective: 7   Section 13.3
    Level: A: Standard   Tag: None

_____ Deltas and alluvial fans are both formed at the mouth of a river.

F

31) Item Type: True-False   Objective: 7   Section 13.3
    Level: A: Standard   Tag: None

_____ Alluvial fans are also called river deltas.

F

32) Item Type: True-False   Objective: 8   Section 13.3
    Level: A: Standard   Tag: None

_____ The volume of water in a stream remains constant from year to year.

F

33) Item Type: True-False   Objective: 8   Section 13.3
    Level: A: Standard   Tag: None

_____ In areas of harsh winters, spring floods are common near headwaters.

T

34) Item Type: True-False   Objective: 8   Section 13.3
    Level: A: Standard   Tag: None

_____ High flow velocity is a major cause of cutting and widening of river channels.

T

35) Item Type: True-False   Objective: 8   Section 13.3
    Level: A: Standard   Tag: None

_____ Human activities contribute to the occurrence of floods.

T

36) Item Type: True-False   Objective: 9   Section 13.3
Level: A: Standard   Tag: None

_____ Artificial levees are resistant to river erosion.

F

37) Item Type: True-False   Objective: 9   Section 13.3
Level: A: Standard   Tag: None

_____ Floodways often prevent streams from overflowing.

T

38) Item Type: True-False   Objective: 9   Section 13.3
Level: A: Standard   Tag: None

_____ Dam building is the most common method of direct flood control.

T

39) Item Type: True-False   Objective: 9   Section 13.3
Level: A: Standard   Tag: None

_____ Artificial levees are used to prevent excess evaporation.

F

40) Item Type: True-False   Objective: 9   Section 13.3
Level: A: Standard   Tag: None

_____ Forest and soil conservation play a major role in flood control.

T

41) Item Type: True-False   Objective: 9   Section 13.3
Level: A: Standard   Tag: None

_____ Floodway construction is primarily a method of erosion control.

F

42) Item Type: True-False   Objective: 9   Section 13.3
Level: A: Standard   Tag: None

_____ Flooding is a natural stage in the development of a stream.

T

43) Item Type: Multiple Choice   Objective: 1   Section 13.1
    Level: A: Standard   Tag: None

The main parts of the water cycle are evapotranspiration, precipitation, and

a) evaporation.   b) condensation.   c) runoff.   d) deposition.

b

44) Item Type: Multiple Choice   Objective: 1   Section 13.1
    Level: A: Standard   Tag: None

Condensation is a change in water from a

a) solid to liquid.   b) liquid to solid.

c) gas to liquid.   d) liquid to gas.

c

45) Item Type: Multiple Choice   Objective: 1   Section 13.1
    Level: B: Enriched   Tag: None

Approximately what percentage of the earth's precipitation falls on the ocean?

a) 5%   b) 25%   c) 75%   d) 99%

c

46) Item Type: Multiple Choice   Objective: 1   Section 13.1
    Level: B: Enriched   Tag: Diagram 1371

Which of the following is represented in the diagram above?

a) the earth's water budget   b) local water budget

c) groundwater movement   d) surface runoff

b

47) Item Type: Multiple Choice   Objective: 1   Section 13.1
    Level: B: Enriched   Tag: Diagram 1371

In the diagram above, the process that must occur before the process labeled **2** can occur is called

a) transpiration.   b) precipitation.   c) condensation.   d) evaporation.

c

48) Item Type: Multiple Choice   Objective: 1   Section 13.1
Level: A: Standard   Tag: Diagram 1371

The process represented by arrow **3** in the diagram above is

a) transpiration.   b) precipitation.   c) condensation.   d) evaporation.

a

49) Item Type: Multiple Choice   Objective: 1   Section 13.1
Level: A: Standard   Tag: Diagram 1372

In the diagram above, the arrow labeled **X** represents

a) absorption.   b) evaporation.   c) rejuvenation.   d) transpiration.

b

50) Item Type: Multiple Choice   Objective: 2   Section 13.1
Level: A: Standard   Tag: None

Which of the following describes the trend in the water movement pattern going from the equator to the poles?

a) condensation decreases   b) evapotranspiration decreases

c) condensation increases   d) evapotranspiration increases

b

51) Item Type: Multiple Choice   Objective: 2   Section 13.1
Level: A: Standard   Tag: None

The amount of precipitation on earth is equal to the amount of evapotranspiration and

a) runoff.   b) headwater.   c) discharge.   d) groundwater.

a

52) Item Type: Multiple Choice   Objective: 3   Section 13.1
Level: A: Standard   Tag: None

Which of the following is an artificial means of producing fresh water from ocean water?

a) desalination   b) saltation   c) evapotranspiration   d) condensation

a

53) Item Type: Multiple Choice   Objective: 3   Section 13.1
Level: A: Standard   Tag: None

The process of removing salt from ocean water is called

a) saltation.   b) condensation.   c) desalination.   d) transpiration.

c

54) Item Type: Multiple Choice Objective: 4 Section 13.2
  Level: B: Enriched   Tag: None

Which of the following is most likely to change drainage patterns in an area?

a) stream piracy   b) floodplain erosion

c) stream deposition   d) meander formation

a

55) Item Type: Multiple Choice Objective: 4 Section 13.2
  Level: B: Enriched   Tag: None

A divide is the imaginary line between two

a) river stages. b) watersheds. c) river loads. d) water gaps.

b

56) Item Type: Multiple Choice Objective: 4 Section 13.2
  Level: A: Standard   Tag: None

What is the term for the main stream and tributaries of a river?

a) river system b) river bed c) drainage basin d) water gap

a

57) Item Type: Multiple Choice Objective: 4 Section 13.2
  Level: A: Standard   Tag: None

The path that a river follows is called its

a) tributary. b) gully. c) channel. d) meander.

c

58) Item Type: Multiple Choice Objective: 5 Section 13.2
  Level: B: Enriched   Tag: None

Channel erosion is quickest in rivers with a

a) high volume of suspended load. b) low volume of suspended load.

c) high volume of bed load.   d) low volume of bed load.

c

59) Item Type: Multiple Choice Objective: 5 Section 13.2
  Level: A: Standard   Tag: None

All of the sediment carried by a river is called the

a) suspended load. b) bed load. c) stream load. d) dissolved load.

c

60) Item Type: Multiple Choice   Objective: 5   Section 13.2
Level: A: Standard   Tag: None

Which of the following creates a water gap?

a)   headward erosion

b)   stream piracy

c)   uplifted land area

d)   increased evaporation

c

61) Item Type: Multiple Choice   Objective: 5   Section 13.2
Level: B: Enriched   Tag: None

The amount of water that flows in a river channel in a given amount of time is called

a) velocity.   b) capacity.   c) discharge.   d) volume.

c

62) Item Type: Multiple Choice   Objective: 5   Section 13.2
Level: A: Standard   Tag: None

One of the principal components of the sediment carried in a bed load is

a) mud.   b) silt.   c) sand.   d) gravel.

d

63) Item Type: Multiple Choice   Objective: 5   Section 13.2
Level: A: Standard   Tag: None

A wind gap is a water gap that has been

a)   deepened.   b)   lengthened.   c)   abandoned.   d)   eroded.

c

64) Item Type: Multiple Choice   Objective: 6   Section 13.2
Level: A: Standard   Tag: Diagram 1373

The wide curves in the river in the diagram are

a)   meanders.   b)   oxbows.   c)   tributaries.   d)   channels.

a

65) Item Type: Multiple Choice   Objective: 6   Section 13.2
Level: B: Enriched   Tag: None

A river can be rejuvenated by

a)   increasing slope.

b)   decreasing velocity.

c)   increasing branching.

d)   decreasing depth.

a

66) Item Type: Multiple Choice   Objective: 6   Section 13.2
Level: A: Standard   Tag: None

As a river passes from a youthful to a mature stage, it

a) shortens.   b) deepens.   c) meanders.   d) narrows.

c

67) Item Type: Multiple Choice   Objective: 6   Section 13.2
Level: A: Standard   Tag: Diagram 1374

In this diagram, the feature labeled **A** represents

a) a meander.   b) an oxbow lake.   c) a pothole.   d) an alluvial fan.

b

68) Item Type: Multiple Choice   Objective: 6   Section 13.2
Level: A: Standard   Tag: None

Meanders typically form during a river's

a) early stage.   b) youthful stage.   c) mature stage.   d) old stage.

c

69) Item Type: Multiple Choice   Objective: 6   Section 13.2
Level: A: Standard   Tag: None

A rejuvenated river typically has a

a) steep gradient.   b) low gradient.   c) high volume.   d) low volume.

a

70) Item Type: Multiple Choice   Objective: 7   Section 13.3
Level: A: Standard   Tag: None

Alluvial fans and deltas have similar

a) causes for deposition.   b) sizes of sediment grains.

c) shapes of deposition.   d) angles of slope.

a

71) Item Type: Multiple Choice   Objective: 7   Section 13.3
Level: A: Standard   Tag: None

One difference between an alluvial fan and a delta is that an alluvial fan is

a) deposited on dry ground.   b) deposited in water.

c) relatively flat.   d) formed by stream deposition.

a

Chapter 13

72) Item Type: Multiple Choice   Objective: 9   Section 13.3
Level: A: Standard   Tag: None

Artificial levees are most commonly used to prevent

a) evaporation.   b) flooding.   c) runoff.   d) gullying.

b

73) Item Type: Multiple Choice   Objective: 9   Section 13.3
Level: A: Standard   Tag: None

Artificial levees are constructed in order to decrease

a) water volume.   b) river erosion.   c) stream velocity.   d) flood danger.

d

74) Item Type: Fill in the Blank   Objective: 1   Section 13.1
Level: A: Standard   Tag: None

When water vapor rises in the atmosphere, it expands and _____.

cools

75) Item Type: Fill in the Blank   Objective: 1   Section 13.1
Level: A: Standard   Tag: None

As water vapor becomes cooler, some of it condenses to form _____.

clouds

76) Item Type: Fill in the Blank   Objective: 2   Section 13.1
Level: A: Standard   Tag: None

The balance between precipitation on one side and evapotranspiration and runoff on the

other in a particular area is called the _____.

local water budget

77) Item Type: Fill in the Blank   Objective: 3   Section 13.1
Level: A: Standard   Tag: None

The process of removing salt from ocean water is called _____.

desalination

78) Item Type: Fill in the Blank   Objective: 4   Section 13.2
Level: B: Enriched   Tag: None

Lengthening and branching of the upper end of a river is called _____.

headward erosion

79) Item Type: Fill in the Blank   Objective: 4   Section 13.2
    Level: A: Standard   Tag: None

The river system begins to form when local precipitation exceeds _____.

evapotranspiration

80) Item Type: Fill in the Blank   Objective: 4   Section 13.2
    Level: A: Standard   Tag: None

Drainage basins are also called _____.

watersheds

81) Item Type: Fill in the Blank   Objective: 4   Section 13.2
    Level: A: Standard   Tag: None

The capture of a river system by a river in a different watershed is called

_____.

stream piracy

82) Item Type: Fill in the Blank   Objective: 5   Section 13.2
    Level: A: Standard   Tag: None

The short jumps by which rocks move in a river bed are called _____.

saltation

83) Item Type: Fill in the Blank   Objective: 5   Section 13.2
    Level: A: Standard   Tag: None

The Cumberland Gap is an example of a(n) _____.

wind gap

84) Item Type: Fill in the Blank   Objective: 6   Section 13.2
    Level: A: Standard   Tag: None

In what stage of a river do large meanders develop? _____

the mature stage

85) Item Type: Fill in the Blank   Objective: 7   Section 13.3
    Level: B: Enriched   Tag: Diagram 1375

The region labeled **X** in the diagram is produced by stream _____.

deposition

Chapter 13

86) Item Type: Fill in the Blank   Objective: 7   Section 13.3
    Level: A: Standard   Tag: None

The accumulation of flood deposits along the banks of a stream eventually produces raised

banks called _____.

| natural levees ■ |

87) Item Type: Fill in the Blank   Objective: 8   Section 13.3
    Level: A: Standard   Tag: None

The part of a valley floor that may be covered with water during a flood is called

_____.

| a floodplain ■ |

88) Item Type: Fill in the Blank   Objective: 9   Section 13.3
    Level: A: Standard   Tag: None

The construction of a water dam usually results in the formation of an artificial

_____.

| lake ■ |

89) Item Type: Essay   Objective: 2   Section 13.1
    Level: B: Enriched   Tag: None

Describe the major factors affecting local water budgets.

_____

_____

_____

_____

_____

Factors that affect a local water budget include temperature, the presence of vegetation, wind, and the amount and duration of rainfall. The presence of vegetation reduces runoff. Winds increase the rate of evapotranspiration. When precipitation exceeds evapotranspiration and runoff, the result is moist soil and possible flooding. When evapotranspiration exceeds precipitation, soil becomes dry.

90) Item Type: Essay   Objective: 2   Section 13.1
     Level: A: Standard   Tag: None

What are the factors that affect the local water budget?

_____

_____

_____

_____

_____

Factors that affect the local water budget include temperature, presence of vegetation, wind, and the amount and duration of precipitation.

91) Item Type: Essay   Objective: 5   Section 13.2
     Level: B: Enriched   Tag: None

Describe the changes in a stream's gradient and velocity as it moves from its headwaters to its mouth.

_____

_____

_____

_____

Near the headwaters, the gradient is steep and the stream generally has a high velocity.
Near the mouth of the stream, the gradient often becomes less steep and velocity decreases.

92) Item Type: Essay   Objective: 7   Section 13.3
     Level: B: Enriched   Tag: None

How does a stream's velocity affect the total load it can carry?

_____

_____

_____

_____

_____

The total load a stream can carry is greatest when a large volume of water is flowing
swiftly. When the velocity of the water decreases, the ability of the stream to carry a load
also decreases. As a result, part of the stream load is deposited as sediment. When
velocity increases again, some or all of the sediment is carried away.

93) Item Type: Essay   Objective: 8   Section 13.3
     Level: B: Enriched   Tag: None

Explain how the Mississippi River can be increasing the land area of the United States.

_____

_____

_____

_____

_____

Stream deposition is the process by which a stream entering a larger body of water
deposits its load. As the Mississippi River enters the Gulf of Mexico, it deposits its load
and slowly adds to the coastline of the United States.

1) Tag Name: Diagram 1371

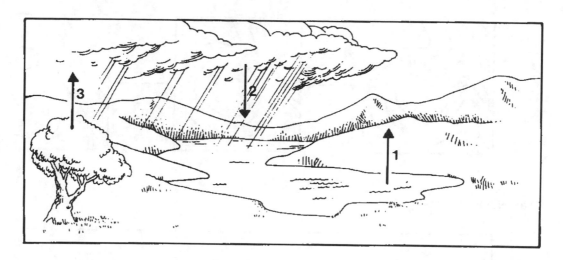

2) Tag Name: Diagram 1372

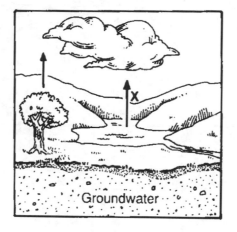

3) Tag Name: Diagram 1373

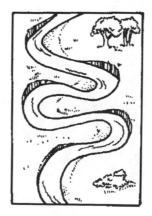

4)   Tag Name: Diagram 1374

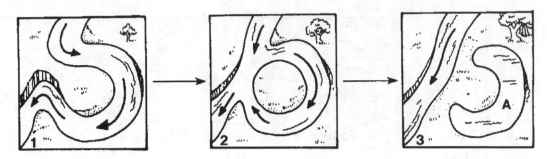

5)   Tag Name: Diagram 1375

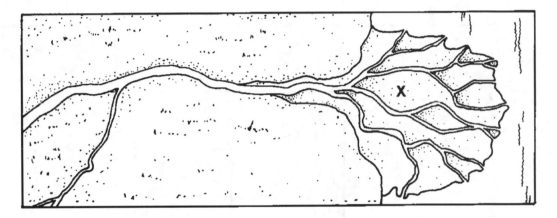

1) Item Type: True-False   Objective: 1   Section 14.1
    Level: B: Enriched   Tag: None

_____ It is possible for a porous rock to be impermeable.

T

2) Item Type: True-False   Objective: 1   Section 14.1
    Level: A: Standard   Tag: None

_____ Porosity is influenced by the way sediment particles are packed together.

T

3) Item Type: True-False   Objective: 1   Section 14.1
    Level: A: Standard   Tag: None

_____ The velocity of groundwater movement increases as the gradient of the water table increases.

T

4) Item Type: True-False   Objective: 1   Section 14.1
    Level: A: Standard   Tag: None

_____ Water flowing on the earth's surface is called groundwater.

F

5) Item Type: True-False   Objective: 1   Section 14.1
    Level: A: Standard   Tag: None

_____ Rocks that do not allow water to pass through them easily are called impermeable.

T

6) Item Type: True-False   Objective: 2   Section 14.1
    Level: B: Enriched   Tag: None

_____ The amount of rainfall in an area has no effect on the level of the area's water table.

F

7) Item Type: True-False   Objective: 2   Section 14.1
    Level: A: Standard   Tag: None

_____ Capillary action is due to the attraction of carbon dioxide molecules to other substances.

F

8) Item Type: True-False   Objective: 3   Section 14.1
Level: A: Standard   Tag: None

_____ The water table tends to parallel the topography of the land.

T

9) Item Type: True-False   Objective: 3   Section 14.1
Level: A: Standard   Tag: None

_____ A spring formed in an area with a perched water table may flow continuously.

T

10) Item Type: True-False   Objective: 4   Section 14.1
Level: A: Standard   Tag: None

_____ Once a groundwater aquifer has been emptied, it will never be productive again.

F

11) Item Type: True-False   Objective: 5   Section 14.2
Level: A: Standard   Tag: None

_____ A cone of depression forms as a result of the collapse of a cavern roof.

F

12) Item Type: True-False   Objective: 5   Section 14.2
Level: B: Enriched   Tag: None

_____ An artesian formation is bounded by two layers of impermeable rock.

T

13) Item Type: True-False   Objective: 5   Section 14.2
Level: B: Enriched   Tag: None

_____ An artesian formation must have a sloping permeable layer that is exposed to the earth's surface.

T

14) Item Type: True-False   Objective: 5   Section 14.2
Level: A: Standard   Tag: None

_____ Many desert oases are formed by artesian springs.

T

15) Item Type: True-False   Objective: 5   Section 14.2
     Level: A: Standard   Tag: None

_____ The layer of cap rock in an artesian formation is impermeable.

```
| T |___|
```

16) Item Type: True-False   Objective: 5   Section 14.2
     Level: A: Standard   Tag: None

_____ Ordinary springs are commonly located where the water table intersects ground surface.

```
| T |___|
```

17) Item Type: True-False   Objective: 5   Section 14.2
     Level: B: Enriched   Tag: None

_____ A well may go dry if the cone of depression drops to the bottom of the well.

```
| T |___|
```

18) Item Type: True-False   Objective: 5   Section 14.2
     Level: A: Standard   Tag: None

_____ A spring can be found when the ground dips below the water table.

```
| T |___|
```

19) Item Type: True-False   Objective: 6   Section 14.2
     Level: A: Standard   Tag: None

_____ Artesian wells erupt from open pools and shoot up sheets of hot water.

```
| F |___|
```

20) Item Type: True-False   Objective: 6   Section 14.2
     Level: A: Standard   Tag: None

_____ Because hot water tends to have fewer minerals dissolved in it than cold water, hot springs often contain very small concentrations of minerals.

```
| F |___|
```

21) Item Type: True-False   Objective: 6   Section 14.2
     Level: A: Standard   Tag: None

_____ Hot springs generally have temperatures at or very near the boiling point of water.

```
| T |___|
```

22) Item Type: True-False   Objective: 6   Section 14.2
Level: A: Standard   Tag: None

_____ Geysers tend to build deposits of the silicate mineral geyserite.

T

23) Item Type: True-False   Objective: 6   Section 14.2
Level: A: Standard   Tag: None

_____ A geyser is a common feature in areas of karst topography.

F

24) Item Type: True-False   Objective: 7   Section 14.3
Level: A: Standard   Tag: None

_____ Caverns often form when carbonic acid dissolves and enlarges cracks formed in limestone.

T

25) Item Type: True-False   Objective: 7   Section 14.3
Level: A: Standard   Tag: None

_____ Soft water contains large amounts of dissolved minerals.

F

26) Item Type: True-False   Objective: 7   Section 14.3
Level: A: Standard   Tag: None

_____ Hot water tends to have more minerals dissolved in it than cold water.

T

27) Item Type: True-False   Objective: 7   Section 14.3
Level: A: Standard   Tag: None

_____ Stalactites and stalagmites may join together to form columns.

T

28) Item Type: True-False   Objective: 8   Section 14.3
Level: A: Standard   Tag: None

_____ Paint pots form when water containing dissolved limestone builds a cone of rock on a cavern floor.

F

29) Item Type: Multiple Choice   Objective: 1   Section 14.1
Level: B: Enriched   Tag: None

An aquifer contains a great deal of water, but the water cannot be removed easily with an ordinary well. This is most likely due to sediment and rock characterized by

a) low porosity.

b) low permeability.

c) poor sorting.

d) high iron content.

b

30) Item Type: Multiple Choice   Objective: 1   Section 14.1
Level: B: Enriched   Tag: None

Which of the following factors most directly influences porosity?

a) grain composition

b) sorting of grains

c) grain size

d) permeability of grains

b

31) Item Type: Multiple Choice   Objective: 1   Section 14.1
Level: B: Enriched   Tag: None

Which of the following is the correct term for a layer of rock that is permeable and has high porosity?

a) aquifer   b) travertine layer   c) natural cavern   d) karst region

a

32) Item Type: Multiple Choice   Objective: 1   Section 14.1
Level: A: Standard   Tag: None

When a sediment is well sorted, its particles are all about the same

a) size.   b) shape.   c) weight.   d) composition.

a

33) Item Type: Multiple Choice   Objective: 1   Section 14.1
Level: B: Enriched   Tag: Diagram 1471

Which of these descriptions best characterizes the sediment sample presented in this diagram?

a) poorly sorted with high porosity

b) poorly sorted with low porosity

c) well sorted with low porosity

d) well sorted with high porosity

b

34) Item Type: Multiple Choice   Objective: 1   Section 14.1
    Level: A: Standard   Tag: None

The attraction of water molecules to other materials, such as soil, is called

a) permeability.   b) gradient.   c) porosity.   d) capillary action.

d

35) Item Type: Multiple Choice   Objective: 2   Section 14.1
    Level: A: Standard   Tag: None

The bottom region of the zone of aeration is known as the

a) aquifer base.   b) capillary fringe.

c) cone of depression.   d) artesian formation.

b

36) Item Type: Multiple Choice   Objective: 2   Section 14.1
    Level: B: Enriched   Tag: Diagram 1472

The groundwater region labeled **X** in the diagram represents a

a) perched water table.   b) zone of saturation.

c) capillary fringe.   d) zone of aeration.

b

37) Item Type: Multiple Choice   Objective: 2   Section 14.1
    Level: A: Standard   Tag: None

The water table forms the upper surface of the

a) zone of saturation.   b) cap rock layer.

c) contour surface.   d) artesian formation.

a

38) Item Type: Multiple Choice   Objective: 2   Section 14.1
    Level: B: Enriched   Tag: None

The rate at which groundwater moves is related to the slope of the water table. The steepness of this slope is called its

a) porosity.   b) capillary fringe.   c) artesian formation.   d) gradient.

d

39) Item Type: Multiple Choice   Objective: 2   Section 14.1
Level: A: Standard   Tag: None

Capillary action in soil results in the upward movement of

a) oxygen.   b) calcite.   c) carbon dioxide.   d) water.

d

40) Item Type: Multiple Choice   Objective: 2   Section 14.1
Level: A: Standard   Tag: None

The capillary fringe lies within the

a) zone of aeration.   b) zone of saturation.

c) water table.   d) artesian formation.

a

41) Item Type: Multiple Choice   Objective: 3   Section 14.1
Level: A: Standard   Tag: None

Which of the following statements about the water table is true?

a) The water table generally follows the topography of the land surface.

b) The water table is found in the zone of aeration.

c) There is only one water table below all land surfaces.

d) The level of the water table remains constant.

a

42) Item Type: Multiple Choice   Objective: 3   Section 14.1
Level: A: Standard   Tag: None

The feature found where a secondary water table is separated from the main water table by a layer of impermeable rock is

a) an artesian formation.   b) a perched water table.

c) a travertine terrace.   d) karst topography.

b

43) Item Type: Multiple Choice   Objective: 3   Section 14.1
Level: A: Standard   Tag: None

A cone of depression in a water table is most likely due to

a) a well.   b) an artesian spring.   c)   a geyser.   d) a cavern.

a

44) Item Type: Multiple Choice   Objective: 4   Section 14.1
    Level: B: Enriched   Tag: None

Which of the following statements about groundwater is true?

a) Groundwater is a nonrenewable resource.

b) Groundwater can be replenished by pumping sea water into the ground.

c) Once groundwater is polluted it can never be used again.

d) Groundwater can be conserved by recycling used water.

45) Item Type: Multiple Choice   Objective: 4   Section 14.1
    Level: A: Standard   Tag: None

Which of the following is most likely to occur when pesticides and fertilizers seep into soil?

a) excessive erosion   b) severe drought

c) acid rain formation   d) groundwater contamination

46) Item Type: Multiple Choice   Objective: 4   Section 14.1
    Level: A: Standard   Tag: None

Contamination of coastal groundwater by salt water is primarily due to

a) highway runoff in winter and spring.

b) lowering of the water table.

c) infiltration of toxic wastes.

d) introduction of hard water.

47) Item Type: Multiple Choice   Objective: 5   Section 14.2
    Level: B: Enriched   Tag: None

A natural flow of water that forms when the water table intersects the ground surface is called

a) a geyser.   b) an ordinary well.

c) an ordinary spring.   d) an artesian well.

Chapter 14

48) Item Type: Multiple Choice   Objective: 5   Section 14.2
Level: B: Enriched   Tag: None

Which of the following statements is true?

a) Artesian formations consist of an impermeable layer of rock sandwiched between two layers of permeable rock.

b) The aquifer in an artesian formation must be sloping and exposed at the surface.

c) When a water table is found on a hill, it is called a perched water table.

d) The position of the water table is fixed within an aquifer.

49) Item Type: Multiple Choice   Objective: 5   Section 14.2
Level: A: Standard   Tag: None

Which of the following is often the source of water in a desert oasis?

a) an ordinary well   b) a geyser

c) an artesian spring   d) a hot spring

50) Item Type: Multiple Choice   Objective: 5   Section 14.2
Level: A: Standard   Tag: None

A sloping layer of permeable rock sandwiched between two layers of impermeable rock and exposed at the surface is

a) mud pot.   b) a hot spring.

c) an ordinary well.   d) an artesian formation.

51) Item Type: Multiple Choice   Objective: 5   Section 14.2
Level: A: Standard   Tag: None

Pumping water from a well creates a lowered area of the water table called a

a) cone of depression.   b) fissure spring.

c) perched water table.   d) paint pot.

a

52) Item Type: Multiple Choice   Objective: 5   Section 14.2
    Level: B: Enriched   Tag: Diagram 1474

In the diagram above, rock layer *I* is called the

a) aquifer.   b) perched water table.

c) cap rock.   d) zone of saturation.

c

53) Item Type: Multiple Choice   Objective: 5   Section 14.2
    Level: B: Enriched   Tag: Diagram 1474

The structure labeled *X* in the diagram above is

a) a geyser.   b) an ordinary spring.   c) a mud pot.   d) an artesian well.

d

54) Item Type: Multiple Choice   Objective: 5   Section 14.2
    Level: A: Standard   Tag: None

A cap rock is the top layer of impermeable rock that is exposed in

a) a travertine deposition.   b) karst topography.

c) an artesian formation.   d) capillary action.

c

55) Item Type: Multiple Choice   Objective: 6   Section 14.2
    Level: B: Enriched   Tag: None

For thousands of years, groundwater has been filling the caverns beneath central Florida, taking the place of limestone that has

a) sunk.   b) washed away with precipitation.

c) dissolved.   d) been contaminated by pollution.

c

56) Item Type: Multiple Choice   Objective: 6   Section 14.2
    Level: B: Enriched   Tag: None

Groundwater is sometimes heated beneath the earth's surface as it passes through areas where there has been recent

a) severe drought.   b) aquifer contamination.

c) mineral deposition.   d) volcanic activity.

d

57) Item Type: Multiple Choice   Objective: 6   Section 14.2
Level: B: Enriched   Tag: None

An area is described as a source of geothermal energy. Which of the following formations would most likely be found there?

a)   a cavern   b)   a sinkhole

c)   a natural bridge   d)   a geyser

d

58) Item Type: Multiple Choice   Objective: 6   Section 14.2
Level: B: Enriched   Tag: Diagram 1475

Which of the following features is represented by this diagram?

a)   travertine terrace   b)   geyser

c)   cap rock   d)   artesian well

b

59) Item Type: Multiple Choice   Objective: 6   Section 14.2
Level: A: Standard   Tag: None

The type of hot spring that is characterized by a crooked tube with several connected underground chambers is called

a)   a geyser.   b)   an ordinary spring.

c)   an artesian spring.   d)   a mud pot.

a

60) Item Type: Multiple Choice   Objective: 7   Section 14.3
Level: B: Enriched   Tag: None

Which of the following processes is most likely to cause caverns to form?

a)   chemical weathering   b)   hydration   c)   hydrolysis   d)   oxidation

a

61) Item Type: Multiple Choice   Objective: 7   Section 14.3
Level: A: Standard   Tag: None

The mineral that forms cone-shaped cavern features is

a)   calcite.   b)   magnesium.   c)   geyserite.   d)   iron.

a

62) Item Type: Multiple Choice   Objective: 7   Section 14.3
    Level: A: Standard   Tag: None

One of the major disadvantages of hard water is that it

a) cannot be used for drinking.

b) inhibits the flow of water up through artesian wells.

c) contaminates groundwater aquifers.

d) damages appliances due to the buildup of mineral deposits.

d

63) Item Type: Multiple Choice   Objective 7   Section 14.3
    Level: A: Standard   Tag: None

For an ordinary well to function properly, it must penetrate deep into rock or sediment that is

a) highly permeable.   b) below two water tables.

c) impermeable.   d) below a layer of cap rock.

a

64) Item Type: Multiple Choice   Objective: 7   Section 14.3
    Level: A: Standard   Tag: None

Most sinkholes are formed by

a) the collapse of a cavern roof.   b) volcanic activity.

c) stream erosion.   d) travertine deposits.

a

65) Item Type: Multiple Choice   Objective: 7   Section 14.3
    Level: A: Standard   Tag: None

Karst topography is a result of

a) geothermal activity.   b) groundwater contamination.

c) chemical weathering.   d) stream erosion.

c

66) Item Type: Multiple Choice   Objective: 8   Section 14.3
    Level: B: Enriched   Tag: Diagram 1476

What feature is represented in the diagram?

a) stalactite   b) column   c) stalagmite   d) natural bridge

c

67) Item Type: Multiple Choice   Objective: 8   Section 14.3
    Level: B: Enriched   Tag: None

Which of the following is most likely to be produced as a result of the process of chemical weathering?

a) caverns   b) hot springs   c) soft water   d) geysers

a

68) Item Type: Multiple Choice   Objective: 8   Section 14.3
    Level: A: Standard   Tag: None

Karst topography occurs in regions that have experienced large amounts of

a) volcanic activity.   b) wind erosion.

c) chemical weathering.   d) groundwater pollution.

c

69) Item Type: Multiple Choice   Objective: 8   Section 14.3
    Level: B: Enriched   Tag: None

Which of the following features would most likely be found in an area that has karst topography?

a) geysers   b) caverns

c) artesian formations   d) travertine terraces

b

70) Item Type: Multiple Choice   Objective: 8   Section 14.3
    Level: A: Standard   Tag: None

The features suspended from the roofs of caverns formed from the deposition of dissolved material are called

a) fissures.   b) natural bridges.

c) stalactites.   d) disappearing streams.

c

71) Item Type: Fill in the Blank   Objective: 1   Section 14.1
    Level: B: Enriched   Tag: None

The percentage of open space in a given volume of rock is referred to as

_____.

porosity

Chapter 14

72) Item Type: Fill in the Blank   Objective: 1   Section 14.1
    Level: A: Standard   Tag: None

Rocks that do not allow water to flow through them are referred to as

_____.

| impermeable |

73) Item Type: Fill in the Blank   Objective: 1   Section 14.1
    Level: A: Standard   Tag: None

The measure of how freely water passes through the open spaces in rock or sediment is

referred to as _____.

| permeability |

74) Item Type: Fill in the Blank   Objective: 1   Section 14.1
    Level: A: Standard   Tag: None

The amount of uniformity in the size of grains in rock is referred to as

_____.

| sorting |

75) Item Type: Fill in the Blank   Objective: 1   Section 14.1
    Level: A: Standard   Tag: None

The ease with which a liquid passes through a rock or sediment is called

_____.

| permeability |

76) Item Type: Fill in the Blank   Objective: 2   Section 14.1
    Level: B: Enriched   Tag: Diagram 1473

In the diagram above, the water in the area labeled **A** is called _____.

| groundwater |

77) Item Type: Fill in the Blank   Objective: 2   Section 14.1
    Level: B: Enriched   Tag: Diagram 1473

In the diagram above, the water level indicated by label **B** is referred to as the

_____.

| water table |

78) Item Type: Fill in the Blank   Objective: 2   Section 14.1
Level: B: Enriched   Tag: None

The attraction of water molecules to other materials, such as soil, is called

_____.

capillary action

79) Item Type: Fill in the Blank   Objective: 2   Section 14.1
Level: A: Standard   Tag: None

The groundwater layer in which all pore spaces are filled with water is called the

_____.

zone of saturation

80) Item Type: Fill in the Blank   Objective: 2   Section 14.1
Level: A: Standard   Tag: None

The upper surface of the zone of saturation is called the _____.

water table

81) Item Type: Fill in the Blank   Objective: 2   Section 14.1
Level: A: Standard   Tag: None

About 90 percent of the earth's unfrozen fresh water is found in _____.

groundwater

82) Item Type: Fill in the Blank   Objective: 2   Section 14.1
Level: A: Standard   Tag: None

Water found below the earth's surface is called _____.

groundwater

83) Item Type: Fill in the Blank   Objective: 2   Section 14.1
Level: A: Standard   Tag: None

Large amounts of water can flow through and be stored in a body of rock called a(n)

_____.

aquifer

84) Item Type: Fill in the Blank   Objective: 3   Section 14.1
    Level: A: Standard   Tag: None

Groundwater flows downhill in response to _____.

gravity

85) Item Type: Fill in the Blank   Objective: 3   Section 14.1
    Level: A: Standard   Tag: None

The steepness of a water table's slope is known as its _____.

gradient

86) Item Type: Fill in the Blank   Objective: 5   Section 14.2
    Level: B: Enriched   Tag: None

When the water table intersects the ground surface, it forms free-flowing bodies of water

called _____.

ordinary springs

87) Item Type: Fill in the Blank   Objective: 6   Section 14.2
    Level: A: Standard   Tag: None

Hot springs that erupt periodically are called _____.

geysers

88) Item Type: Fill in the Blank   Objective: 6   Section 14.2
    Level: A: Standard   Tag: None

Mud pots with highly colored clay are sometimes called _____.

paint pots

89) Item Type: Fill in the Blank   Objective: 7   Section 14.3
    Level: A: Standard   Tag: None

Water that contains relatively large amounts of dissolved minerals is known as

_____.

hard water

90) Item Type: Fill in the Blank   Objective: 7   Section 14.3
    Level: B: Enriched   Tag: None

Most caverns are formed when weak acids break down the rock through the process of

_____.

| chemical weathering ■ |

91) Item Type: Fill in the Blank   Objective: 7   Section 14.3
    Level: A: Standard   Tag: None

Cone-shaped calcite deposits found suspended from the ceiling of a cavern are called

_____.

| stalactites ■ |

92) Item Type: Fill in the Blank   Objective: 8   Section 14.3
    Level: A: Standard   Tag: None

Regions where the effects of chemical weathering due to groundwater are clearly visible at

the surface are said to have _____.

| karst topography ■ |

93) Item Type: Fill in the Blank   Objective: 8   Section 14.3
    Level: A: Standard   Tag: None

The circular depression that forms when the roof of a cavern collapses is called a(n)

_____.

| sinkhole ■ |

94) Item Type: Essay   Objective: 1   Section 14.1
Level: B: Enriched   Tag: None

A scientist discovers that a rock layer is permeable but contains very little groundwater. How can this occur?

_____

_____

_____

_____

_____

The rock layer has high permeability and low porosity. The porosity of rock is influenced by the size and arrangement of the particles forming the rock. If the particles are many different sizes, small particles fill in the spaces between larger ones, making the rock less porous. If the particles are packed tightly together, there are few open spaces to hold water.

95) Item Type: Essay   Objective: 3   Section 14.1
Level: B: Enriched   Tag: None

Describe how the gradient of a water table affects the flow of groundwater?

_____

_____

_____

_____

_____

Gradient is the steepness of any slope, including that of a water table. The velocity of water flow increases as its gradient increases but slows as its gradient decreases.

96) Item Type: Essay   Objective: 4 Section 14.1
  Level: A: Standard   Tag: None

Describe ways in which groundwater may become polluted.

_____

_____

_____

_____

_____

Groundwater may become polluted by waste dumps; leaking underground storage tanks for toxic chemicals; agricultural fertilizers, lawn fertilizers, and pesticides; and leaking sewage systems. If too much groundwater is pumped from an aquifer near the ocean, the water table may drop below sea level, allowing salt water to flow in and contaminate the aquifer.

97) Item Type: Essay   Objective: 6 Section 14.2
  Level: B: Enriched   Tag: None

Explain how mud pots form.

_____

_____

_____

_____

_____

Mud pots form when the rock surrounding a hot spring is chemically weathered by volcanic gases dissolved in the water. The weathered rock mixes with hot water to form a sticky, liquid clay that bubbles at the surface.

98) Item Type: Essay   Objective: 8   Section 14.3
     Level: A: Standard   Tag: None

How do natural bridges form?

_____

_____

_____

_____

_____

Natural bridges form when the roof of a cavern collapses in several places, leaving
stretches of uncollapsed rock between the sinkholes. A natural bridge can also form when
a surface river enters a crack in a rock formation, runs underground, and then reemerges.
The river weathers the rock it passes through and eventually forms a natural bridge.

1)   Tag Name: Diagram 1471

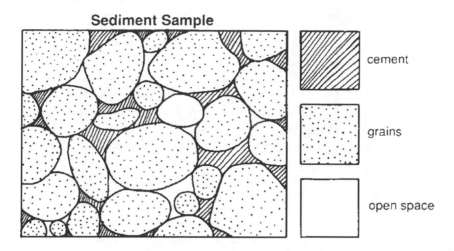

2)   Tag Name: Diagram 1472

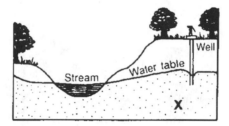

3)   Tag Name: Diagram 1473

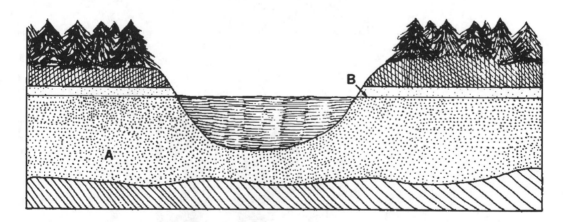

4)   Tag Name: Diagram 1474

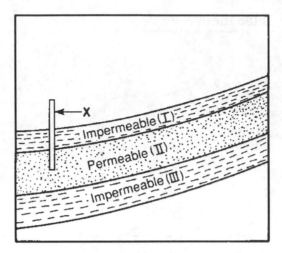

5)   Tag Name: Diagram 1475

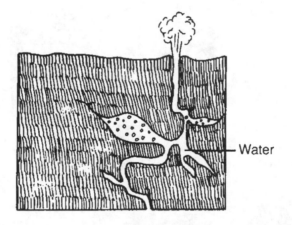

6)   Tag Name: Diagram 1476

1)   Item Type: True-False   Objective: 1   Section 15.1
        Level: A: Standard   Tag: None

     _____ The snowline is the line that marks the farthest advance of a glacier.

F

2)   Item Type: True-False   Objective: 1   Section 15.1
        Level: A: Standard   Tag: None

     _____ A snowfield is a seasonal mass of snow and ice.

F

3)   Item Type: True-False   Objective: 1   Section 15.1
        Level: A: Standard   Tag: None

     _____ Firn is a glacial deposit of unsorted material.

F

4)   Item Type: True-False   Objective: 2   Section 15.1
        Level: A: Standard   Tag: None

     _____ A valley glacier generally forms in mountainous areas.

T

5)   Item Type: True-False   Objective: 2   Section 15.1
        Level: A: Standard   Tag: None

     _____ Greenland is mostly covered by a continental ice sheet.

T

6)   Item Type: True-False   Objective: 2   Section 15.1
        Level: A: Standard   Tag: None

     _____ Valley glaciers generally carve sharp, rugged features.

T

7)   Item Type: True-False   Objective: 3   Section 15.1
        Level: A: Standard   Tag: None

     _____ Friction causes the sides of a valley glacier to move more slowly than the center.

T

8) Item Type: True-False   Objective: 3   Section 15.1
Level: B: Enriched   Tag: None

_____ Basal slip occurs when the weight of a glacier exerts enough pressure to melt the ice where it touches the ground, forming a lubricant.

T

9) Item Type: True-False   Objective: 3   Section 15.1
Level: B: Enriched   Tag: None

_____ An ice shelf forms when an ice sheet moves out over a body of water.

T

10) Item Type: True-False   Objective: 3   Section 15.1
Level: A: Standard   Tag: None

_____ Internal plastic flow in a glacier occurs when water melted by the weight of the glacier acts as a lubricant between the glacier and the underlying rock.

F

11) Item Type: True-False   Objective: 3   Section 15.1
Level: A: Standard   Tag: None

_____ Ice moves faster at the surface of a glacier than at the glacier's base.

T

12) Item Type: True-False   Objective: 4   Section 15.2
Level: A: Standard   Tag: None

_____ An arête is formed by glacial deposition.

F

13) Item Type: True-False   Objective: 4   Section 15.2
Level: A: Standard   Tag: None

_____ Roches moutonnées are rounded knobs of bedrock produced by moving glaciers.

T

14) Item Type: True-False   Objective: 4   Section 15.2
Level: A: Standard   Tag: None

_____ Hanging valleys are caused by glacial meltwater.

F

15) Item Type: True-False   Objective: 4   Section 15.2
     Level: A: Standard   Tag: None

_____ Cirques are depressions left by melted blocks of ice from continental ice sheets.

F

16) Item Type: True-False   Objective: 4   Section 15.2
     Level: A: Standard   Tag: None

_____ Glaciers often form V-shaped valleys.

F

17) Item Type: True-False   Objective: 4   Section 15.2
     Level: A: Standard   Tag: None

_____ The presence of glaciers in the past may be inferred from the presence of V-shaped valleys.

F

18) Item Type: True-False   Objective: 5   Section 15.2
     Level: A: Standard   Tag: None

_____ Stratified drift is material that has been deposited by meltwater flowing from a glacier.

T

19) Item Type: True-False   Objective: 5   Section 15.2
     Level: A: Standard   Tag: None

_____ Eskers resemble raised, winding roadways.

T

20) Item Type: True-False   Objective: 5   Section 15.2
     Level: A: Standard   Tag: None

_____ Lateral moraines are formed at the margins of glaciers.

T

21) Item Type: True-False   Objective: 5   Section 15.2
     Level: A: Standard   Tag: None

_____ Drumlins are rounded projections of rock.

F

22) Item Type: True-False   Objective: 5   Section 15.2
    Level: A: Standard   Tag: None

_____ Till is composed of sorted deposits of rock material.

F

23) Item Type: True-False   Objective: 5   Section 15.2
    Level: A: Standard   Tag: None

_____ An esker is a ridge carved by a valley glacier.

F

24) Item Type: True-False   Objective: 5   Section 15.2
    Level: A: Standard   Tag: None

_____ Drumlins often occur in clusters.

T

25) Item Type: True-False   Objective: 6   Section 15.2
    Level: B: Enriched   Tag: None

_____ Most glacial lakes are formed by deposition rather than by erosion.

T

26) Item Type: True-False   Objective: 6   Section 15.2
    Level: A: Standard   Tag: None

_____ The Great Lakes formed in what were once river valleys.

T

27) Item Type: True-False   Objective: 7   Section 15.3
    Level: A: Standard   Tag: None

_____ Melting of an ice sheet would result in a lowering of sea level.

F

28) Item Type: True-False   Objective: 7   Section 15.3
    Level: A: Standard   Tag: None

_____ Ice ages may begin with a lowering of global temperatures.

T

29) Item Type: True-False   Objective: 7   Section 15.3
     Level: B: Enriched   Tag: None

_____ During the last ice age, ice covered about half of the earth.

F

30) Item Type: True-False   Objective: 7   Section 15.3
     Level: A: Standard   Tag: None

_____ Most of the earth's ice ages have lasted about 1,000 years.

F

31) Item Type: True-False   Objective: 8   Section 15.3
     Level: A: Standard   Tag: None

_____ The Milankovitch theory predicts the rate at which glaciers slide.

F

32) Item Type: True-False   Objective: 8   Section 15.3
     Level: A: Standard   Tag: None

_____ The Milankovitch theory states that ice ages occur when volcanic dust decreases the amount of sunlight reaching the earth's surface.

F

33) Item Type: True-False   Objective: 8   Section 15.3
     Level: A: Standard   Tag: None

_____ The distribution of solar energy reaching the earth's surface depends on the earth's tilt.

T

34) Item Type: Multiple Choice   Objective: 1   Section 15.1
     Level: A: Standard   Tag: None

Which of the following features form because of pressure on the surface of glaciers?

a) crevasses   b) ice shelves

c) kettles   d) roches moutonnées

a

35) Item Type: Multiple Choice   Objective: 1   Section 15.1
    Level: B: Enriched   Tag: None

Firn is formed as a result of

a) the polishing of bedrock by the scraping action of rock particles embedded in glacial ice.

b) changes in the flow rate of glacial meltwater.

c) the partial melting and refreezing of snow crystals into small grains of ice.

d) the deposition of unsorted glacial drift.

c

36) Item Type: Multiple Choice   Objective: 1   Section 15.1
    Level: B: Enriched   Tag: None

As the latitude of snowlines moves from the equator to the poles, the elevation of snowlines

a) remains constant.   b) increases.   c) varies randomly.   d) decreases.

d

37) Item Type: Multiple Choice   Objective: 1   Section 15.1
    Level: A: Standard   Tag: None

Which of the following glacial deposits is formed from stratified drift?

a) drumlin   b) lateral moraine   c) esker   d) medial moraine

c

38) Item Type: Multiple Choice   Objective: 2   Section 15.1
    Level: A: Standard   Tag: None

Today a continental ice sheet is located in

a) Alaska.   b) Antarctica.   c) Canada.   d) Siberia.

b

39) Item Type: Multiple Choice   Objective: 3   Section 15.1
    Level: A: Standard   Tag: None

When a tributary glacier melts, it may leave a suspended formation called

a) a hanging valley.   b) a snowline.   c) a kettle.   d) an erratic.

a

40) Item Type: Multiple Choice   Objective: 3   Section 15.1
Level: A: Standard   Tag: None

Ice in a glacier moves downslope in response to

a) friction.   b) melting.   c) gravity.   d) freezing.

c

41) Item Type: Multiple Choice   Objective: 3   Section 15.1
Level: A: Standard   Tag: Diagram 1571

Which of the following features is represented in the diagram above by number *1*?

a) crevasses   b) ice sheets   c) cirques   d) icebergs

d

42) Item Type: Multiple Choice   Objective: 3   Section 15.1
Level: B: Enriched   Tag: Diagram 1571

An ice shelf is indicated in the diagram above by number

a) *2.*   b) *3.*   c) *4.*   d) *5.*

a

43) Item Type: Multiple Choice   Objective: 4   Section 15.2
Level: B: Enriched   Tag: None

Which of the following features is caused by erosion rather than by deposition?

a) drumlin   b) kettle   c) horn   d) esker

c

44) Item Type: Multiple Choice   Objective: 4   Section 15.2
Level: A: Standard   Tag: None

Which of the following is formed by the joining of arêtes?

a) drumlin   b) till   c) esker   d) horn

d

45) Item Type: Multiple Choice   Objective: 4   Section 15.2
Level: A: Standard   Tag: None

Roches moutonnées are best described as

a) long, low mounds of till.   b) ridges of gravel and coarse sand.

c) rounded rock projections.   d) groups of loose boulders.

c

46) Item Type: Multiple Choice   Objective: 4   Section 15.2
    Level: A: Standard   Tag: None

Glacial meltwater usually appears milky because of the presence in the water of

a) fine rock particles. b) ice crystals.   c) air bubbles.   d) dissolved salts.

a

47) Item Type: Multiple Choice   Objective: 4   Section 15.2
    Level: A: Standard   Tag: None

Cirques are best described as

a) rounded knobs of rock.   b) sharp, curved peaks.

c) bowl-shaped depressions.   d) sharp and jagged ridges.

c

48) Item Type: Multiple Choice   Objective: 5   Section 15.2
    Level: B: Enriched   Tag: None

A glacial moraine is an example of

a) a sorted depositional feature.

b) an unsorted depositional feature.

c) an erosional feature caused by moving ice.

d) an erosional feature caused by moving water.

b

49) Item Type: Multiple Choice   Objective: 5   Section 15.2
    Level: A: Standard   Tag: None

A ground moraine is formed primarily from till deposited

a) at the glacier's leading edge.   b) along the glacier's sides.

c) beneath the glacier.   d) at the glacier's source region.

c

50) Item Type: Multiple Choice   Objective: 5   Section 15.2
    Level: B: Enriched   Tag: Diagram 1572

Which feature in the diagram is formed from stratified drift?

a) *1*   b) *2*   c) *4*   d) *5*

d

51) Item Type: Multiple Choice   Objective: 5   Section 15.2
Level: A: Standard   Tag: Diagram 1572

Which feature in the diagram above may be formed by streams of meltwater flowing in tunnels beneath glacial ice?

a) **2**   b) **3**   c) **4**   d) **5**

d

52) Item Type: Multiple Choice   Objective: 5   Section 15.2
Level: A: Standard   Tag: Diagram 1572

Which feature in the diagram above represents a terminal moraine?

a) **1**   b) **2**   c) **3**   d) **4**

a

53) Item Type: Multiple Choice   Objective: 5   Section 15.2
Level: B: Enriched   Tag: Diagram 1574

Which feature is identified by the number **2** in the diagram above?

a) outwash plain   b) kettle   c) ground moraine   d) arête

a

54) Item Type: Multiple Choice   Objective: 5   Section 15.2
Level: B: Enriched   Tag: Diagram 1574

The features in the diagram above resulted from a glacier that flowed from the

a) north.   b) east.   c) south.   d) west.

d

55) Item Type: Multiple Choice   Objective: 5   Section 15.2
Level: B: Enriched   Tag: Diagram 1574

What type of feature is identified by the number **1** in the diagram above?

a) roches moutonnée   b) esker   c) drumlin   d) moraine

c

56) Item Type: Multiple Choice   Objective: 5   Section 15.2
Level: B: Enriched   Tag: None

Till is best defined as

a) an unsorted deposit of rock material.   b) sorted and layered deposits of sand.

c) sediment sorted by melted ice.   d) unsorted deposits of drift.

a

57) Item Type: Multiple Choice   Objective: 5   Section 15.2
    Level: A: Standard   Tag: None

What type of moraine forms gentle hills and depressions?

a) lateral moraine   b) terminal moraine c) medial moraine   d) ground moraine

d

58) Item Type: Multiple Choice   Objective: 6   Section 15.2
    Level: A: Standard   Tag: None

Which of the following is required to form a salt lake?

a) periods of low temperatures   b) high precipitation rates

c) multiple outlet streams   d) high evaporation rates

d

59) Item Type: Multiple Choice   Objective: 6   Section 15.2
    Level: A: Standard   Tag: None

Today the Great Lakes drain through the

a) Hudson River.   b) Susquehanna River.

c) St. Lawrence River.   d) Mohawk River.

c

60) Item Type: Multiple Choice   Objective: 7   Section 15.3
    Level: B: Enriched   Tag: None

The average length of an ice age on earth is about

a) ten thousand years.   b) fifty thousand years.

c) one hundred thousand years.   d) one million years.

b

61) Item Type: Multiple Choice   Objective: 7   Section 15.3
    Level: A: Standard   Tag: None

Evidence for the cause of ice ages has been found in the

a) mountains.   b) atmosphere.   c) valleys.   d) ocean.

d

62) Item Type: Multiple Choice   Objective: 7   Section 15.3
Level: A: Standard   Tag: None

One condition thought to be necessary to cause an ice age is

a) longer interglacial periods.
b) a rise in sea level.
c) lowered global temperatures.
d) a decrease in precipitation.

c

63) Item Type: Multiple Choice   Objective: 8   Section 15.3
Level: A: Standard   Tag: None

Which of the following is the most commonly accepted cause of ice ages?

a) changes in the amount of radiation produced by the sun
b) blockage of the sun's rays by volcanic dust
c) movements of continents affecting warm ocean currents
d) small changes in the earth's orbit, tilt, and precession

d

64) Item Type: Multiple Choice   Objective: 8   Section 15.3
Level: A: Standard   Tag: None

Changes in the amount of solar energy reaching the earth's surface are most likely to be caused by

a) variation in the tilt of the earth's axis.
b) the advance of continental ice sheets.
c) changes in the position of the continents.
d) periods of increased precipitation.

a

65) Item Type: Fill in the Blank   Objective: 1   Section 15.1
Level: A: Standard   Tag: None

An almost motionless mass of permanent snow and ice is called a(n)

_____.

snowfield

66) Item Type: Fill in the Blank   Objective: 1   Section 15.1
Level: A: Standard   Tag: None

The change from grainy firn to steel-blue ice is caused by _____.

pressure

67) Item Type: Fill in the Blank   Objective: 1   Section 15.1
Level: A: Standard   Tag: None

Masses of moving ice that carve landscapes are called _____.

glaciers

68) Item Type: Fill in the Blank   Objective: 1   Section 15.1
Level: A: Standard   Tag: None

The elevation above which ice and snow remain throughout the year is called the

_____.

snowline

69) Item Type: Fill in the Blank   Objective: 1   Section 15.1
Level: A: Standard   Tag: None

Snowfields form by an accumulation of ice and snow above the _____.

snowline

70) Item Type: Fill in the Blank   Objective: 2   Section 15.1
Level: B: Enriched   Tag: None

The largest ice shelf in Antarctica is the _____.

Ross Ice Shelf

71) Item Type: Fill in the Blank   Objective: 2   Section 15.1
Level: A: Standard   Tag: None

The type of glacier that generally forms in mountainous areas is called a(n)

_____.

valley glacier

72) Item Type: Fill in the Blank   Objective: 2   Section 15.1
Level: A: Standard   Tag: None

The type of glacier that presently covers Greenland is called a(n) _____.

continental ice sheet

73) Item Type: Fill in the Blank   Objective: 2   Section 15.1
Level: A: Standard   Tag: None

The largest continental ice sheet in the world is located in _____.

Antarctica

74) Item Type: Fill in the Blank   Objective: 3   Section 15.1
Level: A: Standard   Tag: None

Travel over the top of a glacier is dangerous because of the presence of deep cracks called

_____.

crevasses

75) Item Type: Fill in the Blank   Objective: 3   Section 15.1
Level: A: Standard   Tag: None

Glacial movement resulting from the slippage of ice crystals over each other is called

_____.

internal plastic flow

76) Item Type: Fill in the Blank   Objective: 3   Section 15.1
Level: A: Standard   Tag: None

A glacier works its way over small barriers in its path by melting and then

_____.

refreezing

77) Item Type: Fill in the Blank   Objective: 4   Section 15.2
Level: B: Enriched   Tag: None

The scraping of solid rock by small rock particles embedded in a glacier is called

_____.

polishing

78) Item Type: Fill in the Blank   Objective: 5   Section 15.2
Level: A: Standard   Tag: Diagram 1573

A lateral moraine is defined in the table by description number _____.

2

79) Item Type: Fill in the Blank   Objective: 5   Section 15.2
Level: A: Standard   Tag: Diagram 1573

A glacial erratic is defined in the table by description number _____.

4

80) Item Type: Fill in the Blank   Objective: 6   Section 15.2
     Level: A: Standard   Tag: None

Lake Winnipeg in Canada is the chief remnant of the former glacial Lake

_____.

Agassiz

81) Item Type: Fill in the Blank   Objective: 6   Section 15.2
     Level: A: Standard   Tag: None

The composition of water in a salt lake depends on the nature of dissolved

_____.

minerals

82) Item Type: Fill in the Blank   Objective: 6   Section 15.2
     Level: A: Standard   Tag: None

The Great Salt Lake in Utah is a remnant of what glacial lake? _____

Lake Bonneville

83) Item Type: Fill in the Blank   Objective: 6   Section 15.2
     Level: A: Standard   Tag: None

Salt lakes generally develop in areas that have high evaporation and low

_____.

precipitation

84) Item Type: Fill in the Blank   Objective: 6   Section 15.2
     Level: A: Standard   Tag: None

Salt lakes lose water only by _____.

evaporation

85) Item Type: Fill in the Blank   Objective: 7   Section 15.3
     Level: B: Enriched   Tag: None

Ice ages probably begin with a drop of about _____ in the earth's
average temperature.

5°C

86) Item Type: Fill in the Blank   Objective: 7   Section 15.3
Level: A: Standard   Tag: None

Ice ages probably begin with a drop in average temperatures combined with an increase in

_____.

snowfall

87) Item Type: Fill in the Blank   Objective: 7   Section 15.3
Level: A: Standard   Tag: None

During ice ages, periods during which glaciers retreat are called _____.

interglacial periods

88) Item Type: Fill in the Blank   Objective: 7   Section 15.3
Level: A: Standard   Tag: None

A long period of climatic cooling during which ice covers much of the earth is called a(n)

_____.

ice age

89) Item Type: Fill in the Blank   Objective: 7   Section 15.3
Level: A: Standard   Tag: None

Periods in which glaciers advance during an ice age are called _____.

glacial periods

90) Item Type: Fill in the Blank   Objective: 8   Section 15.3
Level: B: Enriched   Tag: Diagram 1575

Precession is indicated in the diagram by the arrow labeled _____.

B

91) Item Type: Fill in the Blank   Objective: 8   Section 15.3
Level: B: Enriched   Tag: None

Scientists have discovered that the temperature of the water in which marine animals live is

related to the ratio of two isotopes of _____ in the animals' shells.

oxygen

92) Item Type: Fill in the Blank   Objective: 8   Section 15.3
     Level: A: Standard   Tag: None

A theory that ice ages are caused by changes in the earth's orbit and tilt was proposed by

a scientist named Milutin _____.

> Milankovitch

93) Item Type: Fill in the Blank   Objective: 8   Section 15.3
     Level: A: Standard   Tag: None

Scientists can follow a record of ice ages left in the oceans by studying the ratio of oxygen

isotopes found in marine _____.

> shells

94) Item Type: Essay   Objective: 2   Section 15.1
     Level: B: Enriched   Tag: None

Describe the differences between a valley glacier and a continental ice sheet.

_____

_____

_____

_____

_____

> Valley glaciers and continental glaciers are distinguished by the ways they are formed and
> by their size. Valley glaciers are formed in the snowfields of mountainous areas. These
> glaciers are long, narrow, and wedge-shaped. Continental glaciers cover much greater areas
> of land and are found only in Greenland and Antarctica.

95) Item Type: Essay   Objective: 4   Section 15.2
    Level: B: Enriched   Tag: None

How can a scientist tell if glacier has passed through a particular valley?

_____

_____

_____

_____

_____

As the walls and floor of a valley are scraped away by a glacier, the original stream-cut V-shape of the valley is changed into a glacial U-shape. Because glacial action is the only means by which a valley can acquire a U-shape, scientists can easily tell whether a valley has been glaciated.

96) Item Type: Essay   Objective: 4   Section 15.2
    Level: B: Enriched   Tag: None

Describe how cirques, arêtes, and horns are related.

_____

_____

_____

_____

_____

Cirques are bowl-shaped depressions eroded by glaciers in the upper portions of mountain valleys. The ridges that separate adjacent cirques become sharp and jagged, forming arêtes. When several arêtes are joined, they form a sharp, curved peak called a horn.

97) Item Type: Essay   Objective: 6   Section 15.2
Level: A: Standard   Tag: None

Describe the formation of a salt lake.

_____

_____

_____

_____

_____

> Large lake basins fill with rainfall and meltwater when the glaciers retreat. In lakes that have no outlet streams, water can only leave by evaporation. As water evaporates, salt dissolved in the water is left behind, and the water becomes increasingly salty. This results in the formation of a salt lake.

98) Item Type: Essay   Objective: 7   Section 15.3
Level: B: Enriched   Tag: None

How would an increase in the amount of solar energy reaching the earth's surface most likely affect sea level along the coastline of the United States?

_____

_____

_____

_____

> An increase in solar energy would likely cause the continental ice sheets of the world, such as those in Greenland and Antarctica, to begin melting. This increase in water in the oceans would significantly raise the sea level. Coastlines would then be flooded for some distance inland, depending on the flatness of the coastal region.

99) Item Type: Essay   Objective: 8   Section 15.3
   Level: B: Enriched   Tag: None

What are three kinds of periodic changes that occur in the way the earth moves around the sun?

_____

_____

_____

_____

_____

One change is in the shape of the earth's orbit, which changes back and forth between nearly circular and more elongated. A second change is in the degree of tilt of the earth's axis. A third periodic change results from the precession, or circular motion, of the earth's axis.

1)   Tag Name: Diagram 1571

2)   Tag Name: Diagram 1572

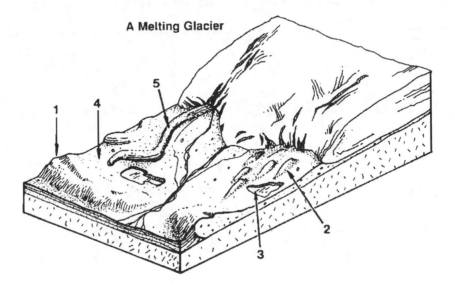

A Melting Glacier

3)   Tag Name: Diagram 1573

GLACIAL DEPOSITS

	Description
1	parallel ridges of till at a glacier's leading edge
2	a ridge of unsorted material along the sides of a glacier
3	material that has been sorted and layered by streams of melted ice
4	a boulder that has been transported and deposited by a glacier
5	long, low mounds composed of sorted till

4)   Tag Name: Diagram 1574

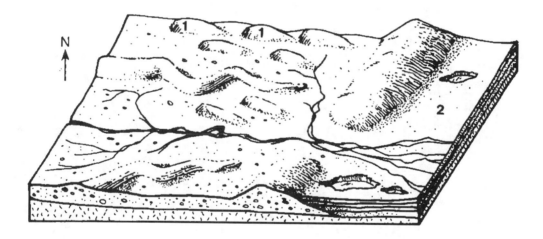

5)   Tag Name: Diagram 1575

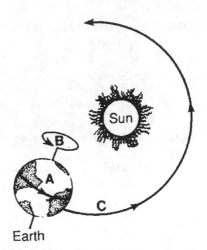

Chapter 16

1) Item Type: True-False   Objective: 1   Section 16.1
   Level: A: Standard   Tag: None

   _____ A ventifact's appearance indicates the direction of the wind that formed it.

   | T |

2) Item Type: True-False   Objective: 1   Section 16.1
   Level: A: Standard   Tag: None

   _____ Desert pavement is hard-packed clay.

   | F |

3) Item Type: True-False   Objective: 1   Section 16.1
   Level: A: Standard   Tag: None

   _____ Most sand grains are made of feldspar.

   | F |

4) Item Type: True-False   Objective: 1   Section 16.1
   Level: A: Standard   Tag: None

   _____ Deflation is a serious problem for farmers.

   | T |

5) Item Type: True-False   Objective: 1   Section 16.1
   Level: A: Standard   Tag: None

   _____ Wind erosion is most effective in moist regions with relatively hard rocks.

   | F |

6) Item Type: True-False   Objective: 2   Section 16.1
   Level: A: Standard   Tag: None

   _____ The windward side of a dune has a steeper slope than the slipface.

   | F |

7) Item Type: True-False   Objective: 2   Section 16.1
   Level: A: Standard   Tag: None

   _____ Ventifacts are rocks that are smoothed by wind abrasion.

   | T |

8) Item Type: True-False   Objective: 2   Section 16.1
    Level: A: Standard   Tag: None

     ____ All the material eroded by the wind is eventually deposited.

T

9) Item Type: True-False   Objective: 2   Section 16.1
    Level: A: Standard   Tag: None

     ____ Dunes usually form where soil is dry and unprotected.

T

10) Item Type: True-False   Objective: 2   Section 16.1
    Level: A: Standard   Tag: None

     ____ Loess deposits are extremely fertile and make good cropland.

T

11) Item Type: True-False   Objective: 3   Section 16.2
    Level: B: Enriched   Tag: None

     ____ Waves striking the shore are sometimes powerful enough to be recorded on seismographs.

T

12) Item Type: True-False   Objective: 3   Section 16.2
    Level: B: Enriched   Tag: None

     ____ Granite shorelines erode more slowly than those made of glacial deposits.

T

13) Item Type: True-False   Objective: 3   Section 16.2
    Level: A: Standard   Tag: None

     ____ Sea cliffs usually erode evenly.

F

14) Item Type: True-False   Objective: 4   Section 16.2
    Level: A: Standard   Tag: None

     ____ The part of a beach most often used by people for recreation is above the berm.

T

15) Item Type: True-False   Objective: 5   Section 16.2
    Level: A: Standard   Tag: None

_____ The berm of a beach is less steep in winter than in summer.

F

16) Item Type: True-False   Objective: 5   Section 16.2
    Level: B: Enriched   Tag: None

_____ When waves strike a beach at an angle, individual sand grains move in a straight line parallel to the shoreline.

F

17) Item Type: True-False   Objective: 5   Section 16.2
    Level: B: Enriched   Tag: None

_____ A longshore current will transport sand in a direction parallel to the shoreline.

T

18) Item Type: True-False   Objective: 5   Section 16.2
    Level: A: Standard   Tag: None

_____ Sand bars are formed when sand carried away from the beach is deposited offshore.

T

19) Item Type: True-False   Objective: 6   Section 16.3
    Level: A: Standard   Tag: None

_____ Scientists believe that the side of a continent near the trailing edge of a moving crustal plate tends to be uplifted.

F

20) Item Type: True-False   Objective: 6   Section 16.3
    Level: B: Enriched   Tag: None

_____ Sea level is currently rising because of melting glaciers.

T

21) Item Type: True-False   Objective: 6   Section 16.3
    Level: A: Standard   Tag: None

_____ An emergent coastline is characterized by the presence of few bays or headlands and many long, wide beaches.

T

Chapter 16

22) Item Type: True-False   Objective: 6   Section 16.3
     Level: A: Standard   Tag: None

____ Changes in exposed coastline can be caused by the rising or sinking of land as well
      as by changes in sea level.

T

23) Item Type: True-False   Objective: 6   Section 16.3
     Level: A: Standard   Tag: None

____ The west coast of the United States has neither risen nor sunk in the past few
      million years.

F

24) Item Type: True-False   Objective: 7   Section 16.3
     Level: A: Standard   Tag: None

____ Storms cause barrier islands to migrate toward the shoreline.

T

25) Item Type: True-False   Objective: 7   Section 16.3
     Level: A: Standard   Tag: None

____ Barrier islands are common along the east coast of the United States.

T

26) Item Type: True-False   Objective: 8   Section 16.3
     Level: A: Standard   Tag: None

____ Coral cannot live in deep water.

T

27) Item Type: True-False   Objective: 9   Section 16.3
     Level: A: Standard   Tag: None

____ Because of recent conservation efforts, coastlines in the United States are no longer
      in danger of being damaged.

F

28) Item Type: Multiple Choice   Objective: 1   Section 16.1
     Level: A: Standard   Tag: None

What are most sand grains made of?

a) coral   b) quartz   c) dust   d) loess

b

29) Item Type: Multiple Choice   Objective: 1   Section 16.1
     Level: A: Standard   Tag: None

The small, closely packed rocks left behind by deflation are called

a) deposits.   b) ventifacts.   c) desert pavement.   d) sedimentary rock.

c

30) Item Type: Multiple Choice   Objective: 1   Section 16.1
     Level: A: Standard   Tag: None

Which type of material is usually moved by saltation?

a) dust   b) sand   c) gravel   d) cobbles

b

31) Item Type: Multiple Choice   Objective: 1   Section 16.1
     Level: B: Enriched   Tag: None

Dust particles have a diameter of less than

a) 0.5 mm.   b) 1 mm.   c) 0.06 mm.   d) 0.3 mm.

c

32) Item Type: Multiple Choice   Objective: 1   Section 16.1
     Level: A: Standard   Tag: None

The process by which fine, very dry soil particles are removed by wind is called

a) deflation.   b) emergence.   c) hollowing.   d) deposition.

a

33) Item Type: Multiple Choice   Objective: 1   Section 16.1
     Level: A: Standard   Tag: None

The most common form of wind erosion is called

a) saltation.   b) deflation.   c) hollowing.   d) deposition.

b

34) Item Type: Multiple Choice   Objective: 2   Section 16.1
     Level: A: Standard   Tag: None

A crescent-shaped dune that has an open side facing the wind is called a

a) transverse dune.   b) barchan.

c) parabolic dune.   d) deflation hollow.

c

35) Item Type: Multiple Choice   Objective: 2   Section 16.1
Level: B: Enriched   Tag: Diagram 1671

What type of dune is pictured in this diagram?

a) longitudinal   b) parabolic   c) transverse   d) barchan

d

36) Item Type: Multiple Choice   Objective: 2   Section 16.1
Level: A: Standard   Tag: None

The downwind side of a dune is called the

a) terrace.   b) slipface.   c) cliff.   d) crescent.

b

37) Item Type: Multiple Choice   Objective: 2   Section 16.1
Level: A: Standard   Tag: None

The fine-grained, yellowish sediment that is soft, easily eroded, and deposited by the wind is called

a) silt.   b) desert pavement.   c) loess.   d) sedimentary rock.

c

38) Item Type: Multiple Choice   Objective: 2   Section 16.1
Level: A: Standard   Tag: None

A mound of windblown sand is called a

a) loess.   b) dune.   c) pavement.   d) spit.

b

39) Item Type: Multiple Choice   Objective: 2   Section 16.1
Level: A: Standard   Tag: None

What type of dune forms at right angles to the wind?

a) barchan   b) longitudinal   c) transverse   d) parabolic

c

40) Item Type: Multiple Choice   Objective: 3   Section 16.2
Level: A: Standard   Tag: None

Which of the following is the result of the greatest amount of wave erosion?

a) sea arch   b) sea cave   c) sea cliff   d) sea stack

d

41) Item Type: Multiple Choice   Objective: 3   Section 16.2
    Level: A: Standard   Tag: None

Isolated columns of rock that result from ocean wave action on rock projections near shore are called sea

a) cliffs.   b) fractures.   c) arches.   d) stacks.

d

42) Item Type: Multiple Choice   Objective: 3   Section 16.2
    Level: A: Standard   Tag: None

A nearly level platform left just below the surface of the water as waves erode the shoreline is called a

a) terrace.   b) berm.   c) beach.   d) bar.

a

43) Item Type: Multiple Choice   Objective: 3   Section 16.2
    Level: A: Standard   Tag: None

Erosion along shorelines is primarily caused by

a) salt.   b) currents.   c) waves.   d) wind.

c

44) Item Type: Multiple Choice   Objective: 3   Section 16.2
    Level: B: Enriched   Tag: None

A sea cliff composed of which of the following would probably erode most slowly?

a) glacial deposits   b) sedimentary rock   c) granite   d) loess

c

45) Item Type: Multiple Choice   Objective: 3   Section 16.2
    Level: A: Standard   Tag: None

Large holes cut into fractured rock at a cliff base are called

a) sea caves.   b) sea stacks.   c) wave-cut terraces.   d) wave-built terraces.

a

46) Item Type: Multiple Choice   Objective: 4   Section 16.2
    Level: A: Standard   Tag: None

Sand that is carried by waves from the berm and deposited offshore usually forms a

a) sand bar.   b) longitudinal dune.   c) spit.   d) cobble.

a

47) Item Type: Multiple Choice   Objective: 4   Section 16.2
     Level: A: Standard   Tag: None

Which of the following is the source of the black sand at beaches?

a) granite   b) volcanic rock   c) coral reefs   d) glacial deposits

b

48) Item Type: Multiple Choice   Objective: 4   Section 16.2
     Level: A: Standard   Tag: None

Many Florida beaches consist of fragments of shells and

a) coral.   b) granite.   c) volcanic rock.   d) glacial sand.

a

49) Item Type: Multiple Choice   Objective: 5   Section 16.2
     Level: A: Standard   Tag: None

Sand deposits that connect islands are called

a) reefs.   b) cobbles.   c) tombolos.   d) spits.

c

50) Item Type: Multiple Choice   Objective: 5   Section 16.2
     Level: A: Standard   Tag: None

Projections of a shore are called

a) estuaries.   b) dunes.   c) headlands.   d) fiords.

c

51) Item Type: Multiple Choice   Objective: 5   Section 16.2
     Level: A: Standard   Tag: None

What is the term for a long, narrow deposit of sand that is connected at one end to the shore?

a) headland   b) berm   c) beach   d) spit

d

52) Item Type: Multiple Choice   Objective: 6   Section 16.3
     Level: A: Standard   Tag: None

The coastal glacial valleys that become flooded as sea level rises are called

a) estuaries.   b) fiords.   c) headlands.   d) berms.

b

53) Item Type: Multiple Choice   Objective: 6   Section 16.3
Level: A: Standard   Tag: None

Sea level is now rising at an average rate of about

a) 1 mm/yr.   b) 5 mm/yr.   c) 15 mm/yr.   d) 25 mm/yr.

a

54) Item Type: Multiple Choice   Objective: 6   Section 16.3
Level: A: Standard   Tag: None

The rising of the lithosphere can be the result of

a) falling sea level.   b) retreating glaciers.

c) eroding sea cliffs.   d) newly forming atolls.

b

55) Item Type: Multiple Choice   Objective: 6   Section 16.3
Level: A: Standard   Tag: None

Large-scale movements of the earth's crust are probably causing the east coast of the United States to

a) rise.   b) sink.

c) move eastward.   d) move southward.

b

56) Item Type: Multiple Choice   Objective: 6   Section 16.3
Level: A: Standard   Tag: None

Many emergent and submergent coastline features are formed by changes in

a) sea level.   b) wind direction.

c) atmospheric pressure.   d) chemical weathering.

a

57) Item Type: Multiple Choice   Objective: 7   Section 16.3
Level: A: Standard   Tag: None

A muddy or sandy part of the shoreline that is visible at low tide and submerged at high tide is a

a) fringing reef.   b) barrier island.

c) tidal flat.   d) deflation hollow.

c

58) Item Type: Multiple Choice   Objective: 7   Section 16.3
     Level: A: Standard   Tag: None

According to some geologists, sand dunes that have become isolated from shoreline by the rising sea level form

a)   tidal lagoons.

b)   barrier islands.

c)   sea cliffs.

d)   atolls.

b

59) Item Type: Multiple Choice   Objective: 7   Section 16.3
     Level: B: Enriched   Tag: Diagram 1675

In which direction will the barrier island in this diagram most likely migrate?

a)   north   b)   south   c)   east   d)   west

d

60) Item Type: Multiple Choice   Objective: 7   Section 16.3
     Level: B: Enriched   Tag: Diagram 1675

In the diagram above, the shallow water in the area labeled **X** is called a

a)   spit.   b)   fiord.   c)   estuary.   d)   lagoon.

d

61) Item Type: Multiple Choice   Objective: 8   Section 16.3
     Level: B: Enriched   Tag: Diagram 1676

What is the term for the area labeled **X** in the diagram above?

a)   barrier island   b)   fringing reef   c)   spit   d)   atoll

d

62) Item Type: Multiple Choice   Objective: 8   Section 16.3
     Level: A: Standard   Tag: Diagram 1676

What is the term for the area labeled **Y** in the diagram above?

a)   lagoon   b)   estuary   c)   tidal flat   d)   submergent coastline

a

Chapter 16

63) Item Type: Multiple Choice   Objective: 8   Section 16.3
Level: A: Standard   Tag: None

What does coral extract from sea water to build its skeleton?

a) sodium chloride   b) quartz

c) calcium carbonate   d) feldspar

c

64) Item Type: Multiple Choice   Objective: 8   Section 16.3
Level: A: Standard   Tag: None

Organisms that build coral reefs always live

a) in cold, deep water.   b) along emergent coastlines.

c) in warm, shallow water.   d) along submergent coastlines.

c

65) Item Type: Multiple Choice   Objective: 8   Section 16.3
Level: A: Standard   Tag: None

What type of structure forms around the borders of a volcanic island in shallow water?

a) sea arch   b) terrace   c) berm   d) fringing reef

d

66) Item Type: Multiple Choice   Objective: 9   Section 16.3
Level: A: Standard   Tag: None

Approximately what percentage of the land in the United States is coastal?

a) 5%   b) 10%   c) 15%   d) 20%

a

67) Item Type: Multiple Choice   Objective: 9   Section 16.3
Level: A: Standard   Tag: None

What have some towns along the New Jersey shoreline used to rebuild eroded beaches?

a) sand   b) limestone   c) clay   d) shale

a

68) Item Type: Multiple Choice   Objective: 9   Section 16.3
Level: A: Standard   Tag: None

One-tenth of the shellfish-producing waters of the United States have been destroyed by

a) erosion.   b) pollution.   c) overfishing.   d) coral.

b

69) Item Type: Fill in the Blank   Objective: 2   Section 16.1
Level: B: Enriched   Tag: None

The type of dune that generally forms when sand is blown from a deflation hollow and

collects around its rim is called a(n) _____.

parabolic dune

70) Item Type: Fill in the Blank   Objective: 3   Section 16.2
Level: A: Standard   Tag: None

A fairly level platform of rock that lies beneath the water at the base of a sea cliff is called

a(n) _____.

wave-cut terrace

71) Item Type: Fill in the Blank   Objective: 3   Section 16.2
Level: A: Standard   Tag: None

Waves often cut into fractures and weak areas along the base of a cliff, forming a large

hole called a(n) _____.

sea cave

72) Item Type: Fill in the Blank   Objective: 3   Section 16.2
Level: A: Standard   Tag: None

When waves cut a hole completely through a shoreline rock projection, the formation that

results is called a(n) _____.

sea arch

73) Item Type: Fill in the Blank   Objective: 3   Section 16.2
Level: A: Standard   Tag: None

Extensions of wave-cut, flat surfaces formed by the deposition of eroded material some

distance from shore are known as _____.

wave-built terraces

74) Item Type: Fill in the Blank   Objective: 4   Section 16.2
Level: A: Standard   Tag: None

Long underwater ridges of sand that form offshore are called _____.

sand bars

75) Item Type: Fill in the Blank Objective: 4 Section 16.2
Level: A: Standard Tag: None

A long, narrow, offshore ridge of sand that may be up to 100 km long and that lies nearly

parallel to the shoreline is called a(n) _____.

barrier island

76) Item Type: Fill in the Blank Objective: 4 Section 16.2
Level: A: Standard Tag: None

Rock fragments that are larger than pebbles but smaller than boulders are called

_____.

cobbles

77) Item Type: Fill in the Blank Objective: 5 Section 16.2
Level: B: Enriched Tag: Diagram 1673

According to the diagram above, the general direction in which sand will be moved along

the beach is _____.

east

78) Item Type: Fill in the Blank Objective: 5 Section 16.2
Level: B: Enriched Tag: Diagram 1673

The term for the type of current that is created when waves approach a beach at the angle

pictured in the diagram above is _____.

longshore

79) Item Type: Fill in the Blank Objective: 5 Section 16.2
Level: A: Standard Tag: None

A resistant rock area that projects out from the shore is called a(n)

_____.

headland

80) Item Type: Fill in the Blank Objective: 5 Section 16.2
Level: B: Enriched Tag: None

The movement of water parallel to the shoreline is called a(n) _____.

longshore current

81) Item Type: Fill in the Blank   Objective: 5   Section 16.2
Level: A: Standard   Tag: None

Sand ridges that connect islands are called _____.

| tombolos ▪ |

82) Item Type: Fill in the Blank   Objective: 6   Section 16.3
Level: A: Standard   Tag: None

Narrow, deep bays with steep walls that result from the flooding of glacial valleys are

called _____.

| fiords ▪ |

83) Item Type: Fill in the Blank   Objective: 6   Section 16.3
Level: B: Enriched   Tag: None

The main cause of the rise in sea level over the last 11,000 years is

_____.

| glacial melting ▪ |

84) Item Type: Fill in the Blank   Objective: 6   Section 16.3
Level: B: Enriched   Tag: None

At what annual rate is the sea level now rising? _____

| 1 mm/yr. ▪ |

85) Item Type: Fill in the Blank   Objective: 7   Section 16.3
Level: A: Standard   Tag: None

The nearly circular reef that forms as a volcanic island disappears is called a(n)

_____.

| atoll ▪ |

86) Item Type: Fill in the Blank   Objective: 7   Section 16.3
Level: A: Standard   Tag: None

The muddy or sandy parts of a lagoon that are visible at low tide are called

_____.

| tidal flats ▪ |

87) Item Type: Essay   Objective: 1   Section 16.1
    Level: B: Enriched   Tag: None

Explain how dunes migrate and how this migration can be slowed or stopped.

_____

_____

_____

_____

_____

Migration occurs as sand is blown off the windward side of a dune over its crest and then builds up on the slipface. In fairly level areas, the dune continues to migrate downwind until a barrier is reached. To prevent dunes from drifting, the planting of grass, trees, or shrubs or the building of fences is often necessary.

88) Item Type: Essay   Objective: 3   Section 16.2
    Level: B: Enriched   Tag: None

How do waves promote the process of chemical weathering of a shoreline?

_____

_____

_____

_____

_____

The waves force salt water and air into small cracks in the rock. Substances in the air and water produce a chemical action that may enlarge the cracks. The enlarged cracks, in turn, provide an increased surface area for physical and chemical weathering.

89) Item Type: Essay   Objective: 3   Section 16.2
    Level: B: Enriched   Tag: None

Explain how the formation of a wave-built terrace helps to slow the rate of sea-cliff erosion.

_____

_____

_____

_____

_____

Eroded material may be deposited some distance from the shore, creating an extension to the wave-cut terrace called a wave-built terrace. The water above the terrace is shallow, and waves lose much of their energy passing through this shallow water. As the energy of a wave lessens over a terrace, the rate of erosion of the cliff is greatly reduced.

90) Item Type: Essay   Objective: 4   Section 16.2
    Level: A: Standard   Tag: None

How do waves form beaches?

_____

_____

_____

_____

_____

As waves wash up on the shore, they move sand and small rock fragments forward. In retreating, the water moves some fragments away from the shore. Beaches form where the amount of rock fragments moving toward the shore is greater than the amount moving away from the shore.

91) Item Type: Essay   Objective: 4   Section 16.2
     Level: B: Enriched   Tag: None

What types of material would most likely compose the beach of a volcanic island that is surrounded by a barrier reef? Why?

_____

_____

_____

_____

_____

Black sand mixed with shell and coral fragments would most likely compose such a beach. Black sand comes from the volcanic rock that is commonly found in Hawaii and on other volcanic islands. A beach located near an offshore coral reef would also consist of fragments of shells and coral that were washed ashore.

92) Item Type: Essay   Objective: 6   Section 16.3
     Level: B: Enriched   Tag: None

Explain how emergent coastlines and their coastal features form.

_____

_____

_____

_____

Emergent coastlines form when the land rises or the sea level falls. If an emergent coastline has a steep slope and is exposed rapidly it will become jagged and dotted with sea cliffs, narrow inlets, and bays. Along some emergent coastlines, a gentle slope is formed when part of the continental slope is lifted and exposed. A relatively smooth coastal plain with few bays or headlands and many long, wide beaches is the result.

93) Item Type: Essay   Objective: 8   Section 16.3
     Level: B: Enriched   Tag: None

Describe the formation of an atoll.

_____

_____

_____

_____

First, a coral reef forms around a tropical volcanic island. Then, as the ocean lithosphere bends under the weight of the volcano, both the volcano and the reef sink. The coral reef builds higher because the animals can live only near the surface. When the volcano is no longer active, the island may sink below the surface. The reef, however, keeps growing upward, and eventually only a nearly circular reef remains above the surface, surrounding a shallow lagoon.

94) Item Type: Essay   Objective: 8   Section 16.3
     Level: B: Enriched   Tag: None

Describe how coral reefs are formed.

_____

_____

_____

_____

Corals reefs are formed by corals, which are small animals that live in warm, shallow waters. A coral extracts calcium carbonate from ocean water and uses it to build a hard skeleton. Corals attach to one another to form large colonies. New corals grow on top of dead ones, forming a coral reef, which is a ridge made up of millions of coral skeletons.

1)   Tag Name: Diagram 1671

2)   Tag Name: Diagram 1673

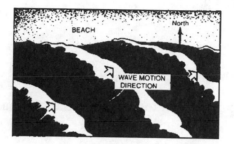

3)   Tag Name: Diagram 1675

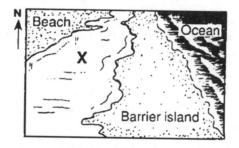

4)   Tag Name: Diagram 1676

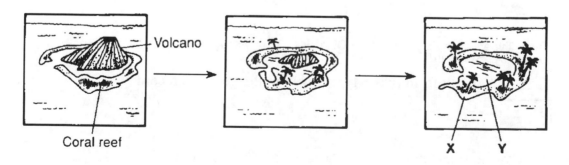

1) Item Type: True-False   Objective: 1   Section 17.1
   Level: A: Standard   Tag: None

_____ The principle of uniformitarianism states that all the earth's geologic features formed at the same time.

F

2) Item Type: True-False   Objective: 1   Section 17.1
   Level: A: Standard   Tag: None

_____ James Hutton stated that the geologic formations on the earth's crust took millions of years to develop.

T

3) Item Type: True-False   Objective: 2   Section 17.1
   Level: A: Standard   Tag: None

_____ Ripple marks can be used to determine the original arrangement of rock layers in an area.

T

4) Item Type: True-False   Objective: 3   Section 17.1
   Level: A: Standard   Tag: None

_____ Nonconformities occur between horizontal rock layers.

T

5) Item Type: True-False   Objective: 3   Section 17.1
   Level: A: Standard   Tag: None

_____ The law of superposition could be used to determine the relative age of a sequence of igneous lava flows.

T

6) Item Type: True-False   Objective: 3   Section 17.1
   Level: A: Standard   Tag: None

_____ A disconformity is a type of gap in the geologic record.

T

7) Item Type: True-False   Objective: 3   Section 17.1
   Level: B: Enriched   Tag: None

_____ Disconformities are produced when folded or tilted rock layers undergo weathering.

F

8) Item Type: True-False   Objective: 3   Section 17.1
   Level: B: Enriched   Tag: None

_____ In an angular unconformity, the bedding planes of older rock layers are not parallel to the bedding planes of younger layers.

T

9) Item Type: True-False   Objective: 4   Section 17.1
   Level: A: Standard   Tag: None

_____ The laws of superposition and crosscutting relationships could be used to determine the relative age of an unconformity.

T

10) Item Type: True-False   Objective: 4   Section 17.1
   Level: A: Standard   Tag: None

_____ According to the law of crosscutting relationships, an intrusion is older than the rocks through which it cuts.

F

11) Item Type: True-False   Objective: 4   Section 17.1
   Level: A: Standard   Tag: None

_____ A fault is always older than the rocks through which it cuts.

F

12) Item Type: True-False   Objective: 5   Section 17.2
   Level: B: Enriched   Tag: None

_____ Measuring the rate of sediment deposition is the most accurate method of determining the absolute age of a geologic feature.

F

13) Item Type: True-False   Objective: 5   Section 17.2
   Level: A: Standard   Tag: None

_____ Erosion rates have been used to establish the age of the Grand Canyon accurately.

F

14) Item Type: True-False   Objective: 6   Section 17.2
   Level: A: Standard   Tag: None

_____ One varve represents two years of deposition of sedimentary particles.

F

15) Item Type: True-False   Objective: 6   Section 17.2
    Level: A: Standard   Tag: None

____ Varves form through the process of sedimentation.

T

16) Item Type: True-False   Objective: 7   Section 17.2
    Level: A: Standard   Tag: None

____ The earth is believed to be about 4.6 billion years old.

T

17) Item Type: True-False   Objective: 7   Section 17.2
    Level: A: Standard   Tag: None

____ The radioactive decay rate is greatly affected by temperature.

F

18) Item Type: True-False   Objective: 7   Section 17.2
    Level: B: Enriched   Tag: None

____ The final product of radioactive decay is usually a nonradioactive element.

T

19) Item Type: True-False   Objective: 7   Section 17.2
    Level: A: Standard   Tag: None

____ Carbon-14 is a radioactive element used to determine the absolute age of rocks.

F

20) Item Type: True-False   Objective: 8   Section 17.3
    Level: A: Standard   Tag: None

____ The hard parts of an organism, such as bones and teeth, are usually the only parts of the body to be preserved in the rock record.

T

21) Item Type: True-False   Objective: 8   Section 17.3
    Level: A: Standard   Tag: None

____ Entire animals can be fossilized in tar pits.

T

22) Item Type: True-False   Objective: 8   Section 17.3
    Level: A: Standard   Tag: None

_____ An organism will undergo mummification only if it is buried by sediments at the bottom of a glacial lake.

F

23) Item Type: True-False   Objective: 8   Section 17.3
    Level: A: Standard   Tag: None

_____ An organism is most likely to be preserved through the process of mummification in the desert.

T

24) Item Type: True-False   Objective: 9   Section 17.3
    Level: A: Standard   Tag: None

_____ Fossil spores and pollen indicate that Antarctica had a much warmer climate 40 million years ago.

T

25) Item Type: True-False   Objective: 9   Section 17.3
    Level: A: Standard   Tag: None

_____ Mold fossils generally provide little information about the internal structures of the original organisms.

T

26) Item Type: True-False   Objective: 9   Section 17.3
    Level: A: Standard   Tag: None

_____ A footprint is an example of a trace fossil.

T

27) Item Type: True-False   Objective: 9   Section 17.3
    Level: A: Standard   Tag: None

_____ Coprolites can be used to study the eating habits of ancient animals.

T

28) Item Type: True-False   Objective: 9   Section 17.3
    Level: A: Standard   Tag: None

_____ A frozen mastodon is an example of a trace fossil.

F

29) Item Type: True-False   Objective: 10   Section 17.3
    Level: A: Standard   Tag: None

_____ Geologists use index fossils to help locate oil and natural gas deposits.

T

30) Item Type: True-False   Objective: 10   Section 17.3
    Level: A: Standard   Tag: None

_____ A species that lived during several different geologic time periods would make a good index fossil.

F

31) Item Type: Multiple Choice   Objective: 1   Section 17.1
    Level: A: Standard   Tag: None

The principle of uniformitarianism states that

a) rocks can be dated by the radioactive elements they contain.

b) geologic processes that occurred in the past are still at work today.

c) rock layers that are buried are older than the layers above them.

d) the major geologic features of the earth all formed at the same time.

b

32) Item Type: Multiple Choice   Objective: 1   Section 17.1
    Level: A: Standard   Tag: None

A period of erosion shows up in layers of rock as

a) an unconformity.   b) a varve.   c) a fault.   d) an intrusion.

a

33) Item Type: Multiple Choice   Objective: 2   Section 17.1
    Level: A: Standard   Tag: None

Which of the following principles would most likely be used to determine the relative age of undisturbed sedimentary strata?

a) principle of uniformitarianism   b) law of superposition

c) principle of intrusion   d) law of crosscutting relationships

b

34) Item Type: Multiple Choice   Objective: 2   Section 17.1
    Level: A: Standard   Tag: None

A nonerosional boundary between two layers of sedimentary rock is called a

a) foliation band.   b) stratification barrier.

c) disconformity line.   d) bedding plane.

d

35) Item Type: Multiple Choice   Objective: 2   Section 17.1
    Level: A: Standard   Tag: None

Sedimentary rocks are generally laid down in nearly horizontal layers called

a) strata.   b) striations.   c) foliates.   d) sheets.

a

36) Item Type: Multiple Choice   Objective: 2   Section 17.1
    Level: B: Enriched   Tag: Diagram 1772

Which feature in the diagram contains the youngest rock?

a) *1*   b) *4*   c) *5*   d) *6*

c

37) Item Type: Multiple Choice   Objective: 2   Section 17.1
    Level: A: Standard   Tag: Diagram 1772

Which rocky layer in the diagram contains the oldest rock?

a) *1*   b) *2*   c) *3*   d) *4*

d

38) Item Type: Multiple Choice   Objective: 3   Section 17.1
    Level: B: Enriched   Tag: None

Which of the following is produced when folded or tilted rock layers undergo erosion and then a new horizontal layer is deposited?

a) a folded intrusion   b) a cross-bedded fault

c) a bedding plane   d) an angular unconformity

d

39) Item Type: Multiple Choice   Objective: 3   Section 17.1
    Level: A: Standard   Tag: None

Environmental conditions during the Mesozoic Era especially favored the survival of

a) reptiles.   b) woolly mammoths.   c) birds.   d) early humans.

a

40) Item Type: Multiple Choice   Objective: 3   Section 17.1
    Level: B: Enriched   Tag: None

Which type of unconformity is the most difficult to identify by sight in rock strata?

a) nonconformity   b) angular unconformity

c) disconformity   d) crossbed intrusion

c

41) Item Type: Multiple Choice   Objective: 3   Section 17.1
    Level: B: Enriched   Tag: None

Which type of unconformity consists of sedimentary layers above metamorphic or igneous rock?

a) a nonconformity   b) an angular unconformity

c) a disconformity   d) a uniformity

a

42) Item Type: Multiple Choice   Objective: 4   Section 17.1
    Level: A: Standard   Tag: Diagram 1772

Which of the following would be used to determine the difference in the relative ages of the rock in layer **6** and layer **3** in the diagram above?

a) law of superposition   b) principle of average erosion rates

c) varve counts   d) law of crosscutting relationships

d

43) Item Type: Multiple Choice   Objective: 4   Section 17.1
    Level: A: Standard   Tag: None

Which of the following would most likely be used to determine the relative age of a fault?

a) law of superposition   b) radioactive decay

c) principle of uniformitarianism   d) law of crosscutting relationships

d

Chapter 17

44) Item Type: Multiple Choice   Objective: 4   Section 17.1
Level: A: Standard   Tag: None

Geologic evidence shows that the first land plants began to grow during which time period or era?

a) Precambrian   b) Cenozoic   c) Paleozoic   d) Mesozoic

c

45) Item Type: Multiple Choice   Objective: 6   Section 17.2
Level: A: Standard   Tag: None

Sedimentary deposits that form in glacial lakes and show definite annual layers are called

a) varves.   b) dikes.   c) sills.   d) crossbeds.

a

46) Item Type: Multiple Choice   Objective: 6   Section 17.2
Level: A: Standard   Tag: Diagram 1773

The sediment layers in the diagram are called

a) molds.   b) imprints.   c) casts.   d) varves.

d

47) Item Type: Multiple Choice   Objective: 6   Section 17.2
Level: A: Standard   Tag: Diagram 1773

How many years of deposition are probably represented in the diagram?

a) 3.5   b) 5   c) 6.5   d) 7

a

48) Item Type: Multiple Choice   Objective: 6   Section 17.2
Level: B: Enriched   Tag: Diagram 1773

The topmost layer in the diagram was most likely deposited during

a) summer.   b) spring.   c) winter.   d) fall.

c

49) Item Type: Multiple Choice   Objective: 7   Section 17.2
Level: A: Standard   Tag: None

Which of the following elements is most appropriate for dating a sample that is between 8,000 and 20,000 years old?

a) uranium-238   b) thorium-234   c) potassium-40   d) carbon-14

d

50) Item Type: Multiple Choice   Objective: 7   Section 17.2
Level: B: Enriched   Tag: None

A rock contains 4 mg of a radioactive element. At the end of 60 years, only 0.5 mg of the element remains in the rock. What is the half-life of the element?

a) 60 years   b) 30 years   c) 20 years   d) 15 years

c

51) Item Type: Multiple Choice   Objective: 7   Section 17.2
Level: A: Standard   Tag: None

Elements that emit atomic particles and energy are called

a) radioactive.   b) stable.   c) magnetic.   d) fluorescent.

a

52) Item Type: Multiple Choice   Objective: 7   Section 17.2
Level: A: Standard   Tag: None

The most accurate method of finding the absolute age of a rock is by

a) varve count dating.   b) erosion rate dating.

c) radioactive decay dating.   d) superposition dating.

c

53) Item Type: Multiple Choice   Objective: 7   Section 17.2
Level: B: Enriched   Tag: None

The absolute age of a rock layer can be determined by studying the

a) superposition of adjacent rock layers.

b) crosscutting relationships within the rock.

c) position of ripple marks on the rock bed.

d) ratio of U-238 to Pb-206 in the rock.

d

54) Item Type: Multiple Choice   Objective: 7   Section 17.2
Level: B: Enriched   Tag: None

Potassium-40 has a half-life of 1.3 billion years. A rock originally contained 8 g of potassium-40, but now contains only 2 g of that element. How many half-lives have passed?

a) 1   b) 2   c) 3   d) 4

b

55) Item Type: Multiple Choice   Objective: 7   Section 17.2
    Level: B: Enriched   Tag: None

Calcium-41 has a half-life 80,000 years. How much calcium-41 would remain in a 240 g sample at the end of 320,000 years?

a) 4 g   b) 15 g   c) 60 g   d) 120 g

b

56) Item Type: Multiple Choice   Objective: 7   Section 17.2
    Level: B: Enriched   Tag: None

Nickel-63 has a half-life of 92 years. How much nickel-63 would remain in a 20 g sample at the end of 184 years?

a) 2 g   b) 5 g   c) 8 g   d) 10 g

b

57) Item Type: Multiple Choice   Objective: 7   Section 17.2
    Level: A: Standard   Tag: None

An element has an atomic number of 29 and a mass number of 63. How many neutrons are there in the element?

a) 29   b) 34   c) 63   d) 92

b

58) Item Type: Multiple Choice   Objective: 7   Section 17.2
    Level: A: Standard   Tag: None

An element has a mass number of 83 and an atomic number of 36. How many electrons are there in the element?

a) 36   b) 47   c) 83   d) 119

a

59) Item Type: Multiple Choice   Objective: 7   Section 17.2
    Level: A: Standard   Tag: None

Which of the following would most likely be dated using C-14?

a) Hawaiian lava   b) varves

c) an ancient skeleton   d) an igneous intrusion

c

60) Item Type: Multiple Choice   Objective: 8   Section 17.3
Level: A: Standard   Tag: None

A fossil animal with its skin and muscles intact was most likely preserved through the process of

a) petrification.   b) freezing.   c) carbonization.   d) casting.

b

61) Item Type: Multiple Choice   Objective: 8   Section 17.3
Level: A: Standard   Tag: None

Which process forms fossils by drying the organic matter in an organism's body?

a) mummification   b) petrification   c) excretion   d) deposition

a

62) Item Type: Multiple Choice   Objective: 8   Section 17.3
Level: A: Standard   Tag: None

Paleontology is defined as the study of

a) rock layers.   b) fossils.   c) erosion.   d) unconformities.

b

63) Item Type: Multiple Choice   Objective: 8   Section 17.3
Level: A: Standard   Tag: None

Tar beds are formed by thick deposits of

a) clay.   b) amber.   c) petroleum.   d) silica.

c

64) Item Type: Multiple Choice   Objective: 8   Section 17.3
Level: A: Standard   Tag: None

Which type of fossil would yield the most information about the internal structure of an organism?

a) a tree imprint   b) a tree cast   c) a tree mold   d) a petrified tree

d

65) Item Type: Multiple Choice   Objective: 9   Section 17.3
Level: A: Standard   Tag: None

Fossils found exclusively in the rock layers of a particular geologic age are called

a) index fossils.   b) gastrolith fossils.

c) metamorphic fossils.   d) complete fossils.

a

66) Item Type: Multiple Choice   Objective: 9   Section 17.3
    Level: A: Standard   Tag: None

Which of the following is an example of a trace fossil?

a)   an intact shark tooth   b)   a petrified log

c)   a dinosaur footprint   d)   an insect in amber

c

67) Item Type: Multiple Choice   Objective: 9   Section 17.3
    Level: A: Standard   Tag: None

Which of the following is a common petrifying mineral?

a)   garnet   b)   diamond   c)   pyrite   d)   feldspar

c

68) Item Type: Multiple Choice   Objective: 9   Section 17.3
    Level: B: Enriched   Tag: None

Which two types of fossils are most likely to be found together in a single rock layer?

a)   graptolites and trilobites   b)   casts and molds

c)   coprolites and petrified plants   d)   imprints and amber drops

b

69) Item Type: Multiple Choice   Objective: 9   Section 17.3
    Level: A: Standard   Tag: None

Gastroliths are fossilized

a)   stones.   b)   footprints.   c)   bones.   d)   plants.

a

70) Item Type: Multiple Choice   Objective: 10   Section 17.3
    Level: B: Enriched   Tag: None

Approximately how old is a layer of rock containing ammonite fossils?

a)   245 million to 570 million years   b)   65 million to 200 million years

c)   100 million to 208 million years   d)   10 million to 65 million years

b

71) Item Type: Multiple Choice   Objective: 10   Section 17.3
    Level: B: Enriched   Tag: Diagram 1774

Which of the following rock layers in the diagram have the same relative age?

a) **1** and **6**   b) **2** and **9**   c) **3** and **7**   d) **5** and **10**

b

72) Item Type: Multiple Choice   Objective: 10   Section 17.3
    Level: B: Enriched   Tag: None

Which of the following would most likely be used to determine the relative age of a rock layer?

a) carbon-14 analysis   b) mineral content analysis

c) crystal shape analysis   d) guide fossil analysis

d

73) Item Type: Fill in the Blank   Objective: 1   Section 17.1
    Level: A: Standard   Tag: None

The principle that states that current geologic processes are the same processes that

occurred in the past is known as the principle of _____.

uniformitarianism

74) Item Type: Fill in the Blank   Objective: 2   Section 17.1
    Level: A: Standard   Tag: None

Layers in sedimentary rock are called _____.

strata

75) Item Type: Fill in the Blank   Objective: 2   Section 17.1
    Level: A: Standard   Tag: None

Small waves formed on the surface of sand by the action of water or wind are called

_____.

ripple marks

76) Item Type: Fill in the Blank   Objective: 2   Section 17.1
    Level: A: Standard   Tag: None

The principle that states that each sedimentary rock layer is older than the layers above it

and younger than the layers below it is called the _____.

law of superposition

77) Item Type: Fill in the Blank   Objective: 2   Section 17.1
Level: A: Standard   Tag: Diagram 1771

The boundary labeled **A** in the diagram above represents a(n) _____.

| bedding plane | ■ |

78) Item Type: Fill in the Blank   Objective: 2   Section 17.1
Level: A: Standard   Tag: Diagram 1771

The principle that can be used to determine the relative age of layers **1** through **3** in the

diagram above is the _____.

| law of superposition | ■ |

79) Item Type: Fill in the Blank   Objective: 3   Section 17.1
Level: B: Enriched   Tag: Diagram 1771

In the diagram above, point **B** represents a type of rock feature called

_____.

| an angular unconformity | ■ |

80) Item Type: Fill in the Blank   Objective: 3   Section 17.1
Level: B: Enriched   Tag: None

An unconformity that consists of stratified sedimentary rock above a horizontal basalt

surface would be classified as a type of _____.

| nonconformity | ■ |

81) Item Type: Fill in the Blank   Objective: 3   Section 17.1
Level: A: Standard   Tag: None

Breaks in the geologic record that make it difficult to determine the relative age of rock

layers are called _____.

| unconformities | ■ |

82) Item Type: Fill in the Blank   Objective: 4   Section 17.1
Level: A: Standard   Tag: None

The relative age of an igneous intrusion would most likely be determined by using the law

of _____.

| crosscutting relationships | ■ |

83) Item Type: Fill in the Blank   Objective: 5   Section 17.2
Level: A: Standard   Tag: None

What did scientists measure to determine the approximate age of Niagara Falls?

_____

erosion rates

84) Item Type: Fill in the Blank   Objective: 5   Section 17.2
Level: A: Standard   Tag: None

The average rate of sediment deposition for rocks such as limestone and sandstone is

_____.

30 cm per 1,000 years

85) Item Type: Fill in the Blank   Objective: 6   Section 17.2
Level: A: Standard   Tag: None

Light and dark banding patterns in the sediments of glacial lakes are called

_____.

varves

86) Item Type: Fill in the Blank   Objective: 8   Section 17.3
Level: A: Standard   Tag: None

Scientists who study fossils to learn about the earth's geologic history are called

_____ .

paleontologists

87) Item Type: Fill in the Blank   Objective: 9   Section 17.3
Level: A: Standard   Tag: None

Fossils formed when mud fills in a mold and hardens are called _____.

casts

88) Item Type: Fill in the Blank   Objective: 10   Section 17.3
Level: A: Standard   Tag: None

Fossils such as trilobites that can be used to establish the age of a rock are called

_____ .

index fossils

89) Item Type: Fill in the Blank   Objective: 10   Section 17.3
    Level: A: Standard   Tag: None

The change in living things over time is called the process of _____.

evolution

90) Item Type: Fill in the Blank   Objective: 10   Section 17.3
    Level: A: Standard   Tag: None

Scientists use index fossils to determine a rock's _____.

relative age

91) Item Type: Fill in the Blank   Objective: 10   Section 17.3
    Level: A: Standard   Tag: None

The index fossils discussed in the text that lived between 245 million and 570 million

years ago are called _____.

trilobites

92) Item Type: Essay   Objective: 2   Section 17.1
    Level: B: Enriched   Tag: None

What geologic features would indicate that several sedimentary rock layers had been
overturned?

_____

_____

_____

_____

_____

Sedimentary layers tend to form with the coarsest-grained rock at the bottom. Another clue to the original position of the rock is found in the bedding plane. Cross-beds are curved at the bottom and cut off at the top of erosion. Ripple marks can also be helpful in determining the order of rock layers. The peaks of ripple marks point upward.

93) Item Type: Essay   Objective: 4   Section 17.1
     Level: B: Enriched   Tag: None

How do geologists use the law of superposition and the law of crosscutting relationships to determine the relative ages of faulted and eroded rock layers?

_____

_____

_____

_____

_____

> According to the law of superposition, all rocks beneath an unconformity are older than those rocks above the unconformity. If a fault or intrusion cuts through rock layers or other rock features, the fault or intrusion is younger than all the rocks and rock features it cuts through.

94) Item Type: Essay   Objective: 5   Section 17.2
     Level: B: Enriched   Tag: None

How can scientists use the average rate of erosion to determine the age of features such as streams? What are the limitations of this method?

_____

_____

_____

_____

_____

> If scientists can measure the rate at which a stream erodes its bed, they can determine the average age of the stream. This method of dating is not dependable for older features because the rates of erosion have varied greatly throughout history.

95) Item Type: Essay   Objective: 7   Section 17.2
Level: B: Enriched   Tag: None

Explain how scientists use carbon dating to determine the age of organic samples.

_____

_____

_____

_____

_____

Carbon-14 and carbon-12 occur naturally in the atmosphere. All living organisms take in C-12 and C-14 in known ratios. When an organism dies, that ratio changes as the radioactive C-14 decays to nitrogen-14, but the stable C-12 remains constant. To determine the age of a sample of organic material, scientists find the ratio of C-12 to C-14, and then use the half-life of C-14 (about 5,730 years) to determine the age of the sample.

96) Item Type: Essay   Objective: 8   Section 17.3
Level: B: Enriched   Tag: None

Describe the process of petrification.

_____

_____

_____

_____

_____

Mineral solutions, such as groundwater, remove the original organic matter in an organism and replace it with mineral matter. Some common petrifying minerals are silica, calcite, and pyrite.

97)   Item Type: Essay   Objective: 10   Section 17.3
       Level: B: Enriched   Tag: None

What are the characteristics of a good index fossil?

_____

_____

_____

_____

_____

> The fossil must be present in rocks scattered over a wide area of the earth's surface and must be clearly distinguishable from other fossils. The organism from which the index fossil formed must have lived for only a short span of geologic time. Finally, index fossils must occur in fairly large numbers within rock layers.

1) Tag Name: Diagram 1771

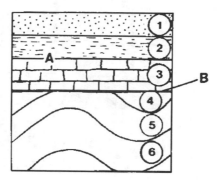

2) Tag Name: Diagram 1772

Diagram of Undisturbed Strata

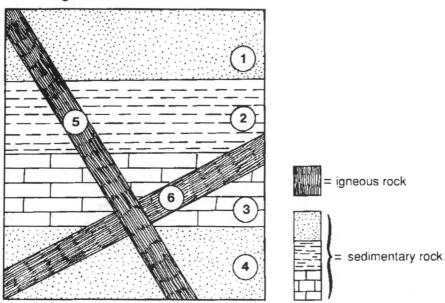

= igneous rock

= sedimentary rock

3) Tag Name: Diagram 1773

**Cross Section of the
Annual Deposition of Sediments
in a Glacial Lake**

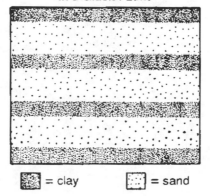

= clay   = sand

4)   Tag Name: Diagram 1774

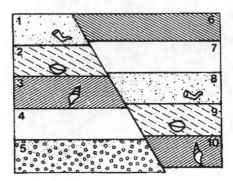

Chapter 18

1) Item Type: True-False   Objective: 1   Section 18.1
    Level: B: Enriched   Tag: None

_____ Using the law of superposition and the study of index fossils, nineteenth-century scientists determined the relative ages of rock layers.

T

2) Item Type: True-False   Objective: 1   Section 18.1
    Level: A: Standard   Tag: None

_____ Scientists use events such as major changes in the earth's surface and climate and the extinction of various species as the basis for dividing the geologic time scale into units.

T

3) Item Type: True-False   Objective: 1   Section 18.1
    Level: A: Standard   Tag: None

_____ Scientists can use the geologic column to estimate the relative ages of rock layers only if the rock contains radioactive minerals.

F

4) Item Type: True-False   Objective: 1   Section 18.1
    Level: A: Standard   Tag: None

_____ Rock layers in the geologic column are distinguished from one another primarily by fossil content and rock type.

T

5) Item Type: True-False   Objective: 1   Section 18.1
    Level: A: Standard   Tag: None

_____ The geologic column is an arrangement of rock layers based on the ages of the layers.

T

6) Item Type: True-False   Objective: 1   Section 18.1
    Level: A: Standard   Tag: None

_____ A few areas on earth contain a complete record of geologic time.

F

7) Item Type: True-False   Objective: 1   Section 18.1
    Level: A: Standard   Tag: None

_____ The geologic column allows scientists to estimate the age of rock layers that contain no radioactive minerals.

T

8) Item Type: True-False   Objective: 2   Section 18.1
    Level: A: Standard   Tag: None

_____ Fossils of mammals are common in Cenozoic rocks.

T

9) Item Type: True-False   Objective: 2   Section 18.1
    Level: A: Standard   Tag: None

_____ Eras are subdivisions of periods.

F

10) Item Type: True-False   Objective: 2   Section 18.1
    Level: A: Standard   Tag: None

_____ Fossil bacteria and algae, thought to be the earth's first life forms, have been found in rocks 4.5 billion years old.

F

11) Item Type: True-False   Objective: 2   Section 18.1
    Level: A: Standard   Tag: None

_____ Scientists have divided earth's history into five eras.

F

12) Item Type: True-False   Objective: 2   Section 18.1
    Level: A: Standard   Tag: None

_____ The Cenozoic Era is subdivided into three periods.

F

13) Item Type: True-False   Objective: 3   Section 18.2
    Level: A: Standard   Tag: None

_____ The theory of evolution states that organisms change over time and that new kinds of organisms are derived from ancestral types.

T

14) Item Type: True-False   Objective: 3   Section 18.2
    Level: A: Standard   Tag: None

_____ The earth's crust solidified during the Cambrian Period.

F

15) Item Type: True-False   Objective: 3   Section 18.2
    Level: A: Standard   Tag: None

_____ Precambrian rocks contain abundant fossils.

F

16) Item Type: True-False   Objective: 3   Section 18.2
    Level: A: Standard   Tag: None

_____ In times of severe environmental change, organisms that cannot adapt become extinct.

T

17) Item Type: True-False   Objective: 3   Section 18.2
    Level: A: Standard   Tag: None

_____ The large areas of exposed Precambrian rocks called shields are found only in Africa.

F

18) Item Type: True-False   Objective: 3   Section 18.2
    Level: A: Standard   Tag: None

_____ In Australia, fossil remains of animals have been found in Precambrian rocks.

T

19) Item Type: True-False   Objective: 3   Section 18.2
    Level: A: Standard   Tag: None

_____ Charles Darwin proposed the theory of evolution by natural selection.

T

20) Item Type: True-False   Objective: 3   Section 18.2
    Level: A: Standard   Tag: None

_____ The presence of stromatolite fossils in Precambrian rocks indicates that shallow seas may have covered much of the earth during some periods of Precambrian time.

T

21) Item Type: True-False   Objective: 4   Section 18.2
     Level: A: Standard   Tag: None

_____ By the end of the Paleozoic Era, the earth's landmasses were grouped together to form the supercontinent Pangaea.

T

22) Item Type: True-False   Objective: 4   Section 18.2
     Level: A: Standard   Tag: None

_____ Land plants first developed during the Silurian Period.

T

23) Item Type: True-False   Objective: 4   Section 18.2
     Level: A: Standard   Tag: None

_____ Brachiopods, a group of shelled animals, are characteristic of the Cambrian Period.

T

24) Item Type: True-False   Objective: 4   Section 18.2
     Level: A: Standard   Tag: None

_____ Fossils are very rare in Paleozoic rocks.

F

25) Item Type: True-False   Objective: 4   Section 18.2
     Level: A: Standard   Tag: None

_____ Fossil evidence indicates that during the Silurian Period a type of fish called the lungfish crawled out of the water to spend short periods of time on land.

F

26) Item Type: True-False   Objective: 5   Section 18.2
     Level: A: Standard   Tag: None

_____ The Mesozoic Era is known as the Age of Fishes.

F

27) Item Type: True-False   Objective: 5   Section 18.2
     Level: A: Standard   Tag: None

_____ The end of the Jurassic Period was marked by mass extinctions.

F

28) Item Type: True-False   Objective: 5   Section 18.2
    Level: A: Standard   Tag: None

_____ The Mesozoic Era is also known as the Age of Reptiles.

T

29) Item Type: True-False   Objective: 5   Section 18.2
    Level: A: Standard   Tag: None

_____ The Jurassic is the third geologic era.

F

30) Item Type: True-False   Objective: 5   Section 18.2
    Level: B: Enriched   Tag: None

_____ Ammonite fossils have worldwide distribution and serve as Mesozoic index fossils.

T

31) Item Type: True-False   Objective: 6   Section 18.2
    Level: A: Standard   Tag: None

_____ Mammals were the dominant animals of the Paleozoic Era.

F

32) Item Type: True-False   Objective: 6   Section 18.2
    Level: A: Standard   Tag: None

_____ Fossils of the ancestors of modern humans date to almost 2 million years ago.

T

33) Item Type: True-False   Objective: 6   Section 18.2
    Level: A: Standard   Tag: None

_____ The Cenozoic Era was a time of drastic climatic change with dropping temperatures and several episodes of extensive glacial activity.

T

34) Item Type: True-False   Objective: 6   Section 18.2
    Level: A: Standard   Tag: None

_____ The Cenozoic Era was a time of increased tectonic activity.

T

35) Item Type: Multiple Choice   Objective: 1   Section 18.1
    Level: A: Standard   Tag: None

What do scientists often use to determine the relative age of a rock layer?

a) mineral content

b) the geologic column

c) rock type

d) the geologic calendar

b

36) Item Type: Multiple Choice   Objective: 1   Section 18.1
    Level: A: Standard   Tag: None

Natural selection is also known as survival of the

a) largest.   b) youngest.   c) strongest.   d) fittest.

d

37) Item Type: Multiple Choice   Objective: 1   Section 18.1
    Level: A: Standard   Tag: None

At the time the geologic column was developed, scientists based their estimates of rock layer ages on factors such as

a) mineral content.

b) rock type.

c) metamorphic temperature.

d) sedimentation rate.

d

38) Item Type: Multiple Choice   Objective: 1   Section 18.1
    Level: A: Standard   Tag: None

The ordered arrangement of rock layers is called the

a) geologic column.

b) radioactive dating.

c) theory of evolution.

d) geologic time scale.

a

39) Item Type: Multiple Choice   Objective: 2   Section 18.1
    Level: A: Standard   Tag: None

The current geologic era is the

a) Mesozoic.   b) Precambrian.   c) Paleozoic.   d) Cenozoic.

d

40) Item Type: Multiple Choice   Objective: 2   Section 18.1
     Level: A: Standard   Tag: None

Some eras are divided into time units characterized by specific fossils and are called

a) epochs.   b) events.   c) periods.   d) eons.

c

41) Item Type: Multiple Choice   Objective: 2   Section 18.1
     Level: A: Standard   Tag: None

The Cenozoic Era contains an extremely detailed fossil record, and its periods may therefore be further divided into

a) epochs.   b) eons.   c) zones.   d) millennia.

a

42) Item Type: Multiple Choice   Objective: 3   Section 18.2
     Level: A: Standard   Tag: None

The presence of cross-bedded sandstone may indicate that the region where it is present was once

a) an ocean.   b) a mountain.   c) a desert.   d) a lake.

c

43) Item Type: Multiple Choice   Objective: 3   Section 18.2
     Level: A: Standard   Tag: None

Shields are large areas of exposed rock of what era or time period?

a) Cenozoic   b) Mesozoic   c) Paleozoic   d) Precambrian

d

44) Item Type: Multiple Choice   Objective: 3   Section 18.2
     Level: A: Standard   Tag: None

Which of the following are rare in Precambrian rocks?

a) fossil remains   b) sandstone   c) minerals   d) metamorphisms

a

45) Item Type: Multiple Choice   Objective: 3   Section 18.2
     Level: B: Enriched   Tag: None

How many billion years did Precambrian time last?

a) 2   b) 4   c) 6   d) 8

b

46) Item Type: Multiple Choice   Objective: 3   Section 18.2
Level: B: Enriched   Tag: None

The reeflike deposits produced by bacteria or algae that are the most common Precambrian fossils are called

a) rhipidistians.   b) plesiosaurs.   c) ceratopsians.   d) stromatolites.

d

47) Item Type: Multiple Choice   Objective: 4   Section 18.2
Level: A: Standard   Tag: None

The Devonian Period is also known as the Age of

a) Birds.   b) Lizards.   c) Mammals.   d) Fishes.

d

48) Item Type: Multiple Choice   Objective: 4   Section 18.2
Level: B: Enriched   Tag: None

The fossil record indicates that which of the following thrived during the Paleozoic Era?

a) reptiles   b) advanced forms of marine life

c) mammals   d) marine algae and bacteria

b

49) Item Type: Multiple Choice   Objective: 4   Section 18.2
Level: B: Enriched   Tag: None

About how many million years ago did the Paleozoic Era begin?

a) 24   b) 144   c) 438   d) 570

d

50) Item Type: Multiple Choice   Objective: 4   Section 18.2
Level: B: Enriched   Tag: None

What animals were common in the Cambrian Period?

a) brachiopods   b) fish   c) lizards   d) eurypterids

a

51) Item Type: Multiple Choice   Objective: 4   Section 18.2
Level: B: Enriched   Tag: None

The final period of the Paleozoic Era is the

a) Silurian.   b) Tertiary.   c) Permian.   d) Triassic.

c

52) Item Type: Multiple Choice   Objective: 5   Section 18.2
Level: A: Standard   Tag: None

What period ended with the extinction of all dinosaurs?

a) Cambrian   b) Pennsylvanian   c) Devonian   d) Cretaceous

d

53) Item Type: Multiple Choice   Objective: 5   Section 18.2
Level: A: Standard   Tag: None

Dinosaurs first appeared during the

a) Triassic Period.   b) Pliocene Epoch.

c) Cretaceous Period.   d) Oligocene Epoch.

a

54) Item Type: Multiple Choice   Objective: 5   Section 18.2
Level: B: Enriched   Tag: Diagram 1874

Which geologic era is represented by the table above?

a) Precambrian   b) Paleozoic   c) Mesozoic   d) Cenozoic

c

55) Item Type: Multiple Choice   Objective: 6   Section 18.2
Level: B: Enriched   Tag: Diagram 1873

In the table above, the missing epoch that belongs in the space labeled *4* is called the

a) Jurassic.   b) Pleistocene.   c) Cretaceous.   d) Eocene.

d

56) Item Type: Multiple Choice   Objective: 6   Section 18.2
Level: B: Enriched   Tag: None

During which epoch did the largest known land mammal, called Baluchitherium, live?

a) Pliocene   b) Miocene   c) Eocene   d) Paleocene

b

57) Item Type: Multiple Choice   Objective: 6   Section 18.2
Level: B: Enriched   Tag: None

About how many million years ago did the Paleocene epoch begin?

a) 245   b) 144   c) 65   d) 58

c

58) Item Type: Multiple Choice   Objective: 6   Section 18.2
Level: B: Enriched   Tag: None

During which time period did the first primates appear?

a) Holocene   b) Miocene   c) Eocene   d) Paleocene

d

59) Item Type: Multiple Choice   Objective: 6   Section 18.2
Level: B: Enriched   Tag: None

What is the second epoch of the Quaternary Period called?

a) Miocene   b) Permian   c) Mississippian   d) Holocene

d

60) Item Type: Multiple Choice   Objective: 6   Section 18.2
Level: A: Standard   Tag: None

Which of the following epochs was called the Golden Age of Mammals?

a) Pliocene   b) Miocene   c) Oligocene   d) Paleocene

b

61) Item Type: Multiple Choice   Objective: 6   Section 18.2
Level: B: Enriched   Tag: None

During which era did mammals become the predominant life form?

a) Precambrian   b) Paleozoic   c) Mesozoic   d) Cenozoic

d

62) Item Type: Fill in the Blank   Objective: 1   Section 18.1
Level: B: Enriched   Tag: Diagram 1871

According to the geologic column in the diagram, the oldest rock layer is

_____.

layer 4

63) Item Type: Fill in the Blank   Objective: 1   Section 18.1
Level: B: Enriched   Tag: None

What is the general term for the method that allows more accurate dating of absolute ages of rock layers than was previously possible based on fossils and rates of deposition?

_____

radioactive dating

64) Item Type: Fill in the Blank   Objective: 3   Section 18.2
Level: B: Enriched   Tag: None

What are the most common Precambrian fossils called? _____

stromatolites

65) Item Type: Fill in the Blank   Objective: 4   Section 18.2
Level: B: Enriched   Tag: None

Rocks of which geologic time period contain nearly half of the world's deposits of

valuable minerals? _____

Precambrian time

66) Item Type: Fill in the Blank   Objective: 4   Section 18.2
Level: B: Enriched   Tag: None

In what period did vertebrates first appear? _____

Ordovician

67) Item Type: Fill in the Blank   Objective: 4   Section 18.2
Level: B: Enriched   Tag: Diagram 1872

Which geologic period is missing from this diagram? _____

Silurian

68) Item Type: Fill in the Blank   Objective: 4   Section 18.2
Level: B: Enriched   Tag: None

During what geologic period did graptolites flourish? _____

Ordovician

69) Item Type: Fill in the Blank   Objective: 4   Section 18.2
Level: B: Enriched   Tag: None

Ichthyostega was the earliest true example of what type of animal? _____

amphibian

70) Item Type: Fill in the Blank   Objective: 4   Section 18.2
Level: B: Enriched   Tag: None

During which period did the first reptiles develop? _____

Pennsylvanian/Carboniferous

71) Item Type: Fill in the Blank   Objective: 4   Section 18.2
Level: B: Enriched   Tag: None

During which period did the Appalachian Mountains form? _____

> Permian

72) Item Type: Fill in the Blank   Objective: 5   Section 18.2
Level: B: Enriched   Tag: None

Scientists have found evidence suggesting that mass extinctions of organisms occurred at about the same time as the earth experienced what dramatic cosmic event?

_____

> meteorite impact

73) Item Type: Fill in the Blank   Objective: 5   Section 18.2
Level: B: Enriched   Tag: None

The extinction of dinosaurs marked the end of what period? _____

> Cretaceous

74) Item Type: Fill in the Blank   Objective: 5   Section 18.2
Level: B: Enriched   Tag: None

What type of animals dominated the Mesozoic Era? _____

> reptiles

75) Item Type: Fill in the Blank   Objective: 5   Section 18.2
Level: A: Standard   Tag: None

During which era did birds first appear? _____

> Mesozoic

76) Item Type: Fill in the Blank   Objective: 5   Section 18.2
Level: B: Enriched   Tag: None

During which period did trees first become abundant? _____

> Cretaceous

77) Item Type: Fill in the Blank   Objective: 6   Section 18.2
Level: B: Enriched   Tag: None

What are the two periods of the Cenozoic Era? _____

> Tertiary and Quaternary

78) Item Type: Fill in the Blank   Objective: 6   Section 18.2
    Level: A: Standard   Tag: None

During which era did glaciers cover nearly one-third of the earth's land area?

_____

Cenozoic

79) Item Type: Fill in the Blank   Objective: 6   Section 18.2
    Level: A: Standard   Tag: None

What sea dried up during the Oligocene Epoch? _____

the Mediterranean

80) Item Type: Fill in the Blank   Objective: 6   Section 18.2
    Level: A: Standard   Tag: None

Which epoch ended with the end of the most recent ice age? _____

Pleistocene

81) Item Type: Fill in the Blank   Objective: 6   Section 18.2
    Level: A: Standard   Tag: None

During which epoch did the first modern horses appear? _____

Pliocene

82) Item Type: Fill in the Blank   Objective: 6   Section 18.2
    Level: A: Standard   Tag: None

Which period of the Cenozoic Era includes the time before the last major ice age?

_____

Tertiary

83) Item Type: Fill in the Blank   Objective: 6   Section 18.2
    Level: B: Enriched   Tag: None

During which epoch did the Great Lakes form? _____

Holocene

84) Item Type: Fill in the Blank   Objective: 6   Section 18.2
   Level: B: Enriched   Tag: None

The periods of the Cenozoic Era have been divided into smaller time spans called

_____.

epochs

85) Item Type: Essay   Objective: 1   Section 18.1
   Level: B: Enriched   Tag: None

Use the theory of evolution by natural selection to explain the extinction of some organisms during times of climatic change.

When environmental changes occur, some organisms can adapt to the changes better than others. The organisms that adapt compete more successfully in the new environment and survive to reproduce, while the organisms that cannot adapt become extinct.

86) Item Type: Essay   Objective: 5   Section 18.2
   Level: B: Enriched   Tag: None

How did climatic conditions in the Mesozoic Era favor reptile survival and development?

The Mesozoic Era was warm and humid. Reptiles are cold-blooded and require external warmth to maintain their body functions at levels sufficient to pursue normal, survival-related activities.

87) Item Type: Essay   Objective: 6   Section 18.2
    Level: B: Enriched   Tag: None

How does the arrival of the ancestors of modern humans in the Pleistocene Epoch help to explain the extinction of some animal species at that time?

_____

_____

_____

_____

_____

The early humans were hunters. Their successful hunting may have contributed to the extinction of some animals.

88) Item Type: Essay   Objective: 6   Section 18.2
    Level: A: Standard   Tag: None

Why can the Cenozoic Era be easily divided into numerous smaller time units?

_____

_____

_____

_____

The fossil record of the Cenozoic Era is very abundant. Smaller time divisions are largely based on changes in the fossil content of rock layers.

Chapter 18

1)   Tag Name: Diagram 1871

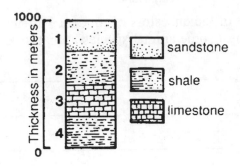

2)   Tag Name: Diagram 1872

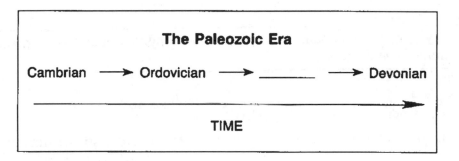

The Paleozoic Era

Cambrian ⟶ Ordovician ⟶ _____ ⟶ Devonian

TIME

3)   Tag Name: Diagram 1873

EPOCH	TIME (millions of years ago)	CHARACTERISTIC ORGANISMS
1. Pliocene	5	carnivores, first modern horses
2. Miocene	24	species of deer, rhinoceros, pig, saber-toothed cat
3. Oligocene	37	species of horses, cats, dogs
4.	58	flying squirrels and bats, whales
5. Paleocene		rodents, lemuroids

4)   Tag Name: Diagram 1874

Period	Years Before Present
	65,000,000
Cretaceous	
	144,000,000
Jurassic	
	208,000,000
Triassic	
	245,000,000

Chapter 19

1) Item Type: True-False   Objective: 1   Section 19.1
    Level: B: Enriched   Tag: None

_____ The part of the supercontinent called Laurasia was located north of the Tethys Sea.

T

2) Item Type: True-False   Objective: 1   Section 19.1
    Level: A: Standard   Tag: None

_____ The lithosphere is made up of plates that ride on the asthenosphere.

T

3) Item Type: True-False   Objective: 1   Section 19.1
    Level: A: Standard   Tag: None

_____ Pangaea was a large supercontinent that formed in Precambrian time.

F

4) Item Type: True-False   Objective: 1   Section 19.1
    Level: A: Standard   Tag: None

_____ Outlines of the continents as they existed in Precambrian time have been determined by fossil evidence.

F

5) Item Type: True-False   Objective: 1   Section 19.1
    Level: A: Standard   Tag: None

_____ Pangaea was initially located near the North Pole.

F

6) Item Type: True-False   Objective: 1   Section 19.1
    Level: A: Standard   Tag: None

_____ The Sahara was once covered by a sheet of ice.

T

7) Item Type: True-False   Objective: 1   Section 19.1
    Level: A: Standard   Tag: None

_____ Australia and Antarctica were once part of a landmass called Laurasia.

F

8) Item Type: True-False   Objective: 2   Section 19.1
    Level: A: Standard   Tag: None

    _____ South America and Africa were part of the same landmass that broke from
    Gondwanaland.

    | T | |

9) Item Type: True-False   Objective: 2   Section 19.1
    Level: A: Standard   Tag: None

    _____ Australia and Antarctica were once part of the same landmass.

    | T | |

10) Item Type: True-False   Objective: 2   Section 19.1
    Level: A: Standard   Tag: None

    _____ The northward movement of Laurasia separated North America from Africa.

    | F | |

11) Item Type: True-False   Objective: 2   Section 19.1
    Level: A: Standard   Tag: None

    _____ The modern Mediterranean Sea formed when the east end of the Tethys Sea was
    closed off.

    | T | |

12) Item Type: True-False   Objective: 2   Section 19.1
    Level: A: Standard   Tag: None

    _____ Australia was once part of Gondwanaland.

    | T | |

13) Item Type: True-False   Objective: 2   Section 19.1
    Level: A: Standard   Tag: None

    _____ Madagascar was once attached to Africa.

    | T | |

14) Item Type: True-False   Objective: 2   Section 19.1
    Level: A: Standard   Tag: None

    _____ The Mediterranean Sea was formed from the Tethys Sea.

    | T | |

15) Item Type: True-False   Objective: 3   Section 19.2
     Level: A: Standard   Tag: None

_____ Precambrian rocks are exposed at the top of the Grand Canyon.

F

16) Item Type: True-False   Objective: 3   Section 19.2
     Level: A: Standard   Tag: None

_____ The ancient landmass around which the present North American continent developed first took shape during the Mesozoic Era.

F

17) Item Type: True-False   Objective: 3   Section 19.2
     Level: A: Standard   Tag: None

_____ Much of North America was once covered by an inland sea.

T

18) Item Type: True-False   Objective: 3   Section 19.2
     Level: A: Standard   Tag: None

_____ Much of what is now the Atlantic coastal area of North America was once attached to the African plate.

T

19) Item Type: True-False   Objective: 3   Section 19.2
     Level: A: Standard   Tag: None

_____ The Canadian Shield is an area of exposed Precambrian rock.

T

20) Item Type: True-False   Objective: 3   Section 19.2
     Level: A: Standard   Tag: None

_____ Salt deposits in Kansas are thought to have formed when a large inland sea dried up.

T

21) Item Type: True-False   Objective: 3   Section 19.2
     Level: A: Standard   Tag: None

_____ A geosyncline is a sediment-filled trough.

T

22) Item Type: True-False   Objective: 3   Section 19.2
     Level: A: Standard   Tag: None

_____ The tectonic activity that accompanied the breakup of Pangaea played a large role in shaping North America.

T

23) Item Type: True-False   Objective: 4   Section 19.2
     Level: B: Enriched   Tag: None

_____ Sediments eroded from the Sierra Nevada eventually formed the Great Plains.

F

24) Item Type: True-False   Objective: 4   Section 19.2
     Level: A: Standard   Tag: None

_____ The Appalachian Mountains existed during Precambrian time.

F

25) Item Type: True-False   Objective: 4   Section 19.2
     Level: A: Standard   Tag: None

_____ The Appalachian Mountains began forming during the Mesozoic Era.

F

26) Item Type: True-False   Objective: 5   Section 19.3
     Level: A: Standard   Tag: None

_____ An outcrop is an exposed rock layer of any type.

T

27) Item Type: True-False   Objective: 5   Section 19.3
     Level: B: Enriched   Tag: None

_____ The Colorado River has probably taken at least 10 million years to erode the Grand Canyon.

T

28) Item Type: True-False   Objective: 5   Section 19.3
     Level: A: Standard   Tag: None

_____ There is evidence that the layers of sedimentary rock in the Grand Canyon have been overturned.

F

29) Item Type: True-False   Objective: 5   Section 19.3
    Level: A: Standard   Tag: None

_____ The Permian Period is not represented by rock layers in the Grand Canyon.

F

30) Item Type: True-False   Objective: 5   Section 19.3
    Level: A: Standard   Tag: None

_____ The Grand Canyon is an artificial outcrop of sedimentary rock.

F

31) Item Type: True-False   Objective: 5   Section 19.3
    Level: A: Standard   Tag: None

_____ The Grand Canyon is located in the Basin and Range Province.

F

32) Item Type: True-False   Objective: 6   Section 19.3
    Level: A: Standard   Tag: None

_____ Radioactive dating is used to estimate the relative age of fossils.

F

33) Item Type: True-False   Objective: 6   Section 19.3
    Level: A: Standard   Tag: None

_____ Sandstone layers in the Grand Canyon often contain fossils of marine organisms.

F

34) Item Type: True-False   Objective: 6   Section 19.3
    Level: A: Standard   Tag: None

_____ Limestone layers in the Grand Canyon often contain fossils of marine organisms.

T

35) Item Type: True-False   Objective: 6   Section 19.3
    Level: A: Standard   Tag: None

_____ Changes in organisms can sometimes be traced by examining their remains in successive rock layers.

T

36) Item Type: True-False   Objective: 6   Section 19.3
     Level: A: Standard   Tag: None

____ Fossils of clams and fishes can be found in some rock layers of the Grand Canyon.

T

37) Item Type: Multiple Choice   Objective: 1   Section 19.1
     Level: A: Standard   Tag: None

Approximately how far do crustal plates move in a year?

a) a few millimeters   b) a few meters

c) a few centimeters   d) a few kilometers

c

38) Item Type: Multiple Choice   Objective: 1   Section 19.1
     Level: A: Standard   Tag: None

The continents existing today contain large areas of deformed Precambrian rocks called

a) cratons.   b) asthenospheres.   c) shields.   d) lithospheres.

a

39) Item Type: Multiple Choice   Objective: 1   Section 19.1
     Level: A: Standard   Tag: Diagram 1971

Which number in the table represents Panthalassa?

a) *1*   b) *2*   c) *3*   d) *4*

a

40) Item Type: Multiple Choice   Objective: 1   Section 19.1
     Level: A: Standard   Tag: Diagram 1971

Which number in the table represents Gondwanaland?

a) *2*   b) *3*   c) *4*   d) *5*

c

41) Item Type: Multiple Choice   Objective: 2   Section 19.1
     Level: B: Enriched   Tag: Diagram 1972

According to the table, which event caused the formation of the South Atlantic Ocean?

a) *1*   b) *2*   c) *4*   d) *5*

c

42) Item Type: Multiple Choice   Objective: 2   Section 19.1
    Level: B: Enriched   Tag: Diagram 1972

According to the table above, event *5* led to the formation of

a) the North Atlantic Ocean.   b) Panthalassa.

c) the South Atlantic Ocean.   d) the Tethys Sea.

a

43) Item Type: Multiple Choice   Objective: 2   Section 19.1
    Level: A: Standard   Tag: None

From which continent did India break away to move northward?

a) Eurasia   b) South America   c) Africa   d) Antarctica

d

44) Item Type: Multiple Choice   Objective: 2   Section 19.1
    Level: A: Standard   Tag: None

Which of the following was once part of Laurasia?

a) South America   b) Antarctica   c) North America   d) Africa

c

45) Item Type: Multiple Choice   Objective: 3   Section 19.2
    Level: B: Enriched   Tag: None

The Paleozoic Era ended approximately how many millions of years ago?

a) 408   b) 245   c) 144   d) 65

b

46) Item Type: Multiple Choice   Objective: 3   Section 19.2
    Level: A: Standard   Tag: None

Parts of the Atlantic coastal plain of North America were once attached to which crustal plate?

a) Australian   b) African   c) South American   d) Indian

b

47) Item Type: Multiple Choice   Objective: 3   Section 19.2
Level: B: Enriched   Tag: None

Which North American mountain range first appeared in the Jurassic Period, was eroded until almost flat, and was then lifted again during the Cenozoic Era?

a) Sierra Nevada

b) Rocky Mountains

c) Appalachian Mountains

d) Cascade Range

a

48) Item Type: Multiple Choice   Objective: 3   Section 19.2
Level: B: Enriched   Tag: None

Approximately how old are fossils found in Precambrian rocks of North America?

a) 2 million years old

b) 200 million years old

c) 1 billion years old

d) 2 billion years old

d

49) Item Type: Multiple Choice   Objective: 3   Section 19.2
Level: A: Standard   Tag: None

Terranes are best described as

a) tilted fault-blocks.

b) exposed portions of cratons.

c) pieces of oceanic crust that are added to continents.

d) sediment-filled troughs.

c

50) Item Type: Multiple Choice   Objective: 3   Section 19.2
Level: B: Enriched   Tag: None

During which time period or era were the Appalachian Mountains formed?

a) Precambrian   b) Paleozoic   c) Mesozoic   d) Cenozoic

b

51) Item Type: Multiple Choice   Objective: 3   Section 19.2
Level: A: Standard   Tag: None

A geosyncline is defined as

a) a large inland sea.

b) a sediment-filled trough.

c) an outcrop of Precambrian rock.

d) an uplifted plateau.

b

52) Item Type: Multiple Choice   Objective: 4   Section 19.2
Level: A: Standard   Tag: None

During the Cenozoic Era, violent movements along plate boundaries in the western part of North America caused the uplift of the

a) Sierra Nevada.

b) Rocky Mountains.

c) Colorado Plateau.

d) Great Plains.

a

53) Item Type: Multiple Choice   Objective: 4   Section 19.2
Level: A: Standard   Tag: None

In North America, volcanic activity during the Cenozoic Era occurred chiefly in the

a) Northwest.   b) Southwest.   c) Northeast.   d) Southeast.

a

54) Item Type: Multiple Choice   Objective: 4   Section 19.2
Level: A: Standard   Tag: None

About 150 million years from now, the portion of California west of the San Andreas fault will probably be located near what present-day region?

a) Mexico   b) Hawaii   c) Oregon   d) Alaska

d

55) Item Type: Multiple Choice   Objective: 4   Section 19.2
Level: A: Standard   Tag: Diagram 1973

Which feature in the table produced the sediments that eventually formed the Great Plains?

a) *1*   b) *2*   c) *3*   d) *4*

d

56) Item Type: Multiple Choice   Objective: 4   Section 19.2
Level: B: Enriched   Tag: Diagram 1973

During which time period did North America exhibit the features listed in the table above?

a) Precambrian   b) Paleozoic   c) Mesozoic   d) Cenozoic

c

57) Item Type: Multiple Choice   Objective: 4   Section 19.2
Level: A: Standard   Tag: None

During the Pleistocene Epoch, a land bridge connected North America and

a) Africa.   b) Australia.   c) Greenland.   d) Eurasia.

d

Chapter 19

58) Item Type: Multiple Choice   Objective: 4   Section 19.2
    Level: B: Enriched   Tag: None

If the continents continue to move as predicted, in 150 million years Africa will collide with

a) South America.   b) Eurasia.   c) India.   d) North America.

b

59) Item Type: Multiple Choice   Objective: 4   Section 19.2
    Level: B: Enriched   Tag: None

Archaeological evidence indicates that the first human inhabitants arrived in North America approximately how many years ago?

a) 10,000 years ago   b) 20,000 years ago

c) 30,000 years ago   d) 40,000 years ago

b

60) Item Type: Multiple Choice   Objective: 4   Section 19.2
    Level: B: Enriched   Tag: None

What North American mountains began to form during the Mesozoic Era?

a) Sierra Nevada   b) Rockies   c) Appalachians   d) Cascades

b

61) Item Type: Multiple Choice   Objective: 5   Section 19.3
    Level: A: Standard   Tag: None

The Colorado River has cut through rock layers to a depth of nearly

a) 1 km   b) 10 km   c) 100 km   d) 1,000 km

a

62) Item Type: Multiple Choice   Objective: 5   Section 19.3
    Level: B: Enriched   Tag: None

From which period have rock layers in the Grand Canyon probably been completely eroded?

a) Cambrian   b) Silurian   c) Mississippian   d) Permian

b

63) Item Type: Multiple Choice   Objective: 5   Section 19.3
      Level: A: Standard   Tag: None

The Mississippian Period is represented in the Grand Canyon by the

a) Redwall limestone.

b) Coconino sandstone.

c) Kaibab limestone.

d) Tapeats sandstone.

a

64) Item Type: Multiple Choice   Objective: 5   Section 19.3
      Level: B: Enriched   Tag: None

Which layer of the Grand Canyon was deposited during the Cambrian Period?

a) Hermit shale

b) Kaibab limestone

c) Bright Angel shale

d) Redwall limestone

c

65) Item Type: Multiple Choice   Objective: 5   Section 19.3
      Level: B: Enriched   Tag: None

During what geologic period was the Bright Angel shale deposited?

a) Cambrian   b) Ordovician   c) Silurian   d) Devonian

a

66) Item Type: Multiple Choice   Objective: 5   Section 19.3
      Level: B: Enriched   Tag: None

Which layer of the Grand Canyon was probably deposited when the area was submerged under a shallow sea?

a) Supai formation

b) Vishnu schist

c) Redwall limestone

d) Coconino sandstone

c

67) Item Type: Fill in the Blank   Objective: 1   Section 19.1
      Level: B: Enriched   Tag: None

Approximately how far do crustal plates drift each year? _____

1 cm

68) Item Type: Fill in the Blank   Objective: 1   Section 19.1
      Level: B: Enriched   Tag: None

During which geologic era did Pangaea exist? _____

Paleozoic

69) Item Type: Fill in the Blank   Objective: 2   Section 19.1
Level: A: Standard   Tag: None

After breaking away from Gondwanaland, India collided with what modern continent?

_____

| Eurasia | ■ |

70) Item Type: Fill in the Blank   Objective: 3   Section 19.2
Level: B: Enriched   Tag: None

In North America, Precambrian rocks contain fossils of what type of organisms?

_____

| bacteria | ■ |

71) Item Type: Fill in the Blank   Objective: 4   Section 19.2
Level: B: Enriched   Tag: None

Most of the major landforms of modern North America were created during which era?

_____

| Cenozoic | ■ |

72) Item Type: Fill in the Blank   Objective: 4   Section 19.2
Level: B: Enriched   Tag: None

During which geologic era did the Rocky Mountains begin to form?

_____

| Mesozoic | ■ |

73) Item Type: Fill in the Blank   Objective: 4   Section 19.2
Level: A: Standard   Tag: None

By the beginning of which geologic era had North America begun to take on its present

shape? _____

| Cenozoic | ■ |

74) Item Type: Fill in the Blank   Objective: 4   Section 19.2
     Level: A: Standard   Tag: None

What is the name of the uplifted area east of the Sierra Nevada through which the

Colorado River flows? _____

Colorado Plateau

75) Item Type: Fill in the Blank   Objective: 4   Section 19.2
     Level: A: Standard   Tag: None

What evidence from Permian layers in Kansas indicate that the area was once under an

inland sea? _____

salt deposits

76) Item Type: Fill in the Blank   Objective: 5   Section 19.3
     Level: B: Enriched   Tag: None

What is the name for the uplifted region that includes the Grand Canyon?

_____

Colorado Plateau

77) Item Type: Fill in the Blank   Objective: 5   Section 19.3
     Level: A: Standard   Tag: None

What river cut the gorge that forms the Grand Canyon? _____

Colorado River

78) Item Type: Fill in the Blank   Objective: 5   Section 19.3
     Level: A: Standard   Tag: None

The lack of deposits from the Ordovician Period in the Grand Canyon is a result of what

process? _____

erosion

79) Item Type: Fill in the Blank   Objective: 5   Section 19.3
     Level: A: Standard   Tag: None

What is the term for any type of exposed rock formation? _____

outcrop

80) Item Type: Fill in the Blank   Objective: 5   Section 19.3
      Level: B: Enriched   Tag: Diagram 1974

In the table above, which geologic time period is represented by *A?*

_____

| Permian Period | ■ |

81) Item Type: Fill in the Blank   Objective: 5   Section 19.3
      Level: A: Standard   Tag: None

What geologic law assumes that young geologic deposits sit on top of older deposits?

_____

| the law of superposition | ■ |

82) Item Type: Fill in the Blank   Objective: 6   Section 19.3
      Level: A: Standard   Tag: None

What method is used to estimate the absolute age of fossils? _____

| radioactive dating | ■ |

83) Item Type: Essay   Objective: 1   Section 19.1
      Level: B: Enriched   Tag: None

How and when did the Alps and the Himalayas form?

_____

_____

_____

_____

_____

| The supercontinent Pangaea broke up in the Mesozoic Era. The new, smaller continents drifted and collided, uplifting these mountains. | ■ |

84) Item Type: Essay   Objective: 1   Section 19.1
     Level: B: Enriched   Tag: None

What effect did northward drift toward the equator have on Pangaea?

_____

_____

_____

_____

_____

Northward drift brought the landmass into warmer climates. Ice covering the continent melted, raising the level of the oceans. This triggered global warming. The northward drift also resulted in an increase in the number of species of land plants and animals.

85) Item Type: Essay   Objective: 3   Section 19.2
     Level: B: Enriched   Tag: None

Describe the Canadian Shield and its location.

_____

_____

_____

_____

_____

The Canadian Shield is a large area in eastern Canada and parts of the northeastern United States where very old Precambrian rocks are exposed. It is the exposed portion of the craton around which the modern continent of North America has been built up.

86) Item Type: Essay   Objective: 4   Section 19.2
     Level: B: Enriched   Tag: None

Describe the configuration of the west coast of North America that scientists predict will exist 150 million years from now.

_____

_____

_____

_____

_____

In North America, Mexico's Baja Peninsula and the portion of California west of the San Andreas fault will have moved to where Alaska is today. If this plate movement occurs as predicted, Los Angeles will one day be located north of where San Francisco is today.

87) Item Type: Essay   Objective: 5   Section 19.3
     Level: B: Enriched   Tag: None

What would most likely happen if the Grand Canyon area were uplifted again?

_____

_____

_____

_____

_____

If the Grand Canyon area were uplifted again, the Colorado River would likely cut deeper into the Colorado Plateau, exposing deeper and deeper rock layers. As a result of this increased erosion, the Grand Canyon would become deeper.

88) Item Type: Essay   Objective: 6   Section 19.3
     Level: A: Standard   Tag: None

What sequence of events can be traced by studying fossils in successive rock layers of the Grand Canyon?

_____

_____

_____

_____

_____

Paleontologists can trace changes in many organisms by examining their remains found in successive rock layers. Analysis of the fossils found in older, lower layers through younger, upper layers suggests that early life-forms gradually became extinct or changed into different forms.

1)   Tag Name: Diagram 1971

	Features of the Paleozoic Earth
1	the single large ocean covering 60% of the earth's surface
2	the triangular body of water that cut into the eastern edge of the supercontinent
3	the northern segment of the giant continent
4	the southern segment of the giant continent
5	the large continental mass covering 40% of the earth's surface

2)   Tag Name: Diagram 1972

	The Breakup of Pangaea
1	closing of the sea in the east
2	opening of an east-west rift
3	opening of the sea in the west
4	opening of a north-south rift
5	clockwise rotation of the northern landmass

3)   Tag Name: Diagram 1973

	Features of the North American Continent
1	Appalachian Mountains
2	sea covering western geosyncline
3	shallow inland sea
4	Rocky Mountains

4)   Tag Name: Diagram 1974

Rock Layers of the Grand Canyon

Time Period	Rock Layers Formed
A	Kaibab limestone Toroweap formation Coconino sandstone **1** Supai formation
B	Redwall limestone
C	Muav limestone **2** Tapeats sandstone Vishnu schist

1) Item Type: True-False   Objective: 1   Section 20.1
   Level: A: Standard   Tag: None

   _____ The world's oceans contain more than 97 percent of the earth's mass.

   F

2) Item Type: True-False   Objective: 1   Section 20.1
   Level: A: Standard   Tag: None

   _____ The global ocean is divided into three principal oceans.

   T

3) Item Type: True-False   Objective: 1   Section 20.1
   Level: A: Standard   Tag: None

   _____ All of the earth's principal oceans are interconnected.

   T

4) Item Type: True-False   Objective: 1   Section 20.1
   Level: A: Standard   Tag: None

   _____ The Pacific Ocean is the largest of the earth's principal oceans.

   T

5) Item Type: True-False   Objective: 1   Section 20.1
   Level: A: Standard   Tag: None

   _____ The earth is the only known planet with a covering of liquid water.

   T

6) Item Type: True-False   Objective: 1   Section 20.1
   Level: A: Standard   Tag: None

   _____ The oceans contain about three-fourths of the earth's water.

   F

7) Item Type: True-False   Objective: 2   Section 20.1
   Level: A: Standard   Tag: None

   _____ Many of the features on the ocean floor were first detected by sonar.

   T

8) Item Type: True-False   Objective: 2   Section 20.1
   Level: A: Standard   Tag: None

   _____ Underwater continental crust makes up part of the deep ocean basin.

   F

9) Item Type: True-False   Objective: 2   Section 20.1
   Level: A: Standard   Tag: None

   _____ Bathyspheres are self-propelled diving vessels that have freedom of movement.

   F

10) Item Type: True-False   Objective: 2   Section 20.1
    Level: A: Standard   Tag: None

    _____ Radar is often used to determine ocean depths.

    F

11) Item Type: True-False   Objective: 2   Section 20.1
    Level: A: Standard   Tag: None

    _____ A bathyscaph is a submersible that is self-propelled and free-moving.

    T

12) Item Type: True-False   Objective: 2   Section 20.1
    Level: B: Enriched   Tag: None

    _____ The *Alvin* is an example of a bathysphere.

    F

13) Item Type: True-False   Objective: 2   Section 20.1
    Level: A: Standard   Tag: None

    _____ When the remains of the *Titanic* were discovered, many of the photographs of the
    inside of the ship were taken with a robot craft.

    T

14) Item Type: True-False   Objective: 3   Section 20.2
    Level: A: Standard   Tag: None

    _____ The shoreline marks the boundary between a continent and the ocean floor.

    F

15) Item Type: True-False   Objective: 3   Section 20.2
     Level: A: Standard   Tag: None

_____ The width of the continental shelf varies depending on its location.

T

16) Item Type: True-False   Objective: 3   Section 20.2
     Level: A: Standard   Tag: None

_____ The continental slope is the portion of the continental margin closest to the shoreline.

F

17) Item Type: True-False   Objective: 3   Section 20.2
     Level: A: Standard   Tag: None

_____ Submarine canyons are most often found on mid-ocean ridges.

F

18) Item Type: True-False   Objective: 3   Section 20.2
     Level: A: Standard   Tag: None

_____ The ocean floor is composed mainly of continental crust.

F

19) Item Type: True-False   Objective: 4   Section 20.2
     Level: A: Standard   Tag: None

_____ Most trenches are located within the mid-ocean ridge system.

F

20) Item Type: True-False   Objective: 4   Section 20.2
     Level: A: Standard   Tag: None

_____ Abyssal plains are the flattest regions on earth.

T

21) Item Type: True-False   Objective: 4   Section 20.2
     Level: A: Standard   Tag: None

_____ Guyots are formed by the erosion and subsidence of seamounts.

T

22) Item Type: True-False   Objective: 5   Section 20.3
     Level: A: Standard   Tag: None

_____ The coarser sediments of the continental margin are usually found close to shore.

T

23) Item Type: True-False   Objective: 5   Section 20.3
     Level: A: Standard   Tag: None

_____ Mineral deposits called nodules probably formed through chemical processes.

T

24) Item Type: True-False   Objective: 5   Section 20.3
     Level: A: Standard   Tag: None

_____ Most deep-ocean sediments come from continental erosion.

F

25) Item Type: True-False   Objective: 5   Section 20.3
     Level: A: Standard   Tag: None

_____ Meteorites contribute to ocean sediments.

T

26) Item Type: True-False   Objective: 5   Section 20.3
     Level: A: Standard   Tag: None

_____ Icebergs commonly deposit ocean sediments on land.

F

27) Item Type: True-False   Objective: 5   Section 20.3
     Level: A: Standard   Tag: None

_____ Turbidity currents are caused by winds and temperature differences.

F

28) Item Type: True-False   Objective: 5   Section 20.3
     Level: A: Standard   Tag: None

_____ Samples of deep-sea sediments are often obtained by taking core samples.

T

29) Item Type: True-False   Objective: 5   Section 20.3
Level: A: Standard   Tag: None

_____ The finest ocean sediments are found closest to shore.

F

30) Item Type: True-False   Objective: 6   Section 20.3
Level: A: Standard   Tag: None

_____ Most material found in oozes on the deep-sea bottom is derived from the continents.

F

31) Item Type: True-False   Objective: 6   Section 20.3
Level: A: Standard   Tag: None

_____ Siliceous ooze is an example of an organic sediment.

T

32) Item Type: True-False   Objective: 6   Section 20.3
Level: A: Standard   Tag: None

_____ Silica and calcium carbonate are commonly found in organic ocean sediment.

T

33) Item Type: True-False   Objective: 6   Section 20.3
Level: A: Standard   Tag: None

_____ Calcareous ooze is found at the ocean's greatest depths.

F

34) Item Type: True-False   Objective: 6   Section 20.3
Level: A: Standard   Tag: None

_____ Most siliceous ooze is found around the continent of Antarctica.

T

35) Item Type: Multiple Choice   Objective: 1   Section 20.1
Level: B: Enriched   Tag: None

The Gulf of Mexico is a part of what ocean?

a) Arctic   b) Indian   c) Atlantic   d) Pacific

c

36) Item Type: Multiple Choice   Objective: 1   Section 20.1
    Level: B: Enriched   Tag: None

The deepest of the earth's principal oceans is the

a) Arctic.   b) Pacific.   c) Atlantic.   d) Indian.

b

37) Item Type: Multiple Choice   Objective: 1   Section 20.1
    Level: B: Enriched   Tag: None

Approximately what percentage of the earth's surface is covered by oceans?

a) 10%   b) 35%   c) 70%   d) 90%

c

38) Item Type: Multiple Choice   Objective: 1   Section 20.1
    Level: A: Standard   Tag: None

Which of the following is a part of three other oceans?

a) Antarctic Ocean   b) Arctic Ocean   c) Indian Ocean   d) Atlantic Ocean

a

39) Item Type: Multiple Choice   Objective: 2   Section 20.1
    Level: B: Enriched   Tag: None

The difference between a bathysphere and a bathyscaph is that a bathysphere

a) carries passengers.   b) remains connected to a ship.

c) uses sonar.   d) is self-propelled.

b

40) Item Type: Multiple Choice   Objective: 2   Section 20.1
    Level: B: Enriched   Tag: None

Which of the following research vessels was recently used to photograph the interior of the *Titanic*?

a) *Argo*   b) *Glomar Challenger*   c) *Alvin*   d) *Jason Jr.*

d

41) Item Type: Multiple Choice   Objective: 2   Section 20.1
    Level: B: Enriched   Tag: None

The research vessel that laid much of the foundation of modern oceanography during the late 19th century was the

a) *Alvin*.   b) H.M.S. *Beagle*.   c) H.M.S. *Challenger*.   d) *Jason Jr.*

c

42) Item Type: Multiple Choice   Objective: 3   Section 20.2
    Level: A: Standard   Tag: None

Submarine canyons most likely formed due to erosion by

a) turbidity currents.   b) volcanic activity.

c) transform faulting.   d) trench formation.

a

43) Item Type: Multiple Choice   Objective: 3   Section 20.2
    Level: A: Standard   Tag: Diagram 2072

Which region in this diagram represents the continental shelf?

a) *1*   b) *2*   c) *3*   d) *4*

b

44) Item Type: Multiple Choice   Objective: 3   Section 20.2
    Level: A: Standard   Tag: None

The boundary between continental crust and oceanic crust occurs at the base of the

a) continental shelf.   b) abyssal plain.   c) continental slope.   d) mid-ocean ridge.

c

45) Item Type: Multiple Choice   Objective: 3   Section 20.2
    Level: A: Standard   Tag: None

Which ocean contains the greatest number of deep-sea trenches?

a) Atlantic   b) Pacific   c) Indian   d) Arctic

b

46) Item Type: Multiple Choice   Objective: 3   Section 20.2
    Level: B: Enriched   Tag: None

Which part of the ocean floor was exposed to increased erosion and weathering during the glacial periods?

a) continental shelf   b) abyssal plain

c) continental rise   d) submarine canyons

a

47) Item Type: Multiple Choice   Objective: 3   Section 20.2
Level: A: Standard   Tag: None

The shallowest portion of the continental margin is the

a) continental shelf.   b) mid-ocean ridge.   c) continental rise.   d) abyssal plain.

a

48) Item Type: Multiple Choice   Objective: 3   Section 20.2
Level: B: Enriched   Tag: Diagram 2073

Where is subduction taking place in the diagram above?

a) *1*   b) *2*   c) *3*   d) *4*

a

49) Item Type: Multiple Choice   Objective: 4   Section 20.2
Level: A: Standard   Tag: Diagram 2073

Which feature in the diagram above represents a guyot?

a) *1*   b) *2*   c) *3*   d) *4*

b

50) Item Type: Multiple Choice   Objective: 4   Section 20.2
Level: A: Standard   Tag: Diagram 2073

Which feature in the diagram above represents a rift zone?

a) *2*   b) *3*   c) *4*   d) *5*

d

51) Item Type: Multiple Choice   Objective: 4   Section 20.2
Level: B: Enriched   Tag: Diagram 2074

Which feature shown in the diagram above is more extensive in the Atlantic Ocean than in the Pacific Ocean?

a) *1*   b) *2*   c) *3*   d) *4*

b

52) Item Type: Multiple Choice   Objective: 4   Section 20.2
Level: A: Standard   Tag: Diagram 2072

Which region in the diagram above is composed of oceanic crust?

a) *2*   b) *3*   c) *4*   d) *5*

d

53) Item Type: Multiple Choice   Objective: 4   Section 20.2
Level: B: Enriched   Tag: None

Which feature is formed where oceanic plates are separating?

a) trench   b) submarine canyon   c) rift   d) abyssal plain

c

54) Item Type: Multiple Choice   Objective: 4   Section 20.2
Level: A: Standard   Tag: None

Which of the following is farthest from the shoreline?

a) continental shelf   b) submarine canyon   c) continental rise   d) abyssal plain

d

55) Item Type: Multiple Choice   Objective: 4   Section 20.2
Level: A: Standard   Tag: None

Which of the following describes a seamount?

a) ocean floor at the edge of a continental margin

b) underwater mountain range

c) sediment piled at the base of the continental slope

d) isolated mid-ocean volcano

d

56) Item Type: Multiple Choice   Objective: 4   Section 20.2
Level: A: Standard   Tag: None

Fracture zones are most often found near

a) trenches.   b) the continental shelf.

c) mid-ocean ridges.   d) the continental slope.

c

57) Item Type: Multiple Choice   Objective: 5   Section 20.3
Level: B: Enriched   Tag: None

Which of the following sediments would most likely be found in a sample from the ocean floor at a depth of 5,000 m?

a) volcanic dust   b) siliceous ooze   c) rock fragments   d) calcareous ooze

b

58) Item Type: Multiple Choice   Objective: 5   Section 20.3
    Level: A: Standard   Tag: None

Scientists obtain deep-ocean core samples in order to analyze

a) deep-ocean basin currents.   b) ocean-floor sediments.

c) deep-ocean temperatures.   d) the composition of sea water.

b

59) Item Type: Multiple Choice   Objective: 5   Section 20.3
    Level: A: Standard   Tag: None

Which of the following is a major source of calcium carbonate in organic sediments on the
ocean floor?

a) radiolaria   b) diatoms   c) foraminifera   d) nodules

c

60) Item Type: Multiple Choice   Objective: 6   Section 20.3
    Level: A: Standard   Tag: None

Which of the following is largely composed of diatoms?

a) red ooze   b) calcareous ooze   c) siliceous ooze   d) lava

c

61) Item Type: Multiple Choice   Objective: 6   Section 20.3
    Level: A: Standard   Tag: None

Which of the following is a common type of ocean-bottom mud?

a) calcareous ooze   b) red clay   c) siliceous ooze   d) volcanic dust

b

62) Item Type: Multiple Choice   Objective: 6   Section 20.3
    Level: A: Standard   Tag: None

Which of the following is composed primarily of diatom skeletons?

a) nodules   b) siliceous ooze   c) red clay   d) calcareous ooze

b

63) Item Type: Fill in the Blank   Objective: 1   Section 20.1
    Level: B: Enriched   Tag: Diagram 2071

In the map above, the ocean labeled **Y** is the _____.

Indian Ocean

64) Item Type: Fill in the Blank   Objective: 1   Section 20.1
Level: A: Standard   Tag: None

A small area of ocean that is partially surrounded by land is called a(n)

_____.

sea

65) Item Type: Fill in the Blank   Objective: 1   Section 20.1
Level: A: Standard   Tag: None

The principal ocean with the smallest area is the _____.

Indian Ocean

66) Item Type: Fill in the Blank   Objective: 1   Section 20.1
Level: A: Standard   Tag: None

The second largest of the earth's principal oceans is the _____.

Atlantic Ocean

67) Item Type: Fill in the Blank   Objective: 2   Section 20.1
Level: B: Enriched   Tag: None

What type of submersible was used to photograph the inside of the remains of the *Titanic?*

_____

bathyscaph

68) Item Type: Fill in the Blank   Objective: 2   Section 20.1
Level: B: Enriched   Tag: None

Strange life-forms were discovered near volcanic vents along mid-ocean ridges by the

bathyscaph named _____.

Alvin

69) Item Type: Fill in the Blank   Objective: 2   Section 20.1
Level: A: Standard   Tag: None

The branch of earth science that includes the study of the physical characteristics of the

ocean is called _____.

oceanography

Chapter 20

70) Item Type: Fill in the Blank   Objective: 2   Section 20.1
Level: A: Standard   Tag: None

The *Alvin* is an example of a submersible called a(n) _____.

| bathyscaph ◼ |

71) Item Type: Fill in the Blank   Objective: 2   Section 20.1
Level: A: Standard   Tag: None

The general name for an underwater research vessel is a(n) _____.

| submersible ◼ |

72) Item Type: Fill in the Blank   Objective: 2   Section 20.1
Level: A: Standard   Tag: None

The technique in which sound waves are used to measure depth is called

_____.

| sonar ◼ |

73) Item Type: Fill in the Blank   Objective: 3   Section 20.2
Level: B: Enriched   Tag: None

The Gulf Stream and the Labrador Current meet at the fishing grounds called the

_____.

| Grand Banks ◼ |

74) Item Type: Fill in the Blank   Objective: 3   Section 20.2
Level: B: Enriched   Tag: None

The boundary between continental crust and oceanic crust at an inactive margin is the

_____.

| base of the continental slope ◼ |

75) Item Type: Fill in the Blank   Objective: 3   Section 20.2
Level: B: Enriched   Tag: None

The portion of the continental margin that is covered by the shallowest water is the

_____.

| continental shelf ◼ |

76) Item Type: Fill in the Blank   Objective: 4   Section 20.2
Level: A: Standard   Tag: None

Isolated volcanic mountains scattered on the ocean floor are called

_____.

| seamounts | ■ |

77) Item Type: Fill in the Blank   Objective: 5   Section 20.3
Level: B: Enriched   Tag: Diagram 2075

In the diagram, the finest sediments are most likely to be found at the point labeled

_____.

| **D** | ■ |

78) Item Type: Fill in the Blank   Objective: 6   Section 20.3
Level: A: Standard   Tag: None

The skeletons of foraminifera and corals are composed of _____.

| calcium carbonate | ■ |

79) Item Type: Fill in the Blank   Objective: 6   Section 20.3
Level: A: Standard   Tag: None

One common type of mud found on the abyssal plains is red _____.

| clay | ■ |

80) Item Type: Fill in the Blank   Objective: 6   Section 20.3
Level: B: Enriched   Tag: None

Red clay and ooze are examples of the general type of sediment called

_____.

| mud | ■ |

81) Item Type: Fill in the Blank   Objective: 6   Section 20.3
Level: A: Standard   Tag: None

Microscopic plants whose remains are composed of silica are called

_____.

| diatoms | ■ |

82) Item Type: Essay   Objective: 2   Section 20.1
     Level: B: Enriched   Tag: None

How is the depth of water determined using sonar?

_____

_____

_____

_____

_____

A sonar system consist of a transmitter and a receiver. The transmitter sends out a
continuous series of sound waves from a ship to the ocean floor. The sound waves,
traveling at a speed of about 1,500 m/sec, bounce off the ocean floor and are reflected up
to the receiver. The time taken by the waves to complete the trip is used to determine the
depth of the water.

83) Item Type: Essay   Objective: 3   Section 20.2
     Level: B: Enriched   Tag: None

How might the number of fish and shellfish be different in the Grand Banks area if it had
a very narrow continental shelf?

_____

_____

_____

_____

With a narrow continental shelf, the currents would probably be weaker. Weaker currents
would reduce the food supply for fish, and the number of fish present would decrease. In
deep waters, no shellfish would exist. They would be restricted to the small area of
continental shelf and therefore would be fewer in number.

84) Item Type: Essay   Objective: 5   Section 20.3
    Level: B: Enriched   Tag: None

How do icebergs provide ocean-floor sediments?

_____

_____

_____

_____

_____

Glaciers pick up great quantities of rock as they move across land. When an iceberg breaks off the glacier and melts, the rock material sinks to the ocean floor.

1)   Tag Name: Diagram 2071

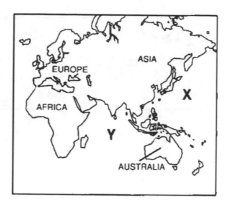

2)   Tag Name: Diagram 2072

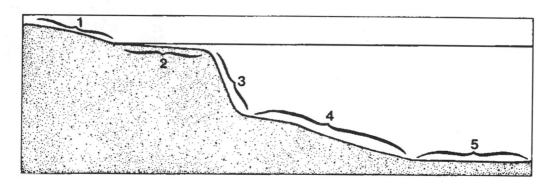

3)   Tag Name: Diagram 2073

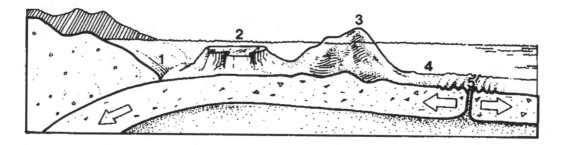

4)   Tag Name: Diagram 2074

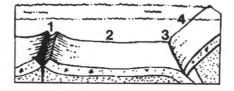

5)   Tag Name: Diagram 2075

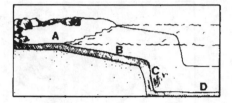

1) Item Type: True-False   Objective: 1   Section 21.1
    Level: B: Enriched   Tag: None

_____ The three principal gases of the atmosphere are also the same three principal gases that dissolve in ocean water.

T

2) Item Type: True-False   Objective: 1   Section 21.1
    Level: B: Enriched   Tag: None

_____ Ocean water is 78 percent pure water.

F

3) Item Type: True-False   Objective: 1   Section 21.1
    Level: A: Standard   Tag: None

_____ The salinity of ocean water tends to be lower near areas of freshwater runoff from rivers.

T

4) Item Type: True-False   Objective: 1   Section 21.1
    Level: A: Standard   Tag: None

_____ The relative amounts of dissolved substances in ocean water have remained almost constant over millions of years.

T

5) Item Type: True-False   Objective: 1   Section 21.1
    Level: B: Enriched   Tag: None

_____ The salinity of ocean water tends to be higher in deep-ocean waters.

F

6) Item Type: True-False   Objective: 1   Section 21.1
    Level: A: Standard   Tag: None

_____ Surface ocean water in colder regions will usually dissolve larger amounts of gases than will surface water in tropical regions.

T

7) Item Type: True-False   Objective: 1   Section 21.1
    Level: A: Standard   Tag: None

_____ The physical properties of ocean water are those characteristics that determine the water's composition and enable the water to dissolve other substances.

F

8) Item Type: True-False   Objective: 2   Section 21.1
Level: B: Enriched   Tag: None

_____ Ocean water becomes less dense as it becomes colder.

F

9) Item Type: True-False   Objective: 2   Section 21.1
Level: B: Enriched   Tag: None

_____ The distribution of life in the ocean depends, to a large extent, on the way life-supporting substances return to the surface from ocean depths.

T

10) Item Type: True-False   Objective: 2   Section 21.1
Level: A: Standard   Tag: None

_____ The Arctic Ocean is covered by pack ice during most of the year.

T

11) Item Type: True-False   Objective: 2   Section 21.1
Level: A: Standard   Tag: None

_____ Pure fresh water is denser than ocean water.

F

12) Item Type: True-False   Objective: 2   Section 21.1
Level: A: Standard   Tag: None

_____ The temperature of ocean water drops sharply not far below the ocean surface.

T

13) Item Type: True-False   Objective: 2   Section 21.1
Level: A: Standard   Tag: None

_____ The total amount of solar energy falling on the surface of the ocean is much greater at the poles than it is at the equator.

F

14) Item Type: True-False   Objective: 2   Section 21.1
Level: A: Standard   Tag: None

_____ The amount of dissolved solids in the water affects the density of ocean water.

T

15) Item Type: True-False   Objective: 2   Section 21.1
     Level: A: Standard   Tag: None

_____ No wavelength of light can penetrate ocean water at depths of more than a few hundred meters.

T

16) Item Type: True-False   Objective: 2   Section 21.1
     Level: A: Standard   Tag: None

_____ Pack ice is often more than 20 m thick on the ocean surface in polar regions.

F

17) Item Type: True-False   Objective: 2   Section 21.1
     Level: A: Standard   Tag: None

_____ The last color of visible light to be absorbed by water is blue.

T

18) Item Type: True-False   Objective: 3   Section 21.2
     Level: A: Standard   Tag: None

_____ Organic remains in ocean water are broken down by bacteria.

T

19) Item Type: True-False   Objective: 3   Section 21.2
     Level: A: Standard   Tag: None

_____ In general, the elements needed by marine life are released at great depths in the ocean.

T

20) Item Type: True-False   Objective: 3   Section 21.2
     Level: A: Standard   Tag: None

_____ Plant growth is present in all ocean environments.

F

21) Item Type: True-False   Objective: 4   Section 21.2
     Level: A: Standard   Tag: None

_____ Zooplankton use the energy from sunlight to carry on photosynthesis.

F

22) Item Type: True-False   Objective: 4   Section 21.2
     Level: A: Standard   Tag: None

_____ Plants that grow on the ocean floor belong to a group of organisms called benthos.

T

23) Item Type: True-False   Objective: 5   Section 21.2
     Level: A: Standard   Tag: None

_____ Squid and octopus live in the hadal zone.

F

24) Item Type: True-False   Objective: 6   Section 21.3
     Level: A: Standard   Tag: None

_____ Evaporating ocean water to desalinate it requires very little heat energy.

F

25) Item Type: True-False   Objective: 6   Section 21.3
     Level: A: Standard   Tag: None

_____ When ocean water freezes, the first ice crystals that form are free of salt.

T

26) Item Type: True-False   Objective: 6   Section 21.3
     Level: A: Standard   Tag: None

_____ Oil spills from offshore drilling are harmless to marine organisms and sea birds.

F

27) Item Type: True-False   Objective: 6   Section 21.3
     Level: A: Standard   Tag: None

_____ An ocean farm could potentially produce more protein-rich food than a land farm of the same size.

T

28) Item Type: True-False   Objective: 6   Section 21.3
     Level: A: Standard   Tag: None

_____ Aquaculture involves developing and raising special breeds of marine animals and plants that yield large amounts of food.

T

29) Item Type: True-False   Objective: 6   Section 21.3
     Level: B: Enriched   Tag: None

_____ Fish and shellfish change their food supply into protein less efficiently than do most land animals.

F

30) Item Type: True-False   Objective: 6   Section 21.3
     Level: B: Enriched   Tag: None

_____ Desalinating ocean water by freezing takes three times the amount of energy needed for distillation.

F

31) Item Type: True-False   Objective: 6   Section 21.3
     Level: A: Standard   Tag: None

_____ Some nations that formerly had abundant supplies of fresh water are now facing shortages.

T

32) Item Type: True-False   Objective: 7   Section 21.3
     Level: A: Standard   Tag: None

_____ Pollutants can be found in measurable amounts everywhere in the oceans.

T

33) Item Type: True-False   Objective: 7   Section 21.3
     Level: A: Standard   Tag: None

_____ Traces of the insecticide DDT have been detected in open ocean.

T

34) Item Type: True-False   Objective: 7   Section 21.3
     Level: A: Standard   Tag: None

_____ Both DDT and lead can cause problems in ocean food chains.

T

35) Item Type: Multiple Choice   Objective: 1   Section 21.1
     Level: B: Enriched   Tag: None

Minerals and salts dissolved in ocean water contain approximately how many known chemical elements?

a) 10   b) 35   c) 75   d) 95

c

36) Item Type: Multiple Choice   Objective: 1   Section 21.1
Level: B: Enriched   Tag: None

Which of the following gases dissolves most easily in ocean water?

a) hydrogen   b) oxygen   c) nitrogen   d) carbon dioxide

d

37) Item Type: Multiple Choice   Objective: 1   Section 21.1
Level: B: Enriched   Tag: None

The most abundant major element dissolved in ocean water is

a) boron.   b) chlorine.   c) calcium.   d) potassium.

b

38) Item Type: Multiple Choice   Objective: 1   Section 21.1
Level: A: Standard   Tag: None

What percent of ocean water is pure water?

a) 3.5%   b) 39. 5%   c) 57.5%   d) 96.5%

d

39) Item Type: Multiple Choice   Objective: 1   Section 21.1
Level: A: Standard   Tag: None

In which of the following environments does salinity tend to be higher?

a) surface temperate waters   b) surface tropical waters

c) deep lake waters   d) deep tropical waters

b

40) Item Type: Multiple Choice   Objective: 1   Section 21.1
Level: A: Standard   Tag: None

Which of the following is one of the three principal gases dissolved in ocean water?

a) hydrogen   b) nitrogen   c) carbon   d) sodium

b

41) Item Type: Multiple Choice   Objective: 2   Section 21.1
Level: B: Enriched   Tag: None

The drop in temperature that begins just below the surface of the ocean continues to a depth of

a) 400 m.   b) 750 m.   c) 1,200 m.   d) 1,500 m.

c

42) Item Type: Multiple Choice   Objective: 2   Section 21.1
Level: A: Standard   Tag: None

Which ocean zone separates warm surface water from the colder deep water?

a) thermocline   b) benthos   c) abyssal plain   d) neritic zone

a

43) Item Type: Multiple Choice   Objective: 3   Section 21.2
Level: A: Standard   Tag: None

Nearly all ocean life is regulated by the

a) life processes of plants.

b) type of sediment on the ocean floor.

c) addition of substances to ocean water.

d) abundance of swimming ocean animals.

a

44) Item Type: Multiple Choice   Objective: 3   Section 21.2
Level: B: Enriched   Tag: None

Which of the following elements is greatly depleted from ocean water by heavy plant growth?

a) nitrogen   b) magnesium   c) oxygen   d) sodium

a

45) Item Type: Multiple Choice   Objective: 3   Section 21.2
Level: A: Standard   Tag: None

The process by which deep, nutrient-rich water moves upward to replace surface water that has blown farther offshore is called

a) thermocline.   b) distilling.   c) desalination.   d) upwelling.

d

46) Item Type: Multiple Choice   Objective: 3   Section 21.2
Level: A: Standard   Tag: None

In general, all the elements necessary to sustain marine life are consumed

a) near the surface of the ocean.   b) near the shoreline.

c) at intermediate ocean depths.   d) at great ocean depths.

a

47)   Item Type: Multiple Choice   Objective: 3   Section 21.2
      Level: A: Standard   Tag: None

Most of the dissolved minerals carried into the ocean by the earth's rivers are

a) salts.   b) ores.   c) silicates.   d) carbonates.

a

48)   Item Type: Multiple Choice   Objective: 4   Section 21.2
      Level: B: Enriched   Tag: None

Plant growth in the ocean is restricted to a depth that extends no more than

a) 50 m below the surface.   b) 100 m below the surface.

c) 200 m below the surface.   d) 300 m below the surface.

b

49)   Item Type: Multiple Choice   Objective: 4   Section 21.2
      Level: A: Standard   Tag: None

Large ocean animals that swim and eat microscopic plants and animals are

a) nekton.   b) zooplankton.   c) aquaculture.   d) anemones.

a

50)   Item Type: Multiple Choice   Objective: 4   Section 21.2
      Level: A: Standard   Tag: None

Which of the following is a characteristic of nekton that enables them to search for food
and avoid some predators?

a) They carry on photosynthesis.   b) They live on the ocean floor.

c) They are able to swim.   d) They are free-floating.

c

51)   Item Type: Multiple Choice   Objective: 4   Section 21.2
      Level: A: Standard   Tag: None

Which of the following types of marine life serve as the first link in the ocean food chain?

a) benthos   b) seaweed   c) nekton   d) plankton

d

52) Item Type: Multiple Choice   Objective: 4   Section 21.2
Level: A: Standard   Tag: None

Organisms that live on the ocean floor are called

a) zooplankton.   b) nekton.   c) benthos.   d) dolphins.

c

53) Item Type: Multiple Choice   Objective: 5   Section 21.2
Level: A: Standard   Tag: Diagram 2172

Which ocean environment in the diagram represents the sublittoral zone?

a) *A*   b) *B*   c) *C*   d) *D*

a

54) Item Type: Multiple Choice   Objective: 5   Section 21.2
Level: A: Standard   Tag: Diagram 2172

Which ocean environment in the diagram represents the hadal zone?

a) *A*   b) *B*   c) *C*   d) *D*

d

55) Item Type: Multiple Choice   Objective: 5   Section 21.2
Level: B: Enriched   Tag: None

Giant clams and blind white crabs have been found in which benthic zone?

a) hadal   b) abyssal   c) sublittoral   d) intertidal

b

56) Item Type: Multiple Choice   Objective: 5   Section 21.2
Level: B: Enriched   Tag: None

In which benthic zone would sea stars and sea lilies be found?

a) hadal   b) abyssal   c) bathyal   d) sublittoral

d

57) Item Type: Multiple Choice   Objective: 5   Section 21.2
Level: B: Enriched   Tag: None

Organisms that live in the bathyal zone include

a) squids and large whales.   b) crabs and mussels.

c) sea anemones and seaweed.   d) sea cucumbers and deep-sea eels.

a

58) Item Type: Multiple Choice   Objective: 5   Section 21.2
Level: B: Enriched   Tag: None

Which benthic environment begins at the end of the continental slope?

a) sublittoral   b) intertidal   c) neritic   d) bathyal

d

59) Item Type: Multiple Choice   Objective: 5   Section 21.2
Level: A: Standard   Tag: None

Organisms that live in the intertidal zone include

a) clams and crabs.   b) squid and octopus.

c) whales and devilfish.   d) sea cucumbers and tube worms.

a

60) Item Type: Multiple Choice   Objective: 5   Section 21.2
Level: A: Standard   Tag: None

Which ocean environment is characterized by breaking waves?

a) neritic   b) benthic   c) intertidal   d) abyssal

c

61) Item Type: Multiple Choice   Objective: 6   Section 21.3
Level: B: Enriched   Tag: None

Nodules are a valuable source of

a) carbon.   b) manganese.   c) silicon.   d) magnesium.

b

62) Item Type: Multiple Choice   Objective: 6   Section 21.3
Level: A: Standard   Tag: None

The most valuable resource taken from the ocean is

a) copper deposited on the ocean floor.   b) calcite found in seashells.

c) petroleum from beneath the ocean floor.   d) gold dissolved in ocean water.

c

63) Item Type: Multiple Choice   Objective: 6   Section 21.3
Level: A: Standard   Tag: None

Manganese from nodules in the ocean is used to manufacture certain types of

a) plastic.   b) steel.   c) fertilizer.   d) glass.

b

64) Item Type: Multiple Choice   Objective: 6   Section 21.3
    Level: B: Enriched   Tag: None

Which of the following is the most valuable resource currently taken from the ocean?

a) petroleum   b) iron   c) calcium   d) sodium

a ∎

65) Item Type: Multiple Choice   Objective: 6   Section 21.3
    Level: B: Enriched   Tag: None

The method used to remove salt by evaporating ocean water is called

a) freezing.   b) reverse osmosis. c) distillation.   d) electrolysis.

c ∎

66) Item Type: Multiple Choice   Objective: 6   Section 21.3
    Level: B: Enriched   Tag: None

Freezing ocean water can create fresh water because the salt

a) crystallizes within the ice.   b) remains in pockets of liquid water.

c) freezes faster than water.   d) evaporates as the water is frozen.

b ∎

67) Item Type: Multiple Choice   Objective: 6   Section 21.3
    Level: A: Standard   Tag: None

Which of the following is extracted easily from the ocean and refined to produce metal?

a) magnesium   b) phosphate   c) gold   d) sulfur

a ∎

68) Item Type: Multiple Choice   Objective: 7   Section 21.3
    Level: A: Standard   Tag: None

What part of the ocean is in the greatest danger of being polluted?

a) polar waters   b) coastal waters

c) deep marine waters   d) intermediate ocean depths

b ∎

69) Item Type: Fill in the Blank   Objective: 1   Section 21.1
Level: A: Standard   Tag: None

What is defined by the number of grams of dissolved salt in 1 kg of ocean water?

_____

salinity

70) Item Type: Fill in the Blank   Objective: 1   Section 21.1
Level: A: Standard   Tag: None

Of the three principal gases dissolved in ocean water, the one that dissolves most easily is

_____.

carbon dioxide

71) Item Type: Fill in the Blank   Objective: 1   Section 21.1
Level: B: Enriched   Tag: None

The most abundant dissolved elements in ocean water are sodium and

_____.

chlorine

72) Item Type: Fill in the Blank   Objective: 2   Section 21.1
Level: B: Enriched   Tag: None

What major physical property of ocean water is affected by water temperature and the

amount of dissolved solids in the water? _____

density

73) Item Type: Fill in the Blank   Objective: 2   Section 21.1
Level: A: Standard   Tag: None

Which color wavelengths of light tend to be reflected by ocean water?

_____

blue

74) Item Type: Fill in the Blank   Objective: 2   Section 21.1
     Level: A: Standard   Tag: None

What factor, besides the amount of dissolved solids, affects the density of water?

_____

| temperature |

75) Item Type: Fill in the Blank   Objective: 3   Section 21.2
     Level: A: Standard   Tag: None

What is the process in which deep water moves upward to replace surface water?

_____

| upwelling |

76) Item Type: Fill in the Blank   Objective: 3   Section 21.2
     Level: A: Standard   Tag: None

Organisms that live on the ocean floor, such as oysters and sea stars, are examples of the

general group of organisms called _____.

| benthos |

77) Item Type: Fill in the Blank   Objective: 3   Section 21.2
     Level: A: Standard   Tag: None

Nutrients are restored to surface waters from deep ocean depths through the process called

_____.

| upwelling |

78) Item Type: Fill in the Blank   Objective: 4   Section 21.2
     Level: B: Enriched   Tag: None

There is not enough light to sustain plants in the ocean waters below an approximate depth

of _____.

| 80 m |

79) Item Type: Fill in the Blank   Objective: 4   Section 21.2
    Level: A: Standard   Tag: None

What is the name for the larger swimming ocean animals that eat microscopic animals and

plants? _____

| nekton |

80) Item Type: Fill in the Blank   Objective: 4   Section 21.2
    Level: A: Standard   Tag: None

What are the free-floating, microscopic plants found in most regions of the near-surface

ocean called? _____

| phytoplankton |

81) Item Type: Fill in the Blank   Objective: 4   Section 21.2
    Level: A: Standard   Tag: None

The general name for free-floating microscopic plants and animals is

_____.

| plankton |

82) Item Type: Fill in the Blank   Objective: 4   Section 21.2
    Level: A: Standard   Tag: None

The process by which microscopic plants produce food from sunlight is

_____.

| photosynthesis |

83) Item Type: Fill in the Blank   Objective: 5   Section 21.2
    Level: B: Enriched   Tag: None

Which pelagic zone is the source of much of the seafood that people eat?

_____

| neritic zone |

84) Item Type: Fill in the Blank   Objective: 5   Section 21.2
      Level: A: Standard   Tag: None

Which benthic zone is a relatively unstable environment for marine life but is populated by

crabs, clams, and seaweed? _____

intertidal

85) Item Type: Fill in the Blank   Objective: 5   Section 21.2
      Level: A: Standard   Tag: None

Which ocean environment extends seaward beyond the continental shelf?

_____

pelagic

86) Item Type: Fill in the Blank   Objective: 5   Section 21.2
      Level: B: Enriched   Tag: Diagram 2173

The pelagic zone labeled X in the diagram is the _____.

neritic zone

87) Item Type: Fill in the Blank   Objective: 5   Section 21.2
      Level: B: Enriched   Tag: None

Which ocean environment is characterized by areas that are rich in minerals and an ocean

floor marked by volcanic vents? _____

abyssal zone

88) Item Type: Fill in the Blank   Objective: 5   Section 21.2
      Level: A: Standard   Tag: None

The benthic zone characterized by exposure at low tides and breaking waves is the

_____.

intertidal zone

89) Item Type: Fill in the Blank   Objective: 7   Section 21.3
      Level: A: Standard   Tag: None

The presence of what substance found in gasoline has increased tenfold in the Pacific

Ocean in the last 50 years? _____

lead

90) Item Type: Essay   Objective: 1   Section 21.1
    Level: A: Standard   Tag: None

How does the process of evaporation affect the salinity of ocean water?

_____

_____

_____

_____

_____

Evaporation increases the salinity of ocean water. During evaporation, only water
molecules are removed. Dissolved salts and other solids remain in the ocean. When the
rate of evaporation is high, the relative amount of dissolved solids in surface water
increases.

91) Item Type: Essay   Objective: 2   Section 21.1
    Level: B: Enriched   Tag: None

Describe a thermocline in terms of where it occurs and why.

_____

_____

_____

_____

_____

Because the sun cannot directly heat ocean water below its uppermost regions, the
temperature of the water drops sharply as the depth increases. In most places in the ocean,
this sudden temperature drop begins not far below the surface. This zone of rapid
temperature change is called the thermocline. The warm uppermost water cannot mix easily
with the cold, dense water below; a thermocline marks the distinct separation between the
warm surface water and the colder deep water.

92) Item Type: Essay   Objective: 5   Section 21.2
     Level: B: Enriched   Tag: None

How might a dramatic drop in sea level affect the current position of the ocean environments?

_____

_____

_____

_____

_____

As the sea level dropped, the zones would have to move farther toward the open ocean than they currently are. Sunlight would most likely penetrate more of the shallower ocean, and plants would grow in the expanded neritic zone. The three deepest bottom zones, the bathyal, abyssal, and hadal zones, would probably occupy a smaller area because the oceans would be generally more shallow.

93) Item Type: Essay   Objective: 6   Section 21.3
     Level: A: Standard   Tag: None

Describe the practice of aquaculture.

_____

_____

_____

_____

_____

Aquaculture is the farming of the ocean. Aquaculture involves developing and raising special breeds of marine animals and plants that yield large amounts of food. Aquaculture has been used successfully to grow catfish and salmon on large aquatic farms.

94) Item Type: Essay   Objective: 7   Section 21.3
       Level: B: Enriched   Tag: None

How has the addition of lead to gasoline affected marine life in the Pacific Ocean?

_____

_____

_____

_____

_____

Lead can cause problems in ocean food webs. As animals eat other animals, the amount of lead builds up in the bodies of the predators. In some areas, the lead concentration in fish has made the fish inedible.

1)   Tag Name: Diagram 2172

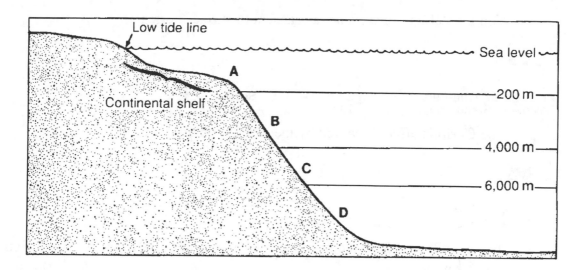

2)   Tag Name: Diagram 2173

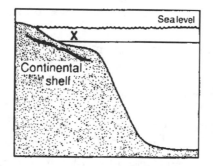

1) Item Type: True-False   Objective: 1   Section 22.1
   Level: A: Standard   Tag: None

   _____ Continental land masses act as barriers to surface currents.

   T

2) Item Type: True-False   Objective: 1   Section 22.1
   Level: A: Standard   Tag: None

   _____ The Coriolis effect is caused by the earth's rotation.

   T

3) Item Type: True-False   Objective: 1   Section 22.1
   Level: A: Standard   Tag: None

   _____ The overall pattern of the movement of surface currents in the Northern Hemisphere
   is clockwise.

   T

4) Item Type: True-False   Objective: 1   Section 22.1
   Level: A: Standard   Tag: None

   _____ The Gulf Stream is a warm current.

   T

5) Item Type: True-False   Objective: 1   Section 22.1
   Level: A: Standard   Tag: None

   _____ The winds in the Indian Ocean that change direction with each season are called
   typhoons.

   F

6) Item Type: True-False   Objective: 1   Section 22.1
   Level: B: Enriched   Tag: None

   _____ All surface currents north of the equator flow in a counterclockwise direction.

   F

7) Item Type: True-False   Objective: 1   Section 22.1
   Level: B: Enriched   Tag: None

   _____ Near the equator, the trade winds are deflected to the west because of the Coriolis
   effect.

   T

8) Item Type: True-False   Objective: 1   Section 22.1
     Level: A: Standard   Tag: None

_____ The trade winds are located in the middle latitudes of the Northern Hemisphere.

F

9) Item Type: True-False   Objective: 1   Section 22.1
     Level: A: Standard   Tag: None

_____ The Sargasso Sea lies in the center of a circle formed by the Gulf Stream, the North Atlantic Drift, and the North Equatorial Current.

T

10) Item Type: True-False   Objective: 1   Section 22.1
     Level: A: Standard   Tag: None

_____ Wind produces both surface currents and waves.

T

11) Item Type: True-False   Objective: 2   Section 22.1
     Level: B: Enriched   Tag: None

_____ Water in the Mediterranean Sea is more saline than water in the Atlantic Ocean is.

T

12) Item Type: True-False   Objective: 2   Section 22.1
     Level: A: Standard   Tag: None

_____ Deep currents tend to have cold, highly saline water in them.

T

13) Item Type: True-False   Objective: 2   Section 22.1
     Level: A: Standard   Tag: None

_____ Deep currents are produced by density differences within the water.

T

14) Item Type: True-False   Objective: 2   Section 22.1
     Level: A: Standard   Tag: None

_____ The deep currents of the Atlantic Ocean tend to flow in the same direction as the Atlantic surface currents do.

F

15) Item Type: True-False   Objective: 3   Section 22.2
     Level: A: Standard   Tag: None

_____ Energy is transferred from wind to water to form waves.

| T | |

16) Item Type: True-False   Objective: 3   Section 22.2
     Level: A: Standard   Tag: None

_____ The lowest point between two wave crests is a trough.

| T | |

17) Item Type: True-False   Objective: 4   Section 22.2
     Level: B: Enriched   Tag: None

_____ If the ocean shore slopes gently, waves generally will break with great force.

| F | |

18) Item Type: True-False   Objective: 4   Section 22.2
     Level: A: Standard   Tag: None

_____ An undertow is generally a very strong current.

| F | |

19) Item Type: True-False   Objective: 4   Section 22.2
     Level: A: Standard   Tag: None

_____ The depth of the water relative to wavelength determines when a wave breaks.

| T | |

20) Item Type: True-False   Objective: 4   Section 22.2
     Level: A: Standard   Tag: None

_____ Longshore currents create a zigzag-shaped shoreline.

| F | |

21) Item Type: True-False   Objective: 4   Section 22.2
     Level: A: Standard   Tag: None

_____ Longshore currents occur when waves approach a beach at an angle.

| T | |

22) Item Type: True-False   Objective: 4   Section 22.2
     Level: B: Enriched   Tag: None

____ Waves break primarily as a result of friction.

T

23) Item Type: True-False   Objective: 4   Section 22.2
     Level: A: Standard   Tag: None

____ Rip currents form when water returns to the ocean through channels in underwater sandbars.

T

24) Item Type: True-False   Objective: 5   Section 22.3
     Level: B: Enriched   Tag: None

____ High tides occur at the same time each day.

F

25) Item Type: True-False   Objective: 5   Section 22.3
     Level: A: Standard   Tag: None

____ The position of the sun has the greatest influence on the tides.

F

26) Item Type: True-False   Objective: 5   Section 22.3
     Level: A: Standard   Tag: None

____ The tidal range is the difference in time between two successive high tides.

F

27) Item Type: True-False   Objective: 5   Section 22.3
     Level: B: Enriched   Tag: None

____ According to Newton's law of gravitation, the gravitational pull of the moon on the earth and its waters causes the rise and fall of the tides.

T

28) Item Type: True-False   Objective: 5   Section 22.3
     Level: A: Standard   Tag: None

____ There are always two tidal bulges somewhere on the earth's surface.

T

29) Item Type: True-False   Objective: 5   Section 22.3
    Level: A: Standard   Tag: None

____ Because of the frictional pull of tides on the ocean floor, the earth's rotational speed is slowing.

T

30) Item Type: True-False   Objective: 5   Section 22.3
    Level: A: Standard   Tag: None

____ Tides occur on the sides of the earth that face toward and away from the moon.

T

31) Item Type: True-False   Objective: 5   Section 22.3
    Level: A: Standard   Tag: None

____ The difference between the water level at high tide and the water level at low tide for a particular location is called the tidal bore.

F

32) Item Type: True-False   Objective: 5   Section 22.3
    Level: A: Standard   Tag: None

____ Tidal patterns are greatly influenced by the size, shape, and location of the ocean basin in which they occur.

T

33) Item Type: True-False   Objective: 5   Section 22.3
    Level: A: Standard   Tag: None

____ The effects of tidal oscillations are most apparent along straight coastlines.

F

34) Item Type: True-False   Objective: 6   Section 22.3
    Level: A: Standard   Tag: None

____ The tidal movement toward the coast is called the ebb tide.

F

35) Item Type: True-False   Objective: 6   Section 22.3
    Level: A: Standard   Tag: None

____ Tidal flats are largely unaffected by tidal currents.

F

36) Item Type: True-False   Objective: 6   Section 22.3
Level: A: Standard   Tag: None

_____ Tides create rapid currents in the open ocean.

F

37) Item Type: True-False   Objective: 6   Section 22.3
Level: A: Standard   Tag: None

_____ The energy of tidal movements has been harnessed to generate electricity.

T

38) Item Type: True-False   Objective: 6   Section 22.3
Level: B: Enriched   Tag: None

_____ Tidal oscillations increase the tidal range in the Bay of Fundy.

T

39) Item Type: Multiple Choice   Objective: 1   Section 22.1
Level: B: Enriched   Tag: None

The global winds that are located just north and south of the equator are called

a) rip currents.   b) westerlies.   c) trade winds.   d) monsoons.

c

40) Item Type: Multiple Choice   Objective: 1   Section 22.1
Level: B: Enriched   Tag: None

The flow of ocean currents in the Indian Ocean is divided into two circular patterns called

a) gyres.   b) drifts.   c) refractions.   d) westerlies.

a

41) Item Type: Multiple Choice   Objective: 1   Section 22.1
Level: B: Enriched   Tag: None

From April to October, the surface waters of the northern Indian Ocean flow eastward along the southern coast of Asia, forming the

a) Antarctic Bottom Current.   b) Monsoon Current.

c) North Equatorial Current.   d) Gulf Stream.

b

42) Item Type: Multiple Choice Objective: 1 Section 22.1
     Level: B: Enriched Tag: Diagram 2271

Which number in the diagram represents the location of the winds called the westerlies?

a) **1**   b) **2**   c) **3**   d) **4**

b ▪

43) Item Type: Multiple Choice Objective: 1 Section 22.1
     Level: B: Enriched Tag: Diagram 2272

What is the name of the ocean current in this diagram?

a) Labrador Current   b) North Atlantic Drift

c) Gulf Stream   d) North Equatorial Current

c ▪

44) Item Type: Multiple Choice Objective: 1 Section 22.1
     Level: A: Standard Tag: None

Which type of current results from underwater landslides?

a) rip current   b) turbidity current   c) longshore current   d) tidal current

b ▪

45) Item Type: Multiple Choice Objective: 1 Section 22.1
     Level: A: Standard Tag: None

The currents in the northern Indian Ocean are governed by winds called

a) monsoons.   b) tsunamis.   c) trade winds.   d) tombolos.

a ▪

46) Item Type: Multiple Choice Objective: 2 Section 22.1
     Level: A: Standard Tag: None

Which of the following best describes the water found in the deep currents of the Atlantic Ocean?

a) cold with high salinity   b) cold with low salinity

c) warm with high salinity   d) warm with low salinity

a ▪

Chapter 22

47) Item Type: Multiple Choice   Objective: 2   Section 22.1
    Level: A: Standard   Tag: None

Deep Atlantic currents are formed largely as a result of the

a) erosion by surface winds.   b) heating of surface water.

c) rotation of the earth.   d) sinking of denser water.

d

48) Item Type: Multiple Choice   Objective: 2   Section 22.1
    Level: A: Standard   Tag: None

Turbidity currents probably form as a result of

a) underwater landslides.   b) temperature differences.

c) surface winds.   d) salinity differences.

a

49) Item Type: Multiple Choice   Objective: 2   Section 22.1
    Level: A: Standard   Tag: None

Which of the following is the deepest current of the major oceans?

a) Equatorial Countercurrent   b) Antarctic Bottom Water

c) South Equatorial Current   d) North Atlantic Drift

b

50) Item Type: Multiple Choice   Objective: 3   Section 22.2
    Level: B: Enriched   Tag: None

Which of these factors is used to calculate wave speed?

a) wind speed   b) wave period   c) water depth   d) wave height

b

51) Item Type: Multiple Choice   Objective: 3   Section 22.2
    Level: B: Enriched   Tag: None

The speed at which a wave moves is calculated by

a) multiplying the wavelength by the period. b) dividing the wavelength by the period.

c) multiplying the period by the height.   d) dividing the period by the height.

b

52) Item Type: Multiple Choice   Objective: 3   Section 22.2
    Level: A: Standard   Tag: None

The distance that wind can blow across open water is called the

a) wavelength.   b) fetch.   c) wave period.   d) swell.

b

53) Item Type: Multiple Choice   Objective: 3   Section 22.2
    Level: A: Standard   Tag: None

Which of the following factors influences the size of a wave?

a) wind speed   b) air temperature

c) deep-current speed   d) water density

a

54) Item Type: Multiple Choice   Objective: 3   Section 22.2
    Level: A: Standard   Tag: None

The direction of water particles in a wave in deep water can best be described as moving

a) forward.   b) in a long ellipse.

c) up and down.   d) in a circle.

d

55) Item Type: Multiple Choice   Objective: 3   Section 22.2
    Level: B: Enriched   Tag: None

What is calculated by dividing the wavelength of a wave by the wave's period?

a) wave speed   b) size of wave crest

c) wave height   d) size of wave trough

a

56) Item Type: Multiple Choice   Objective: 3   Section 22.2
    Level: A: Standard   Tag: None

Which of the following is one of a group of long, rolling waves that are all the same size?

a) drift   b) fetch   c) swell   d) crest

c

57) Item Type: Multiple Choice   Objective: 3   Section 22.2
Level: B: Enriched   Tag: Diagram 2273

Which number in the diagram above indicates a wavelength?

a) **1**   b) **2**   c) **3**   d) **4**

b

58) Item Type: Multiple Choice   Objective: 3   Section 22.2
Level: B: Enriched   Tag: Diagram 2273

Which number in the diagram above indicates a wave crest?

a) **1**   b) **2**   c) **3**   d) **4**

a

59) Item Type: Multiple Choice   Objective: 4   Section 22.2
Level: A: Standard   Tag: None

When water returns to the ocean through channels in underwater sandbars, it may produce a current known as a

a) turbidity current.   b) bottom current.   c) rip current.   d) surface current.

c

60) Item Type: Multiple Choice   Objective: 4   Section 22.2
Level: A: Standard   Tag: None

The relatively weak, irregular current produced when the water from breaking waves is pulled back to deep water is called

a) drift.   b) a longshore current.   c) an undertow.   d) a rip current.

c

61) Item Type: Multiple Choice   Objective: 4   Section 22.2
Level: B: Enriched   Tag: None

Which of the following is most likely to be created by longshore currents?

a) breakers   b) sandbars   c) whitecaps   d) landslides

b

62) Item Type: Multiple Choice   Objective: 4   Section 22.2
Level: B: Enriched   Tag: Diagram 2274

In the diagram above, the movement of water in the direction of arrow **X** represents

a) a tidal current.   b) an undertow.

c) a longshore current.   d) a breaker.

c

63) Item Type: Multiple Choice   Objective: 4   Section 22.2
Level: B: Enriched   Tag: Diagram 2275

Which of the following occurs when waves approach a shoreline, as shown in this diagram?

a) The wavelength increases.

b) The wave period increases.

c) The wave height decreases.

d) The wave height increases.

d

64) Item Type: Multiple Choice   Objective: 4   Section 22.2
Level: A: Standard   Tag: None

In the open ocean, a tsunami is characterized by

a) a short wavelength.

b) little wave energy.

c) a long wave period.

d) a large wave height.

c

65) Item Type: Multiple Choice   Objective: 4   Section 22.2
Level: A: Standard   Tag: None

Whitecaps are caused by

a) wind during storms.

b) longshore currents.

c) refraction of waves.

d) tidal currents.

a

66) Item Type: Multiple Choice   Objective: 5   Section 22.3
Level: A: Standard   Tag: None

Which of the following occurs only during the full moon and new moon phases?

a) spring tides   b) tidal oscillations   c) neap tides   d) tidal bores

a

67) Item Type: Multiple Choice   Objective: 5   Section 22.3
Level: A: Standard   Tag: None

Spring tides occur twice every

a) day.   b) week.   c) month.   d) year.

c

68) Item Type: Multiple Choice Objective: 5 Section 22.3
   Level: A: Standard Tag: None

What happens as a result of friction caused by the movement of tidal bulges around the earth?

a) The earth's rotation is slowed.   b) The ocean floor is eroded.

c) The tidal currents are slowed.   d) The shoreline is eroded.

a

69) Item Type: Multiple Choice Objective: 5 Section 22.3
   Level: B: Enriched Tag: None

The time between two daily tidal cycles is

a) 6 hours and 30 minutes.   b) 12 hours and 50 minutes.

c) 20 hours and 30 minutes.   d) 24 hours and 50 minutes.

d

70) Item Type: Multiple Choice Objective: 5 Section 22.3
   Level: A: Standard Tag: None

Tides are primarily due to

a) density differences. b) currents. c) salinity differences. d) gravitational attraction.

d

71) Item Type: Multiple Choice Objective: 6 Section 22.3
   Level: A: Standard Tag: None

Where would a tidal bore most likely occur?

a) along a straight shoreline   b) where an enclosed sea connects to the ocean

c) along a steeply sloped shoreline   d) where a river empties into a long bay

d

72) Item Type: Multiple Choice Objective: 6 Section 22.3
   Level: A: Standard Tag: None

The flow of the tidal current toward the ocean is called

a) an ebb current.   b) a rip current.

c) a flood current.   d) a deep current.

a

73) Item Type: Multiple Choice   Objective: 6   Section 22.3
Level: A: Standard   Tag: None

When a river enters the ocean through a long bay, the ocean tide may rush into the river, creating a surge of water called a tidal

a) bore.　　　　b) oscillation.　　　c) range.　　　d) flat.

a

74) Item Type: Multiple Choice   Objective: 6   Section 22.3
Level: B: Enriched   Tag: None

Where are tidal currents generally strongest?

a) along straight coastlines　　　　b) in the open ocean

c) along irregular coastlines　　　　d) in a narrow bay

d

75) Item Type: Fill in the Blank   Objective: 1   Section 22.1
Level: B: Enriched   Tag: None

From which direction do the trade winds blow in the Northern Hemisphere?

northeast

76) Item Type: Fill in the Blank   Objective: 2   Section 22.1
Level: A: Standard   Tag: None

What deep-water current travels north along the ocean floor for thousands of kilometers?

Antarctic Bottom Water

77) Item Type: Fill in the Blank   Objective: 2   Section 22.1
Level: A: Standard   Tag: None

What short-lived current is produced by underwater landslides? _____

turbidity current

78) Item Type: Fill in the Blank   Objective: 3   Section 22.2
Level: A: Standard   Tag: None

What is the term for the time it takes for one complete wavelength to pass a fixed point?

wave period

Chapter 22

79) Item Type: Fill in the Blank Objective: 3 Section 22.2
Level: A: Standard Tag: None
What is formed when the crest is blown off a wave? _____

a whitecap

80) Item Type: Fill in the Blank Objective: 3 Section 22.2
Level: B: Enriched Tag: None
What is the term used to describe the distance that wind has traveled across open water?

_____

fetch

81) Item Type: Fill in the Blank Objective: 3 Section 22.2
Level: A: Standard Tag: None
What is the name of the highest part of a wave? _____

crest

82) Item Type: Fill in the Blank Objective: 3 Section 22.2
Level: A: Standard Tag: None
What is the term used to describe the distance between two consecutive wave troughs?

_____

wavelength

83) Item Type: Fill in the Blank Objective: 4 Section 22.2
Level: A: Standard Tag: None
What is the bending of waves called? _____

refraction

84) Item Type: Fill in the Blank Objective: 5 Section 22.3
Level: B: Enriched Tag: Diagram 2276
What type of tide is produced when the earth, sun, and moon are aligned as they are in

this diagram? _____

spring tide

85) Item Type: Fill in the Blank   Objective: 6   Section 22.3
Level: B: Enriched   Tag: None

In which body of water does the tidal range exceed 15 m due to tidal oscillations?

_____

Bay of Fundy ◼

86) Item Type: Fill in the Blank   Objective: 6   Section 22.3
Level: A: Standard   Tag: None

What is the term for a tidal current that flows toward the coast? _____

flood tide ◼

87) Item Type: Essay   Objective: 2   Section 22.1
Level: A: Standard   Tag: None

Explain the reason for the difference in density between warm water and cold water.

_____

_____

_____

_____

_____

When water is cooled, it contracts and the water molecules move closer together. This contraction makes the water more dense and, as a result, the cold water sinks. When water is warmed, it expands and the water molecules move farther apart. Warm water is thus less dense and remains above the cold water. ◼

88) Item Type: Essay   Objective: 5   Section 22.3
     Level: B: Enriched   Tag: None

What causes a tide to occur on the side of the earth opposite the moon?

_____

_____

_____

_____

_____

> The tidal bulge that occurs on the side of the earth opposite the moon is caused by the mutual revolution of the earth and moon around a common center of gravity. Their spinning around each other generates outward forces that cause the ocean to bulge out from the earth's surface.

89) Item Type: Essay   Objective: 5   Section 22.3
     Level: B: Enriched   Tag: None

Which is probably a better time to plan the site of an oceanside building, during a period of spring tides or neap tides? Explain your answer.

_____

_____

_____

_____

_____

> During spring tide, the daily tidal range is largest. This is the time of the highest tide of the month and, therefore, it will indicate the maximum distance up the shore that the water will travel.

Chapter 22

90) Item Type: Essay   Objective: 6   Section 22.3
     Level: B: Enriched   Tag: None

How do the tidal patterns on the Atlantic Coast and the Pacific Coast of the United States differ?

_____

_____

_____

_____

_____

Along the Atlantic Coast, tides follow a semidiurnal, or twice daily, pattern. There are two high tides and two low tides each day, with a fairly regular tidal range. Along the Pacific Coast, tides follow a mixed pattern, with an irregular tidal range.

1) Tag Name: Diagram 2271

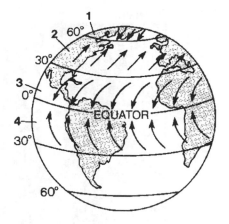

2) Tag Name: Diagram 2272

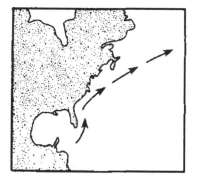

3) Tag Name: Diagram 2273

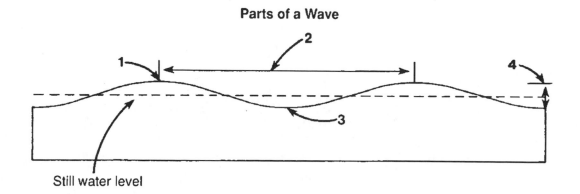

Parts of a Wave

4)   Tag Name: Diagram 2274

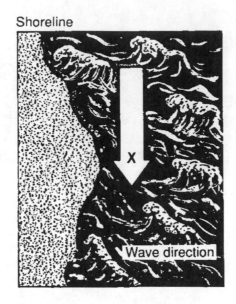

Shoreline

X

Wave direction

5)   Tag Name: Diagram 2275

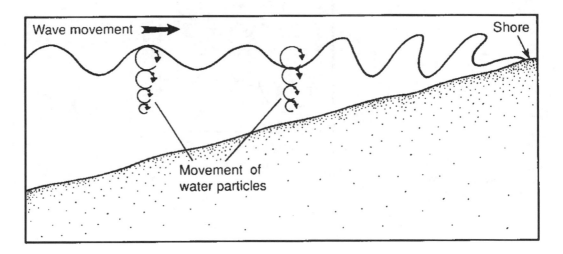

Wave movement

Shore

Movement of
water particles

6)   Tag Name: Diagram 2276

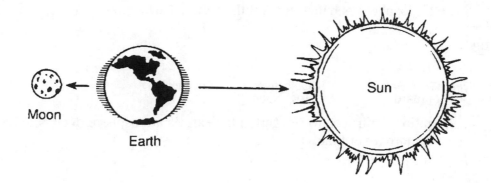

1) Item Type: True-False   Objective: 1   Section 23.1
   Level: A: Standard   Tag: None

   _____ The atmosphere is pulled toward the earth's surface by gravity.

   T

2) Item Type: True-False   Objective: 1   Section 23.1
   Level: A: Standard   Tag: None

   _____ The ratio of nitrogen to oxygen in the earth's atmosphere does not change significantly over time.

   T

3) Item Type: True-False   Objective: 2   Section 23.1
   Level: A: Standard   Tag: None

   _____ Standard atmospheric pressure is equal to a reading of 760 mm of mercury on a barometer.

   T

4) Item Type: True-False   Objective: 2   Section 23.1
   Level: A: Standard   Tag: None

   _____ Air pressure increases with elevation.

   F

5) Item Type: True-False   Objective: 2   Section 23.1
   Level: A: Standard   Tag: None

   _____ An aneroid barometer can be used as an altimeter.

   T

6) Item Type: True-False   Objective: 2   Section 23.1
   Level: A: Standard   Tag: None

   _____ An aneroid barometer uses mercury to measure atmospheric pressure.

   F

7) Item Type: True-False   Objective: 2   Section 23.1
   Level: A: Standard   Tag: None

   _____ Standard atmospheric pressure is equivalent to one atmosphere.

   T

8) Item Type: True-False   Objective: 3   Section 23.1
   Level: A: Standard   Tag: None

   _____ Most of the ozone in the earth's atmosphere is located in the ionosphere.

   F

9) Item Type: True-False   Objective: 3   Section 23.1
   Level: B: Enriched   Tag: None

   _____ The altitude of the tropopause changes with latitude and with the season of the year.

   T

10) Item Type: True-False   Objective: 3   Section 23.1
    Level: A: Standard   Tag: None

    _____ Most weather changes take place in the stratosphere.

    F

11) Item Type: True-False   Objective: 4   Section 23.1
    Level: A: Standard   Tag: None

    _____ Auto emissions are a major contributor of air pollution.

    T

12) Item Type: True-False   Objective: 4   Section 23.1
    Level: A: Standard   Tag: None

    _____ Most air pollution comes from the burning of coal and petroleum fuels.

    T

13) Item Type: True-False   Objective: 4   Section 23.1
    Level: A: Standard   Tag: None

    _____ Temperature inversions occur when warm air is trapped under cold air.

    F

14) Item Type: True-False   Objective: 5   Section 23.2
    Level: B: Enriched   Tag: None

    _____ About 50 percent of the solar energy reaching the earth's atmosphere is absorbed by the earth's surface.

    T

15) Item Type: True-False   Objective: 5   Section 23.2
    Level: B: Enriched   Tag: None

____ Almost all of the solar energy reaching the earth is reflected back into space.

F

16) Item Type: True-False   Objective: 5   Section 23.2
    Level: A: Standard   Tag: None

____ Scattering is responsible for making the sky appear blue.

T

17) Item Type: True-False   Objective: 5   Section 23.2
    Level: A: Standard   Tag: None

____ Radiation from the sun travels through space in the form of waves.

T

18) Item Type: True-False   Objective: 5   Section 23.2
    Level: A: Standard   Tag: None

____ Sunburns are caused by infrared radiation.

F

19) Item Type: True-False   Objective: 5   Section 23.2
    Level: A: Standard   Tag: None

____ A mirage is created by the refraction of light rays.

T

20) Item Type: True-False   Objective: 6   Section 23.2
    Level: A: Standard   Tag: None

____ The process by which the earth's atmosphere traps infrared rays is called the greenhouse effect.

T

21) Item Type: True-False   Objective: 6   Section 23.2
    Level: A: Standard   Tag: None

____ Desert temperatures usually change dramatically between day and night.

T

22) Item Type: True-False   Objective: 6   Section 23.2
Level: A: Standard   Tag: None

_____ Increases in atmospheric levels of carbon dioxide would probably cause the atmosphere to become warmer.

T

23) Item Type: True-False   Objective: 7   Section 23.2
Level: B: Enriched   Tag: None

_____ Atmospheric pressure is generally higher under a body of warm air than under a body of cool air.

F

24) Item Type: True-False   Objective: 7   Section 23.2
Level: A: Standard   Tag: None

_____ Air is a good conductor of heat.

F

25) Item Type: True-False   Objective: 7   Section 23.2
Level: A: Standard   Tag: None

_____ Solid materials are good conductors of heat.

T

26) Item Type: True-False   Objective: 8   Section 23.3
Level: B: Enriched   Tag: None

_____ Areas close to the equator have low atmospheric pressure due to ascending air.

T

27) Item Type: True-False   Objective: 8   Section 23.3
Level: A: Standard   Tag: None

_____ Polar easterlies are generally strong winds.

F

28) Item Type: True-False   Objective: 8   Section 23.3
Level: A: Standard   Tag: None

_____ The positions of the global wind belts shift during the year.

T

29) Item Type: True-False   Objective: 8   Section 23.3
     Level: A: Standard   Tag: None

_____ In the Northern Hemisphere, the Coriolis effect deflects surface winds to the left.

F

30) Item Type: True-False   Objective: 8   Section 23.3
     Level: A: Standard   Tag: None

_____ Movement of air is primarily a result of pressure differences in the atmosphere.

T

31) Item Type: True-False   Objective: 9   Section 23.3
     Level: A: Standard   Tag: None

_____ Valley breezes blow from mountains into valleys.

F

32) Item Type: True-False   Objective: 9   Section 23.3
     Level: A: Standard   Tag: None

_____ The Darrieus turbine generates power from the wind.

T

33) Item Type: True-False   Objective: 9   Section 23.3
     Level: A: Standard   Tag: None

_____ A sea breeze generally occurs at night.

F

34) Item Type: True-False   Objective: 9   Section 23.3
     Level: A: Standard   Tag: None

_____ A mountain breeze is created when cooler air descends from mountain peaks.

T

35) Item Type: True-False   Objective: 9   Section 23.3
     Level: A: Standard   Tag: None

_____ During daylight hours, there is little temperature difference between a body of water and the land along its shore.

F

36) Item Type: Multiple Choice   Objective: 1   Section 23.1
Level: A: Standard   Tag: None

Ozone is a form of

a) nitrogen.   b) carbon dioxide.   c) oxygen.   d) water vapor.

c

37) Item Type: Multiple Choice   Objective: 1   Section 23.1
Level: B: Enriched   Tag: None

Water vapor is added to the atmosphere primarily by

a) convection.   b) evaporation.   c) condensation.   d) precipitation.

b

38) Item Type: Multiple Choice   Objective: 1   Section 23.1
Level: A: Standard   Tag: None

The most abundant element in the earth's atmosphere is

a) oxygen.   b) argon.   c) hydrogen.   d) nitrogen.

d

39) Item Type: Multiple Choice   Objective: 1   Section 23.1
Level: A: Standard   Tag: None

Which of the following adds oxygen to the atmosphere?

a) forest fires   b) weathering of rocks

c) photosynthesis   d) life processes of animals

c

40) Item Type: Multiple Choice   Objective: 1   Section 23.1
Level: A: Standard   Tag: None

What would be the long-range effect on the earth's surface if the ozone layer were destroyed?

a) decreased infrared radiation   b) increased acid precipitation

c) increased ultraviolet radiation   d) decreased atmospheric temperature

c

41) Item Type: Multiple Choice   Objective: 2   Section 23.1
Level: B: Enriched   Tag: None

Which of the following is a unit used to measure atmospheric pressure?

a) millibar   b) newton   c) milligram   d) degree

a

42) Item Type: Multiple Choice   Objective: 2   Section 23.1
Level: A: Standard   Tag: Diagram 2371

Which of the following meteorological instruments does this illustration represent?

a) altimeter   b) mercurial barometer   c) thermometer   d) aneroid barometer

b

43) Item Type: Multiple Choice   Objective: 3   Section 23.1
Level: A: Standard   Tag: None

The layer of the atmosphere in which weather change occurs is the

a) stratopause.   b) troposphere.   c) mesopause.   d) thermosphere.

b

44) Item Type: Multiple Choice   Objective: 3   Section 23.1
Level: B: Enriched   Tag: None

The coldest layer of the atmosphere is the

a) mesosphere.   b) exosphere.   c) stratosphere.   d) ionosphere.

a

45) Item Type: Multiple Choice   Objective: 3   Section 23.1
Level: A: Standard   Tag: None

In which layer of the earth's atmosphere is most ozone found?

a) troposphere   b) thermosphere   c) stratosphere   d) mesosphere

c

46) Item Type: Multiple Choice   Objective: 3   Section 23.1
Level: A: Standard   Tag: None

The atmospheric layer closest to the earth's surface is called the

a) troposphere.   b) stratosphere.   c) ionosphere.   d) mesosphere.

a

47) Item Type: Multiple Choice   Objective: 3   Section 23.1
Level: A: Standard   Tag: None

The ionosphere is the lower portion of the

a) thermosphere.   b) troposphere.   c) mesosphere.   d) stratosphere.

a

48) Item Type: Multiple Choice   Objective: 4   Section 23.1
Level: A: Standard   Tag: None

Acid rain usually results when water vapor in the atmosphere combines with

a) ash blown from forest fires.

b) products formed from the breakdown of ozone.

c) gases from aerosol cans.

d) gases emitted by the burning of fossil fuels.

d

49) Item Type: Multiple Choice   Objective: 4   Section 23.1
Level: A: Standard   Tag: None

Which of the following is an air pollutant?

a) carbon dioxide   b) oxygen   c) sulfur dioxide   d) nitrogen

c

50) Item Type: Multiple Choice   Objective: 5   Section 23.2
Level: A: Standard   Tag: None

The albedo of the earth's surface is the fraction of solar radiation that is

a) scattered.   b) reflected.   c) inverted.   d) absorbed.

b

51) Item Type: Multiple Choice   Objective: 5   Section 23.2
Level: A: Standard   Tag: None

Which of the following phenomena is responsible for red sunsets?

a) reflection   b) albedo   c) convection   d) scattering

d

52) Item Type: Multiple Choice   Objective: 5   Section 23.2
    Level: A: Standard   Tag: None

The lower boundary of the stratosphere extends upward from the

a) tropopause.   b) ionosphere.   c) mesopause.   d) thermosphere.

a

53) Item Type: Multiple Choice   Objective: 5   Section 23.2
    Level: A: Standard   Tag: None

The white color of a foggy sky is produced by the phenomenon called

a) reflection.   b) scattering.   c) refraction.   d) albedo.

b

54) Item Type: Multiple Choice   Objective: 6   Section 23.2
    Level: B: Enriched   Tag: Diagram 2375

In the diagram above, the jet streams usually form as a result of interaction between zones *4* and

a) *6.*   b) *5.*   c) *3.*   d) *2.*

b

55) Item Type: Multiple Choice   Objective: 6   Section 23.2
    Level: B: Enriched   Tag: None

At what time does the highest temperature of the day typically occur?

a) 10:00 A.M.   b) 12:00 noon   c) 3:00 P.M.   d) 6.00 P.M.

c

56) Item Type: Multiple Choice   Objective: 6   Section 23.2
    Level: A: Standard   Tag: None

What type of radiation is trapped on the earth's surface by the greenhouse effect?

a) ultraviolet rays   b) X rays   c) gamma rays   d) infrared rays

d

57) Item Type: Multiple Choice   Objective: 6   Section 23.2
    Level: A: Standard   Tag: None

Which of the following gases is responsible for the greenhouse effect?

a) carbon dioxide   b) ozone   c) water vapor   d) nitrogen

a

58) Item Type: Multiple Choice   Objective: 7   Section 23.2
    Level: B: Enriched   Tag: None

Which of the following processes is primarily responsible for heating air in the lower part of the atmosphere?

a) refraction   b) convection   c) reflection   d) conduction

d

59) Item Type: Multiple Choice   Objective: 7   Section 23.2
    Level: A: Standard   Tag: None

The vertical movement of air due to uneven heating is called

a) convection.   b) refraction.   c) conduction.   d) reflection.

a

60) Item Type: Multiple Choice   Objective: 7   Section 23.2
    Level: A: Standard   Tag: None

The transfer of heat that takes place when gases are unevenly heated is called

a) reflection.   b) conduction.   c) radiation.   d) convection.

d

61) Item Type: Multiple Choice   Objective: 8   Section 23.3
    Level: A: Standard   Tag: None

How many convection cells are in the Southern Hemisphere?

a) 3   b) 4   c) 6   d) 8

a

62) Item Type: Multiple Choice   Objective: 8   Section 23.3
    Level: A: Standard   Tag: None

Near the equator, the narrow zone formed by the convergence of the trade winds is called the

a) horse latitudes. b) jet stream.   c) prevailing westerlies. d) doldrums.

d

63) Item Type: Multiple Choice   Objective: 8   Section 23.3
    Level: A: Standard   Tag: Diagram 2373

In the diagram above, which number represents the location of the subpolar lows?

a) *1*   b) *2*   c) *3*   d) *4*

d

64) Item Type: Multiple Choice   Objective: 8   Section 23.3
    Level: A: Standard   Tag: Diagram 2374

In the diagram above, which number represents the horse latitudes?

a) **1**   b) **2**   c) **3**   d) **4**

b

65) Item Type: Multiple Choice   Objective: 8   Section 23.3
    Level: A: Standard   Tag: Diagram 2375

In the diagram above, which number represents a convection cell?

a) **1**   b) **3**   c) **5**   d) **6**

d

66) Item Type: Multiple Choice   Objective: 8   Section 23.3
    Level: B: Enriched   Tag: Diagram 2375

In the diagram above, which winds are represented by the arrows labeled **4**?

a) southeast trades   b) westerlies   c) northeast trades   d) polar easterlies

b

67) Item Type: Multiple Choice   Objective: 8   Section 23.3
    Level: B: Enriched   Tag: None

The general direction of flow of the jet stream is toward the

a) north.   b) south.   c) east.   d) west.

c

68) Item Type: Multiple Choice   Objective: 8   Section 23.3
    Level: A: Standard   Tag: None

In what region do the northeast and southeast trade winds meet?

a) horse latitudes   b) doldrums   c) subpolar laws   d) jet streams

b

69) Item Type: Multiple Choice   Objective: 9   Section 23.3
    Level: B: Enriched   Tag: None

When is a land breeze most likely to occur?

a) morning   b) noon   c) late afternoon   d) night

d

70) Item Type: Fill in the Blank   Objective: 1   Section 23.1
    Level: A: Standard   Tag: None

The general weather conditions over many years is called _____.

climate

71) Item Type: Fill in the Blank   Objective: 1   Section 23.1
    Level: A: Standard   Tag: None

The most abundant compounds in the earth's atmosphere are water vapor and

_____.

carbon dioxide

72) Item Type: Fill in the Blank   Objective: 1   Section 23.1
    Level: A: Standard   Tag: None

The general condition of the atmosphere at a particular time and place is called

_____.

weather

73) Item Type: Fill in the Blank   Objective: 1   Section 23.1
    Level: A: Standard   Tag: None

The mixture of gases and particles that surrounds the earth is called the

_____.

atmosphere

74) Item Type: Fill in the Blank   Objective: 2   Section 23.1
    Level: A: Standard   Tag: None

The units used on weather maps to indicate atmospheric pressure are called

_____.

millibars

75) Item Type: Fill in the Blank   Objective: 2   Section 23.1
    Level: A: Standard   Tag: None

The ratio of the weight of air to the area on which it presses is called

_____.

atmospheric pressure

76) Item Type: Fill in the Blank   Objective: 2   Section 23.1
Level: A: Standard   Tag: None

An instrument that measures atmospheric pressure is called a(n) _____.

barometer

77) Item Type: Fill in the Blank   Objective: 3   Section 23.1
Level: A: Standard   Tag: None

The outermost layer of the earth's atmosphere is the _____.

exosphere

78) Item Type: Fill in the Blank   Objective: 3   Section 23.1
Level: A: Standard   Tag: None

The layer of the atmosphere closest to the earth's surface is the _____.

troposphere

79) Item Type: Fill In the Blank   Objective: 3   Section 23.1
Level: A: Standard   Tag: None

The layer of the atmosphere that can reflect radio waves back to the earth is the

_____.

ionosphere

80) Item Type: Fill in the Blank   Objective: 3   Section 23.1
Level: A: Standard   Tag: None

The coldest layer of the earth's atmosphere is the _____.

mesosphere

81) Item Type: Fill in the Blank   Objective: 4   Section 23.1
Level: A: Standard   Tag: None

Acid precipitation is produced primarily by the burning of _____.

fossil fuels

82) Item Type: Fill in the Blank   Objective: 4   Section 23.1
Level: A: Standard   Tag: None

The condition in which cold air is trapped under warm air is called a(n)

_____.

temperature inversion

83) Item Type: Fill in the Blank   Objective: 4   Section 23.1
     Level: A: Standard   Tag: None

Any substance in the atmosphere that is harmful to organisms or property is called a(n)

_____.

air pollutant

84) Item Type: Fill in the Blank   Objective: 5   Section 23.2
     Level: A: Standard   Tag: None

The fraction of solar radiation reflected by a surface is called the surface's

_____.

albedo

85) Item Type: Fill in the Blank   Objective: 5   Section 23.2
     Level: A: Standard   Tag: None

The sun appears red at sunset because of the process called _____.

scattering

86) Item Type: Fill in the Blank   Objective: 5   Section 23.2
     Level: A: Standard   Tag: Diagram 2372

The entire range of radiation represented by this diagram is called the

_____.

electromagnetic spectrum

87) Item Type: Fill in the Blank   Objective: 6   Section 23.2
     Level: A: Standard   Tag: None

Objects on earth are heated when they absorb the sun's visible light rays and its

_____.

infrared rays

88) Item Type: Fill in the Blank   Objective: 6   Section 23.2
     Level: A: Standard   Tag: None

The name of the process by which the atmosphere traps infrared rays is the

_____.

greenhouse effect

89) Item Type: Fill in the Blank   Objective: 6   Section 23.2
Level: A: Standard   Tag: None

Which rays in the electromagnetic spectrum are trapped by the greenhouse effect?

_____

infrared

90) Item Type: Fill in the Blank   Objective: 7   Section 23.2
Level: B: Enriched   Tag: None

The movement of air due to uneven heating is called _____.

convection

91) Item Type: Fill in the Blank   Objective: 7   Section 23.2
Level: B: Enriched   Tag: None

The form of heat transfer in which fast-moving molecules cause other molecules to also move

faster is called _____.

conduction

92) Item Type: Fill in the Blank   Objective: 9   Section 23.3
Level: A: Standard   Tag: None

Gentle winds that extend over distances of less than 100 km are called

_____.

breezes

93) Item Type: Fill in the Blank   Objective: 9   Section 23.3
Level: A: Standard   Tag: None

What type of breeze blows downslope at night? _____

mountain breeze

94) Item Type: Essay   Objective: 4   Section 23.1
     Level: A: Standard   Tag: None

What are two harmful effects of air pollution on humans?

_____

_____

_____

_____

_____

> On a few occasions, severe air pollution has caused a large number of deaths. Also, studies have shown that long-term exposure to pollution reduces a person's ability to resist many illnesses.

95) Item Type: Essay   Objective: 5   Section 23.2
     Level: B: Enriched   Tag: None

From where does the light of the full moon originate, and by what process is this light directed toward the earth?

_____

_____

_____

_____

> The source of the moon's light is the sun. The light is reflected by the moon's surface toward the earth.

96) Item Type: Essay   Objective: 6   Section 23.2
    Level: A: Standard   Tag: None

Why is the hottest part of the day usually at mid-afternoon rather than at noon, when the sun is highest in the sky?

_____

_____

_____

_____

_____

Because air is such a poor conductor of heat, it takes several hours for heat to be absorbed and reradiated from the ground to warm the air. Although the sun is highest at noon, because of the time lag, it is mid-afternoon before the heat warms the air.

97) Item Type: Essay   Objective: 6   Section 23.2
    Level: B: Enriched   Tag: None

What causes a sunset to be red?

_____

_____

_____

_____

Short-wavelength rays, such as blue light rays, are more easily scattered than long-wavelength rays are. The sun appears red when it is low in the sky because more blue light rays are being scattered by the atmosphere. Thus, more of the longer-wavelength red light rays reach the earth's surface, giving the setting sun its red color.

98)   Item Type: Essay   Objective: 7   Section 23.2
      Level: B: Enriched   Tag: None

Describe how the process of convection helps warm the earth's surface.

_____

_____

_____

_____

_____

As some of the air is heated, it becomes less dense and tends to rise. Nearby cooler air, which is more dense, tends to sink. As it sinks, the cooler air pushes the warm air up. The cold air, in turn, is warmed and then rises. The continuous cycle helps warm the earth's surface.

1)   Tag Name: Diagram 2371

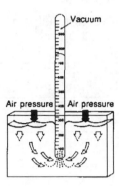

2)   Tag Name: Diagram 2372

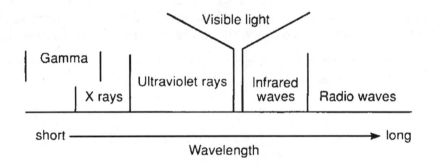

3)   Tag Name: Diagram 2373

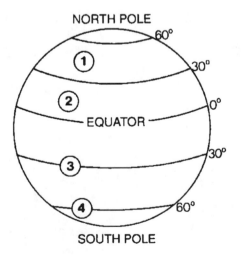

4)   Tag Name: Diagram 2374

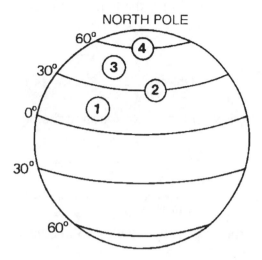

5)   Tag Name: Diagram 2375

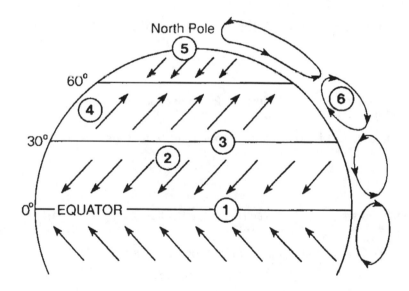

Chapter 24

1) Item Type: True-False   Objective: 1   Section 24.1
   Level: B: Enriched   Tag: None

   _____ Sublimation occurs only when the temperature is above freezing.

   F

2) Item Type: True-False   Objective: 1   Section 24.1
   Level: A: Standard   Tag: None

   _____ Most evaporation occurs in regions around the equator.

   T

3) Item Type: True-False   Objective: 1   Section 24.1
   Level: A: Standard   Tag: None

   _____ Sublimation is the process by which a solid changes to a liquid.

   F

4) Item Type: True-False   Objective: 1   Section 24.1
   Level: A: Standard   Tag: None

   _____ The heat energy necessary for evaporation of ocean water is absorbed from the sun.

   T

5) Item Type: True-False   Objective: 2   Section 24.1
   Level: A: Standard   Tag: None

   _____ Cold air can hold more water vapor than warm air.

   F

6) Item Type: True-False   Objective: 2   Section 24.1
   Level: A: Standard   Tag: None

   _____ Air at 60°C can hold more water vapor than air at 40°C.

   T

7) Item Type: True-False   Objective: 2   Section 24.1
   Level: B: Enriched   Tag: None

   _____ The amount of water vapor that a volume of air can hold decreases as air temperature rises.

   F

8) Item Type: True-False   Objective: 2   Section 24.1
    Level: A: Standard   Tag: None

_____ Specific humidity compares the amount of water vapor present in the air to the amount it can hold at saturation.

F

9) Item Type: True-False   Objective: 3   Section 24.1
    Level: A: Standard   Tag: None

_____ The dew point temperature depends on the amount of water vapor in the air.

T

10) Item Type: True-False   Objective: 3   Section 24.1
    Level: A: Standard   Tag: None

_____ Dew forms only when the dew point falls below 30°C.

F

11) Item Type: True-False   Objective: 3   Section 24.1
    Level: A: Standard   Tag: None

_____ Humid air usually has a higher dew point than dry air.

T

12) Item Type: True-False   Objective: 3   Section 24.1
    Level: A: Standard   Tag: None

_____ Water usually condenses when air temperature falls below dew point.

T

13) Item Type: True-False   Objective: 4   Section 24.2
    Level: A: Standard   Tag: None

_____ Air must be saturated before clouds can form.

T

14) Item Type: True-False   Objective: 4   Section 24.2
    Level: B: Enriched   Tag: None

_____ Condensing water vapor slows the rate of adiabatic temperature change.

T

Chapter 24

15) Item Type: True-False   Objective: 4   Section 24.2
    Level: A: Standard   Tag: None

_____   As air rises and expands, it undergoes advective cooling.

F

16) Item Type: True-False   Objective: 4   Section 24.2
    Level: A: Standard   Tag: None

_____   Clouds require condensation nuclei in order to form.

T

17) Item Type: True-False   Objective: 5   Section 24.2
    Level: A: Standard   Tag: None

_____   The lowest clouds in the sky are cirrus clouds.

F

18) Item Type: True-False   Objective: 6   Section 24.2
    Level: A: Standard   Tag: None

_____   Fog does not require condensation nuclei in order to form.

F

19) Item Type: True-False   Objective: 6   Section 24.2
    Level: A: Standard   Tag: None

_____   Fog is usually heaviest in cities because of low ground temperature.

F

20) Item Type: True-False   Objective: 7   Section 24.3
    Level: A: Standard   Tag: None

_____   Hail forms when rain falls through a layer of freezing air.

F

21) Item Type: True-False   Objective: 7   Section 24.3
    Level: A: Standard   Tag: None

_____   Snowflakes are usually larger in 30°C air than in 10°C air.

T

22) Item Type: True-False   Objective: 7   Section 24.3
      Level: A: Standard   Tag: None

_____ Raindrops are generally larger than 5 mm in diameter.

F

23) Item Type: True-False   Objective: 7   Section 24.3
      Level: A: Standard   Tag: None

_____ Hail is formed by rain that freezes as it strikes the ground.

F

24) Item Type: True-False   Objective: 7   Section 24.3
      Level: A: Standard   Tag: None

_____ Glaze ice forms when rain freezes as it strikes the ground.

T

25) Item Type: True-False   Objective: 8   Section 24.3
      Level: A: Standard   Tag: None

_____ In tropical regions, rain commonly forms by the process of coalescence.

T

26) Item Type: True-False   Objective: 8   Section 24.3
      Level: B: Enriched   Tag: None

_____ Water vapor is the most common form of water found in supercooled clouds.

F

27) Item Type: True-False   Objective: 8   Section 24.3
      Level: A: Standard   Tag: None

_____ The size of cloud droplets depends on the size of the condensation nuclei on which they form.

T

28) Item Type: True-False   Objective: 8   Section 24.3
      Level: A: Standard   Tag: None

_____ Cloud droplets must increase in size in order to fall as precipitation.

T

29) Item Type: True-False   Objective: 9   Section 24.3
     Level: A: Standard   Tag: None

_____ Silver-iodide vapor and powdered dry ice have been used to cause or increase precipitation.

T

30) Item Type: True-False   Objective: 9   Section 24.3
     Level: A: Standard   Tag: None

_____ Silver iodide is used in cloud seeding because it resembles ice crystals.

T

31) Item Type: True-False   Objective: 9   Section 24.3
     Level: A: Standard   Tag: None

_____ Cloud seeding may eventually be used to control the severity of storms.

T

32) Item Type: True-False   Objective: 10   Section 24.3
     Level: A: Standard   Tag: None

_____ Rain gauges measure the precipitation that falls over a large region.

F

33) Item Type: True-False   Objective: 10   Section 24.3
     Level: A: Standard   Tag: None

_____ A rain gauge can be constructed using a funnel and a bucket.

T

34) Item Type: True-False   Objective: 10   Section 24.3
     Level: A: Standard   Tag: None

_____ On average, the melting of 10 cm of snow will yield 1 cm of water.

T

35) Item Type: Multiple Choice   Objective: 1   Section 24.1
     Level: A: Standard   Tag: None

Ice changes directly into water vapor by the process of

a) evaporation.   b) sublimation.   c) condensation.   d) saltation.

b

36) Item Type: Multiple Choice   Objective: 1   Section 24.1
    Level: A: Standard   Tag: None

Most water enters the atmosphere by the process of

a) evaporation.   b) sublimation.   c) condensation.   d) precipitation.

a

37) Item Type: Multiple Choice   Objective: 2   Section 24.1
    Level: B: Enriched   Tag: None

What is the relative humidity when there are 7 g/m^3 of water vapor in air with a saturation point of 14 g/m^3?

a) 7%   b) 14%   c) 50%   d) 98%

c

38) Item Type: Multiple Choice   Objective: 2   Section 24.1
    Level: A: Standard   Tag: None

A psychrometer is used to measure

a) freezing point.   b) specific humidity.

c) dew point   d) relative humidity.

d

39) Item Type: Multiple Choice   Objective: 2   Section 24.1
    Level: A: Standard   Tag: None

The mass of water vapor in a sample of air compared to the mass of water vapor the air can hold at saturation is called the

a) dew point.   b) relative humidity.

c) condensation level.   d) specific humidity.

b

40) Item Type: Multiple Choice   Objective: 3   Section 24.1
    Level: A: Standard   Tag: None

What forms when the dew point is below 0°C?

a) dew   b) fog   c) frost   d) drizzle

c

41) Item Type: Multiple Choice   Objective: 3   Section 24.1
Level: B: Enriched   Tag: None

Air becomes saturated when its temperature equals its

a) condensation level.

b) dew point.

c) specific humidity.

d) adiabatic temperature.

b

42) Item Type: Multiple Choice   Objective: 3   Section 24.1
Level: A: Standard   Tag: None

Frost forms by the process of

a) advective cooling.   b) sublimation.   c) supercooling.   d) precipitation.

b

43) Item Type: Multiple Choice   Objective: 4   Section 24.2
Level: A: Standard   Tag: None

As air rises and expands, it undergoes

a) advective heating.

b) advective cooling.

c) convective heating.

d) convective cooling.

d

44) Item Type: Multiple Choice   Objective: 4   Section 24.2
Level: B: Enriched   Tag: None

Which of the following changes as air temperature changes?

a) dew point

b) relative humidity

c) freezing point

d) specific humidity

b

45) Item Type: Multiple Choice   Objective: 4   Section 24.2
Level: A: Standard   Tag: None

Forceful lifting of an air mass can occur as the air meets a

a) body of water.   b) mountain range.   c) cloud bank.   d) valley.

b

46) Item Type: Multiple Choice   Objective: 4   Section 24.2
     Level: B: Enriched   Tag: Diagram 2472

Which type of cooling is most likely to occur as the air mass in the diagram above moves over the water?

a) condensational   b) adiabatic   c) convective   d) advective

d

47) Item Type: Multiple Choice   Objective: 5   Section 24.2
     Level: A: Standard   Tag: Diagram 2473

What type of cloud is pictured in the diagram above?

a) cumulus   b) stratus   c) cirrus   d) nimbus

c

48) Item Type: Multiple Choice   Objective: 5   Section 24.2
     Level: B: Enriched   Tag: Diagram 2473

At what altitude would you expect to find the type of cloud pictured above?

a) 1,000 m   b) 2,000 m   c) 4,000 m   d) 6,000 m

d

49) Item Type: Multiple Choice   Objective: 5   Section 24.2
     Level: B: Enriched   Tag: None

What type of cloud forms when warm, moist air overrides a layer of cool air?

a) stratus   b) cumulonimbus   c) nimbostratus   d) altonimbus

a

50) Item Type: Multiple Choice   Objective: 5   Section 24.2
     Level: A: Standard   Tag: None

Which of the following cloud types occurs at the lowest altitude?

a) cirrus   b) cumulus   c) stratus   d) nimbus

c

51) Item Type: Multiple Choice   Objective: 5   Section 24.2
     Level: A: Standard   Tag: None

Which type of cloud usually forms in fair weather?

a) nimbostratus   b) cumulus   c) cirrocumulus   d) nimbus

b

52) **Item Type:** Multiple Choice   **Objective:** 5   Section 24.2
    Level: A: Standard   **Tag:** None

Which of the following cloud types occurs at the highest altitude?

a) cumulus   b) stratus   c) cirrus   d) nimbus

c

53) **Item Type:** Multiple Choice   **Objective:** 5   Section 24.2
    Level: B: Enriched   **Tag:** None

Which type of cloud is produced by the rising and cooling of large air masses?

a) nimbus   b) stratus   c) cirrus   d) cumulus

d

54) **Item Type:** Multiple Choice   **Objective:** 5   Section 24.2
    Level: A: Standard   **Tag:** None

Which type of cloud is composed entirely of ice crystals?

a) stratus   b) nimbus   c) cirrus   d) cumulus

c

55) **Item Type:** Multiple Choice   **Objective:** 6   Section 24.2
    Level: A: Standard   **Tag:** None

Which type of fog usually forms over inland rivers and lakes?

a) radiation   b) steam   c) upslope   d) advection

b

56) **Item Type:** Multiple Choice   **Objective:** 6   Section 24.2
    Level: A: Standard   **Tag:** None

Fog differs from clouds in that fog

a) forms closer to the ground.

b) requires no condensation nuclei.

c) forms in saturated air.

d) contains many ice crystals.

a

57) Item Type: Multiple Choice   Objective: 6   Section 24.2
Level: A: Standard   Tag: None

Which type of fog occurs as warm moist air moves across a cold surface?

a) radiation   b) upslope   c) advection   d) steam

c

58) Item Type: Multiple Choice   Objective: 7   Section 24.3
Level: A: Standard   Tag: None

The most common form of solid precipitation is

a) glaze ice.   b) hail.   c) sleet.   d) snow.

d

59) Item Type: Multiple Choice   Objective: 7   Section 24.3
Level: A: Standard   Tag: None

Which type of precipitation is most likely to occur when a layer of warm air overlies a layer of air that is below the freezing point?

a) rain   b) sleet   c) drizzle   d) snow

b

60) Item Type: Multiple Choice   Objective: 7   Section 24.3
Level: B: Enriched   Tag: None

In which type of cloud is hail usually formed?

a) cumulonimbus   b) cirrostratus   c) nimbostratus   d) cirrocumulus

a

61) Item Type: Multiple Choice   Objective: 7   Section 24.3
Level: B: Enriched   Tag: None

Precipitation consisting of drops smaller than 0.5 mm in diameter is called

a) drizzle.   b) rain.   c) sleet.   d) snow.

a

62) Item Type: Multiple Choice   Objective: 7   Section 24.3
Level: A: Standard   Tag: None

Which type of precipitation forms as rain falls through a layer of freezing air?

a) rain   b) snow   c) drizzle   d) sleet

d

63) Item Type: Multiple Choice   Objective: 8   Section 24.3
Level: B: Enriched   Tag: None

Most of the water in supercooled clouds exists as

a) snowflakes   b) water droplets.   c) ice crystals.   d) water vapor.

b

64) Item Type: Multiple Choice   Objective: 8   Section 24.3
Level: A: Standard   Tag: None

The collision and combination of large and small cloud droplets are described as

a) supercooling.   b) coalescence.   c) precipitation.   d) convection.

b

65) Item Type: Multiple Choice   Objective: 8   Section 24.3
Level: A: Standard   Tag: None

Snow forms in supercooled clouds by the

a) combination of supercooled droplets.

b) reaction between dry ice and water.

c) condensation of water onto ice crystals.

d) aggregation of ice crystals.

c

66) Item Type: Multiple Choice   Objective: 8   Section 24.3
Level: A: Standard   Tag: None

Rain in tropical regions is commonly formed by the process of

a) supercooling.   b) convection.   c) sublimation.   d) coalescence.

d

67) Item Type: Multiple Choice   Objective: 9   Section 24.3
Level: B: Enriched   Tag: None

Precipitation increases after cloud seeding because clouds contain more

a) cloud droplets.   b) water vapor.

c) coalesced droplets.   d) condensation nuclei.

d

68) Item Type: Multiple Choice   Objective: 9   Section 24.3
Level: A: Standard   Tag: None

Cloud seeding may eventually be used to control

a) winds.   b) evaporation.   c) drought.   d) humidity.

c

69) Item Type: Multiple Choice   Objective: 10   Section 24.3
Level: A: Standard   Tag: None

A funnel and a cylindrical container could be used to measure the

a) amount of rainfall.

b) specific humidity.

c) adiabatic temperature change.

d) dew point.

a

70) Item Type: Multiple Choice   Objective: 10   Section 24.3
Level: B: Enriched   Tag: Diagram 2474

How much liquid water would most likely result from the melting of snow in the diagram?

a) 1 cm   b) 2 cm   c) 10 cm   d) 20 cm

b

71) Item Type: Multiple Choice   Objective: 10   Section 24.3
Level: A: Standard   Tag: None

Which of the following properties of snow is commonly determined by melting a sample of snow and measuring the result?

a) water content   b) depth   c) crystal size   d) density

a

72) Item Type: Fill in the Blank   Objective: 1   Section 24.1
Level: A: Standard   Tag: None

Evaporation is greatest near the equator because this area receives large amounts of

_____.

solar energy

73) Item Type: Fill in the Blank   Objective: 1   Section 24.1
    Level: B: Enriched   Tag: None

The heat released when water condenses and changes to liquid form is called

_____.

latent heat

74) Item Type: Fill in the Blank   Objective: 1   Section 24.1
    Level: A: Standard   Tag: None

By the process of sublimation, ice changes into _____.

water vapor

75) Item Type: Fill in the Blank   Objective: 1   Section 24.1
    Level: A: Standard   Tag: None

Most atmospheric water is found in the form of _____.

water vapor

76) Item Type: Fill in the Blank   Objective: 1   Section 24.1
    Level: A: Standard   Tag: None

The principal source of atmospheric moisture is _____.

ocean water

77) Item Type: Fill in the Blank   Objective: 2   Section 24.1
    Level: A: Standard   Tag: None

The actual amount of water vapor in a volume of air is expressed as specific

_____.

humidity

78) Item Type: Fill in the Blank   Objective: 2   Section 24.1
    Level: B: Enriched   Tag: None

What instrument package is carried in a balloon to high altitudes in order to measure

temperature and humidity? _____

radiosonde

Chapter 24

79) Item Type: Fill in the Blank   Objective: 2   Section 24.1
    Level: B: Enriched   Tag: None

As the temperature of an air mass falls below the dew point, water vapor will

_____ .

condense ▪

80) Item Type: Fill in the Blank   Objective: 2   Section 24.1
    Level: A: Standard   Tag: None

The instrument that uses wet-bulb and dry-bulb thermometers to measure relative humidity

is called a(n) _____ .

psychrometer ▪

81) Item Type: Fill in the Blank   Objective: 2   Section 24.1
    Level: A: Standard   Tag: None

What term is used to express the actual amount of moisture in the air?

_____

specific humidity ▪

82) Item Type: Fill in the Blank   Objective: 2   Section 24.1
    Level: B: Enriched   Tag: Diagram 2471

What is the relative humidity of air sample 2 in the table? _____

41% ▪

83) Item Type: Fill in the Blank   Objective: 3   Section 24.1
    Level: A: Standard   Tag: None

When air contacts a cold surface, the air may cool to its dew point by the process of

_____ .

conduction ▪

84) Item Type: Fill in the Blank   Objective: 3   Section 24.1
    Level: A: Standard   Tag: None

What forms on surfaces when the dew point is below the freezing temperature of water?

_____

frost ▪

85) Item Type: Fill in the Blank   Objective: 3   Section 24.1
Level: A: Standard   Tag: None

The temperature to which air must be cooled to reach saturation is called the

_____.

dew point

86) Item Type: Fill in the Blank   Objective: 4   Section 24.2
Level: B: Enriched   Tag: None

Before clouds can form, air must become _____.

saturated

87) Item Type: Fill in the Blank   Objective: 4   Section 24.2
Level: A: Standard   Tag: None

Water vapor condenses on suspended particles called _____.

condensation nuclei

88) Item Type: Fill in the Blank   Objective: 4   Section 24.2
Level: A: Standard   Tag: None

The condensation level is marked by the base of the _____.

cloud layer

89) Item Type: Fill in the Blank   Objective: 4   Section 24.2
Level: A: Standard   Tag: None

What level is marked by the base of a cloud layer? _____

condensation level

90) Item Type: Fill in the Blank   Objective: 4   Section 24.2
Level: A: Standard   Tag: None

Temperature changes that result solely from the expansion and compression of air are

called _____.

adiabatic

91) Item Type: Fill in the Blank   Objective: 5   Section 24.2
     Level: A: Standard   Tag: None

Clouds produced by the rising and cooling of large bodies of air are called

_____.

cumulus clouds

92) Item Type: Fill in the Blank   Objective: 5   Section 24.2
     Level: A: Standard   Tag: None

Sheetlike, layered clouds are called _____.

stratus clouds

93) Item Type: Fill in the Blank   Objective: 5   Section 24.2
     Level: A: Standard   Tag: None

What type of cloud creates a halo effect around the sun? _____

cirrostratus

94) Item Type: Fill in the Blank   Objective: 5   Section 24.2
     Level: A: Standard   Tag: None

What is another name for cumulonimbus clouds? _____

thunderheads

95) Item Type: Fill in the Blank   Objective: 6   Section 24.2
     Level: A: Standard   Tag: None

Radiation fog is also called _____.

ground fog

96) Item Type: Fill in the Blank   Objective: 6   Section 24.2
     Level: A: Standard   Tag: None

The type of fog that is formed by the lifting and adiabatic cooling of air along hillsides is

called _____.

upslope fog

97) Item Type: Fill in the Blank   Objective: 6   Section 24.2
     Level: A: Standard   Tag: None

What type of fog usually forms over inland rivers and lakes? _____

steam fog

98) Item Type: Fill in the Blank   Objective: 6   Section 24.2
    Level: A: Standard   Tag: None

What type of fog usually forms on calm, clear nights and results from the nightly cooling

of the earth? _____

| radiation fog   ▪ |

99) Item Type: Fill in the Blank   Objective: 7   Section 24.3
    Level: A: Standard   Tag: None

Snowflakes vary in size depending on the air _____.

| temperature   ▪ |

100) Item Type: Fill in the Blank   Objective: 7   Section 24.3
     Level: A: Standard   Tag: None

Raindrops carried aloft by convective currents in cumulonimbus clouds form

_____.

| hailstones   ▪ |

101) Item Type: Fill in the Blank   Objective: 8   Section 24.3
     Level: A: Standard   Tag: None

The type of condensation nucleus that has a crystalline structure similar to ice is called

_____.

| silver iodide   ▪ |

102) Item Type: Fill in the Blank   Objective: 8   Section 24.3
     Level: A: Standard   Tag: None

In tropical regions, rain usually results from the process of _____.

| coalescence   ▪ |

103) Item Type: Fill in the Blank   Objective: 9   Section 24.3
     Level: A: Standard   Tag: None

Droughts may eventually be ended using a rain-producing method called

_____.

| cloud seeding   ▪ |

104) Item Type: Fill in the Blank   Objective: 10   Section 24.3
Level: A: Standard   Tag: None

Snow is measured by both the depth of accumulation and the _____.

water content ▪

105) Item Type: Fill in the Blank   Objective: 10   Section 24.3
Level: A: Standard   Tag: None

An instrument used to measure the amount of rainfall is called a(n)

_____.

rain gauge ▪

106) Item Type: Essay   Objective: 1   Section 24.1
Level: A: Standard   Tag: None

Describe the differences in molecular motion found in ice, liquid water, and water vapor.

_____

_____

_____

_____

_____

In ice, water molecules are held almost stationary in a definite crystalline arrangement. As heat is added, the molecules begin to move more rapidly and the ice melts. In liquid water, the molecules move around each other but stay close together. As water is heated, molecules move even faster and collide. Collisions cause molecules to move rapidly enough to evaporate and become water vapor. ▪

107) Item Type: Essay   Objective: 2   Section 24.1
     Level: B: Enriched   Tag: None

Describe how a psychrometer works.

_____

_____

_____

_____

_____

A psychrometer consists of a dry-bulb thermometer and a wet-bulb thermometer. The two thermometers are whirled to circulate air around the bulbs. As water on the wet-bulb evaporates, heat is withdrawn from the thermometer. The temperature drop in the wet-bulb thermometer depends on the humidity of the air. A table is used to translate the temperature difference into relative humidity.

108) Item Type: Essay   Objective: 3   Section 24.1
     Level: B: Enriched   Tag: None

Explain why drops of water may form on the outside of a glass filled with ice water.

_____

_____

_____

_____

_____

Water from the warmer air surrounding the glass cools to its dew point and condenses on the cooled surface of the glass. This is also the way dew forms on many surfaces at night.

109) Item Type: Essay   Objective: 8 Section 24.3
Level: B: Enriched   Tag: None

Explain how precipitation forms in supercooled clouds.

_____

_____

_____

_____

_____

In supercooled clouds, most of the water exists as supercooled water droplets. There are few freezing nuclei. Precipitation forms as water molecules evaporate from supercooled water droplets and condense onto ice crystals. The ice crystals increase in size until they are large enough to fall as snow and rain.

110) Item Type: Essay   Objective: 9 Section 24.3
Level: B: Enriched   Tag: None

How does the process of cloud seeding work?

_____

_____

_____

_____

_____

Cloud seeding involves the release of silver-iodide vapor or powdered dry ice into clouds in an attempt to cause or increase precipitation. In supercooled clouds, precipitation forms by the evaporation of water from supercooled cloud droplets and its condensation onto ice crystals. However, there are few ice crystals because there are few freezing nuclei. Cloud seeding adds freezing nuclei to the supercooled clouds.

1)   Tag Name: Diagram 2471

Sample	T °C	Mass of Water Vapor	Mass of Water Vapor at Saturation
1	20°	12 g	17 g
2	32°	14 g	34 g
3	41°	20 g	56 g

2)   Tag Name: Diagram 2472

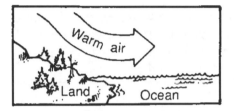

3)   Tag Name: Diagram 2473

4) Tag Name: Diagram 2474

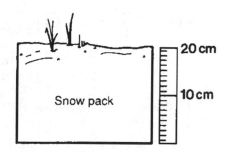

1) Item Type: True-False   Objective: 1   Section 25.1
Level: A: Standard   Tag: None

_____ A large body of air with highly variable temperature and moisture content is called an air mass.

F

2) Item Type: True-False   Objective: 1   Section 25.1
Level: A: Standard   Tag: None

_____ An air mass can be thousands of kilometers in diameter.

T

3) Item Type: True-False   Objective: 1   Section 25.1
Level: A: Standard   Tag: None

_____ The temperature and humidity of an air mass depend primarily on the source region of the air mass.

T

4) Item Type: True-False   Objective: 1   Section 25.1
Level: A: Standard   Tag: None

_____ Differences in air pressure and temperature cause convection cells to form.

T

5) Item Type: True-False   Objective: 2   Section 25.1
Level: A: Standard   Tag: None

_____ Three polar air masses influence weather in North America.

T

6) Item Type: True-False   Objective: 2   Section 25.1
Level: A: Standard   Tag: None

_____ An air mass that forms over Mexico is classified as an mP air mass.

F

7) Item Type: True-False   Objective: 2   Section 25.1
Level: A: Standard   Tag: None

_____ In summer, maritime polar Pacific air masses bring cool, foggy weather to the Pacific Northwest of the United States.

T

Chapter 25

8) Item Type: True-False   Objective: 2   Section 25.1
   Level: A: Standard   Tag: None

_____ The letters *cT* designate a cold, turbulent air mass.

F

9) Item Type: True-False   Objective: 2   Section 25.1
   Level: A: Standard   Tag: None

_____ Polar Atlantic air masses usually move eastward toward Europe.

T

10) Item Type: True-False   Objective: 3   Section 25.2
   Level: A: Standard   Tag: None

_____ A squall line is a long line of heavy thunderstorms.

T

11) Item Type: True-False   Objective: 3   Section 25.2
   Level: A: Standard   Tag: None

_____ Storms usually form along fronts.

T

12) Item Type: True-False   Objective: 3   Section 25.2
   Level: A: Standard   Tag: None

_____ Fronts form when two air masses with identical characteristics merge.

F

13) Item Type: True-False   Objective: 4   Section 25.2
   Level: A: Standard   Tag: None

_____ In North America, wave cyclones generally move in a westerly direction and spin clockwise.

F

14) Item Type: True-False   Objective: 4   Section 25.2
   Level: A: Standard   Tag: None

_____ Anticyclones generally spin counterclockwise in the Northern Hemisphere.

F

Chapter 25

15) Item Type: True-False   Objective: 5   Section 25.2
     Level: A: Standard   Tag: None

_____ The World Meteorological Organization can predict precisely when precipitation will occur and can forecast accurately the amount of precipitation that will fall.

F

16) Item Type: True-False   Objective: 5   Section 25.2
     Level: A: Standard   Tag: None

_____ Hurricanes are more powerful than wave cyclones.

T

17) Item Type: True-False   Objective: 5   Section 25.2
     Level: A: Standard   Tag: None

_____ In order for lightning to occur, a cloud must have areas with distinct electrical charges.

T

18) Item Type: True-False   Objective: 6   Section 25.3
     Level: A: Standard   Tag: None

_____ An anemometer measures wind direction.

F

19) Item Type: True-False   Objective: 6   Section 25.3
     Level: A: Standard   Tag: None

_____ A bimetal thermometer translates changes in electric current into temperature readings.

F

20) Item Type: True-False   Objective: 6   Section 25.3
     Level: A: Standard   Tag: None

_____ Wind direction may be described by one of 32 points on the compass.

T

21) Item Type: True-False   Objective: 7   Section 25.3
     Level: A: Standard   Tag: None

_____ Radar is often used to detect cloud droplets and dust particles.

F

Chapter 25

22) Item Type: True-False   Objective: 8   Section 25.4
    Level: A: Standard   Tag: None

_____ Station models summarize weather conditions at individual weather observation
      stations.

> T

23) Item Type: True-False   Objective: 8   Section 25.4
    Level: A: Standard   Tag: None

_____ The dew point is a measure of air pressure.

> F

24) Item Type: True-False   Objective: 9   Section 25.4
    Level: A: Standard   Tag: None

_____ Lines on a weather map connecting points of equal humidity are isobars.

> F

25) Item Type: True-False   Objective: 9   Section 25.4
    Level: A: Standard   Tag: None

_____ To forecast the weather, meteorologists compare the most recent weather map with
      maps from the previous 24 hours.

> T

26) Item Type: True-False   Objective: 9   Section 25.4
    Level: A: Standard   Tag: None

_____ Meteorological technicians must have advanced college degrees in meteorology.

> F

27) Item Type: True-False   Objective: 9   Section 25.4
    Level: A: Standard   Tag: None

_____ Extended forecasts usually predict precipitation very accurately.

> F

28) Item Type: True-False   Objective: 9   Section 25.4
    Level: A: Standard   Tag: None

_____ Hurricanes have been successfully controlled using cloud seeding.

> F

29) Item Type: True-False   Objective: 9   Section 25.4
Level: A: Standard   Tag: None

_____ The World Meteorological Organization was created to promote the rapid exchange of weather information.

T

30) Item Type: Multiple Choice   Objective: 1   Section 25.1
Level: A: Standard   Tag: None

What is the general name for a large body of air having uniform temperature and moisture content?

a) wind cell   b) air mass   c) stationary front   d) wave cyclone

b

31) Item Type: Multiple Choice   Objective: 2   Section 25.1
Level: B: Enriched   Tag: None

Which letter designation is used for an air mass that forms over the southwestern United States?

a) mT   b) mP   c) cP   d) cT

d

32) Item Type: Multiple Choice   Objective: 2   Section 25.1
Level: B: Enriched   Tag: Diagram 2571

What type of air mass usually forms in Region **A** and moves along the path of the arrow on the map?

a) continental tropical Mexican   b) maritime tropical gulf

c) maritime tropical Atlantic   d) continental tropical desert

b

33) Item Type: Multiple Choice   Objective: 2   Section 25.1
Level: A: Standard   Tag: None

During which season do continental tropical air masses flow over North America?

a) autumn   b) spring   c) summer   d) winter

c

34) Item Type: Multiple Choice   Objective: 2   Section 25.1
Level: B: Enriched   Tag: None

What letters would be used to designate an air mass that forms over Antarctica?

a) cP   b) cT   c) mP   d) mT

a

35) Item Type: Multiple Choice   Objective: 2   Section 25.1
Level: B: Enriched   Tag: None

Which area is a source region for mT air masses?

a) southwestern United States   b) northern Canada

c) New England coast   d) southern Pacific Ocean

d

36) Item Type: Multiple Choice   Objective: 2   Section 25.1
Level: A: Standard   Tag: None

Which type of air mass causes heavy precipitation in Oregon and Washington?

a) maritime polar Atlantic   b) continental tropical

c) maritime polar Pacific   d) maritime tropical gulf

c

37) Item Type: Multiple Choice   Objective: 3   Section 25.2
Level: A: Standard   Tag: None

During the summers, air masses that form over the southwestern United States usually bring weather that is

a) hot and clear.   b) rainy and cold.

c) cool and clear.   d) hot and humid.

a

38) Item Type: Multiple Choice   Objective: 3   Section 25.2
Level: A: Standard   Tag: None

Which type of front forms when a warm air mass overtakes a cold air mass?

a) cold   b) warm   c) stationary   d) occluded

b

39) Item Type: Multiple Choice   Objective: 3   Section 25.2
Level: B: Enriched   Tag: Diagram 2572

Which type of front is shown in this diagram?

a) stationary   b) cold   c) warm   d) occluded

a

40) Item Type: Multiple Choice   Objective: 3   Section 25.2
     Level: A: Standard   Tag: None

Which type of front generally has a gradual slope and produces heavy precipitation over large areas?

a) occluded   b) stationary   c) cold   d) warm

d

41) Item Type: Multiple Choice   Objective: 4   Section 25.2
     Level: A: Standard   Tag: None

Large, low-pressure storm centers that form along polar fronts and influence weather in the middle latitudes are called

a) Coriolis formations.   b) jet-stream systems.

c) wave cyclones.   d) polar storm fronts.

c

42) Item Type: Multiple Choice   Objective: 5   Section 25.2
     Level: A: Standard   Tag: None

Lightning causes a rapid expansion and collapse of the air that produces

a) thunder.   b) hail.   c) cyclones.   d) rain.

a

43) Item Type: Multiple Choice   Objective: 5   Section 25.2
     Level: A: Standard   Tag: None

Which of the following types of storms is the shortest-lived and has the highest wind-speed?

a) typhoon   b) thunderstorm   c) hurricane   d) tornado

d

44) Item Type: Multiple Choice   Objective: 6   Section 25.3
     Level: A: Standard   Tag: None

Which of the following instruments is used to measure temperature changes?

a) barometer   b) hygrometer   c) psychrometer   d) thermograph

d

45) Item Type: Multiple Choice   Objective: 6   Section 25.3
    Level: A: Standard   Tag: None

When a weather map shows a wind symbol at 270° it indicates that the wind is coming from the

a) north.   b) south.   c) east.   d) west.

d

46) Item Type: Multiple Choice   Objective: 6   Section 25.3
    Level: A: Standard   Tag: None

A barometer is used to measure

a) wind speed.   b) precipitation rate.

c) humidity level.   d) air pressure.

d

47) Item Type: Multiple Choice   Objective: 6   Section 25.3
    Level: A: Standard   Tag: Diagram 2573

What type of thermometer is shown in the diagram?

a) bimetal   b) liquid   c) thermograph   d) electrical

b

48) Item Type: Multiple Choice   Objective: 6   Section 25.3
    Level: A: Standard   Tag: None

Which instrument measures wind direction?

a) wind scale   b) wind compass   c) wind gauge   d) wind vane

d

49) Item Type: Multiple Choice   Objective: 7   Section 25.3
    Level: A: Standard   Tag: None

Satellite images made using infrared energy primarily reveal information about atmospheric

a) temperature.   b) humidity.   c) pollution.   d) wind speed.

a

50) Item Type: Multiple Choice   Objective: 7   Section 25.3
    Level: A: Standard   Tag: None

Which type of thermometer contains a bar made of brass and iron that curves when it is heated?

a) liquid   b) bimetal   c) thermograph   d) electrical

b

51) Item Type: Multiple Choice   Objective: 7   Section 25.3
    Level: A: Standard   Tag: None

Which weather instrument carries mounted cameras designed to photograph the tops of clouds in the uppermost layers of the atmosphere?

a) radiosonde   b) radar detector   c) satellite   d) weather balloon

c

52) Item Type: Multiple Choice   Objective: 8   Section 25.4
    Level: A: Standard   Tag: None

On a weather map, station models are used to indicate weather conditions

a) near an observation center.   b) around a storm system.

c) over the oceans.   d) in the upper atmosphere.

a

53) Item Type: Multiple Choice   Objective: 8   Section 25.4
    Level: A: Standard   Tag: Diagram 2575

What feature is shown on the weather map?

a) warm front   b) hurricane   c) cold front   d) typhoon

c

54) Item Type: Multiple Choice   Objective: 8   Section 25.4
    Level: A: Standard   Tag: Diagram 2575

The feature on the map is moving toward the

a) northwest   b) northeast   c) southwest   d) southeast

d

55) Item Type: Multiple Choice   Objective: 9   Section 25.4
    Level: A: Standard   Tag: None

The process of adding freezing nuclei to supercooled clouds is called

a) ionizing.   b) seeding.   c) jetting.   d) linking.

b

56) Item Type: Fill in the Blank   Objective: 1   Section 25.1
    Level: A: Standard   Tag: None

What type of front is generally preceded by cirrus and cirrostratus clouds?

_____

warm front

57) Item Type: Fill in the Blank   Objective: 1   Section 25.1
    Level: A: Standard   Tag: None

The deflection of wind caused by the earth's rotation is called _____.

| the Coriolis effect |

58) Item Type: Fill in the Blank   Objective: 2   Section 25.1
    Level: B: Enriched   Tag: None

The type of air mass that forms over land is called a(n) _____.

| continental air mass |

59) Item Type: Fill in the Blank   Objective: 2   Section 25.1
    Level: B: Enriched   Tag: None

How many tropical air masses influence the weather in North America?

_____

| four |

60) Item Type: Fill in the Blank   Objective: 2   Section 25.1
    Level: B: Enriched   Tag: None

What letter designation is used for air masses that form over northern Canada?

_____

| cP |

61) Item Type: Fill in the Blank   Objective: 2   Section 25.1
    Level: A: Standard   Tag: None

What type of air mass forms over the northern Pacific and southwestern Alaska?

_____

| maritime polar Pacific |

62) Item Type: Fill in the Blank   Objective: 3   Section 25.2
    Level: A: Standard   Tag: None

What type of front forms when two air masses meet but neither is displaced?

_____

| stationary |

63) Item Type: Fill in the Blank   Objective: 3   Section 25.2
Level: A: Standard   Tag: None

What type of storm is usually found along a squall line? _____

thunderstorm

64) Item Type: Fill in the Blank   Objective: 3   Section 25.2
Level: A: Standard   Tag: None

What type of front is produced when a cold front overtakes and lifts warm air completely

off the ground? _____

occluded front

65) Item Type: Fill in the Blank   Objective: 4   Section 25.2
Level: A: Standard   Tag: None

In which direction do wave cyclones spin in the Northern Hemisphere?

_____

counterclockwise

66) Item Type: Fill in the Blank   Objective: 4   Section 25.2
Level: A: Standard   Tag: None

What is the name of the boundary at which cold polar air meets the warmer air of the

middle latitudes? _____

polar front

67) Item Type: Fill in the Blank   Objective: 4   Section 25.2
Level: A: Standard   Tag: None

What causes the clockwise circulation of air around a center of high pressure in the

Northern Hemisphere? _____

the Coriolis effect

68) Item Type: Fill in the Blank   Objective: 5   Section 25.2
Level: A: Standard   Tag: None

Hurricanes that form over the western Pacific Ocean are called _____.

typhoons

69) Item Type: Fill in the Blank   Objective: 5   Section 25.2
Level: A: Standard   Tag: None

What is the name for a small storm that forms when a thunderstorm meets high-altitude,

horizontal winds? _____

tornado

70) Item Type: Fill in the Blank   Objective: 5   Section 25.2
Level: A: Standard   Tag: None

What term is used to describe the center of a hurricane? _____

eye

71) Item Type: Fill in the Blank   Objective: 6   Section 25.3
Level: A: Standard   Tag: None

How is a south wind indicated in degrees? _____

180°

72) Item Type: Fill in the Blank   Objective: 6   Section 25.3
Level: A: Standard   Tag: None

What two temperature scales are commonly used in the United States?

_____

Fahrenheit and Celsius

73) Item Type: Fill in the Blank   Objective: 7   Section 25.3
Level: A: Standard   Tag: None

What type of meteorological instrument package is carried into the atmosphere by

helium-filled balloons, to investigate weather conditions? _____

radiosonde

74) Item Type: Fill in the Blank   Objective: 7   Section 25.3
Level: A: Standard   Tag: None

To take images of cloud cover over all of North America, cameras are mounted on

_____.

satellites

75) Item Type: Fill in the Blank   Objective: 7   Section 25.3
Level: A: Standard   Tag: None

Which instrument do scientists use to measure wind speed? _____

anemometer

76) Item Type: Fill in the Blank   Objective: 7   Section 25.3
Level: A: Standard   Tag: None

What information do scientists gather by tracking the path of a helium-filled balloon?

_____

wind speed and direction

77) Item Type: Fill in the Blank   Objective: 8   Section 25.4
Level: A: Standard   Tag: None

On a weather map, what is a line connecting the points of equal atmospheric pressure

called? _____

isobar

78) Item Type: Fill in the Blank   Objective: 8   Section 25.4
Level: A: Standard   Tag: None

What is the general term for the temperature to which air must be cooled to become

saturated with moisture? _____

dew point

79) Item Type: Fill in the Blank   Objective: 8   Section 25.4
Level: A: Standard   Tag: Diagram 2574

What temperature is indicated by the station model above? _____

4°C

80) Item Type: Fill in the Blank   Objective: 8   Section 25.4
Level: A: Standard   Tag: Diagram 2574

The dew point indicated by the station model above is _____.

1°C

81) Item Type: Fill in the Blank   Objective: 8   Section 25.4
Level: A: Standard   Tag: Diagram 2576

What cloud conditions are indicated by the symbol labeled *1* in the diagram above?

_____

overcast (100% cloud coverage) ▮

82) Item Type: Fill in the Blank   Objective: 8   Section 25.4
Level: A: Standard   Tag: Diagram 2576

What is the wind direction indicated by the symbol labeled *2* in the diagram above?

_____

north (0°) ▮

83) Item Type: Fill in the Blank   Objective: 9   Section 25.4
Level: A: Standard   Tag: None

The release of large quantities of ions near the ground can modify the electrical properties

of small cumulus clouds to help control _____.

lightning ▮

84) Item Type: Fill in the Blank   Objective: 9   Section 25.4
Level: A: Standard   Tag: None

What do scientists try to produce by adding freezing nuclei to supercooled clouds?

_____

precipitation ▮

85) Item Type: Fill in the Blank   Objective: 9   Section 25.4
Level: A: Standard   Tag: None

Daily forecasts usually predict weather conditions for how many consecutive hours?

_____

48 hours ▮

86) Item Type: Essay   Objective: 1   Section 25.1
     Level: B: Enriched   Tag: None

Explain how convection cells form in the Northern and Southern Hemispheres.

_____

_____

_____

_____

_____

Cold, dense polar air creates high-pressure zones at the poles. Warmer, less-dense air creates low-pressure zones at the equator. As cold polar air sinks, it pushes toward the equator along the earth's surface. Warm equatorial air rises and moves toward the poles, where it cools and sinks. These conditions create three convection cells in both the Northern and Southern Hemispheres.

87) Item Type: Essay   Objective: 5   Section 25.2
     Level: A: Standard   Tag: None

How do tornadoes develop?

_____

_____

_____

_____

_____

A thunderstorm meets high-altitude, horizontal winds. These winds cause the rising air to rotate. A narrow, funnel-shaped, rapidly spinning extension may develop below one of the storm clouds. This extension reaches downward and may or may not touch the ground.

1)   Tag Name: Diagram 2571

2)   Tag Name: Diagram 2572

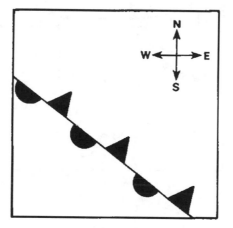

3)   Tag Name: Diagram 2573

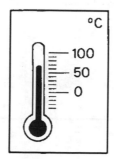

4) Tag Name: Diagram 2574

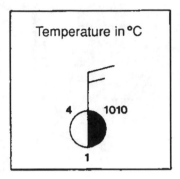

5) Tag Name: Diagram 2575

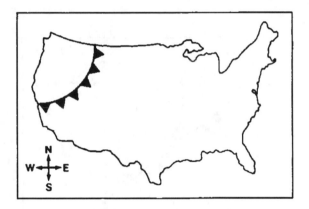

6) Tag Name: Diagram 2576

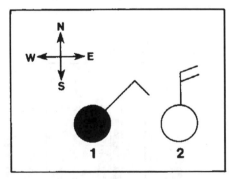

Chapter 26

1) Item Type: True-False   Objective: 1   Section 26.1
   Level: B: Enriched   Tag: None

   _____ Wave cyclones frequently develop in the subtropical highs.

   F

2) Item Type: True-False   Objective: 1   Section 26.1
   Level: A: Standard   Tag: None

   _____ The average weather condition that occurs in an area over a long period of time is
   called climate.

   T

3) Item Type: True-False   Objective: 1   Section 26.1
   Level: A: Standard   Tag: None

   _____ The equatorial region is characterized by dry seasons and little temperature
   fluctuation.

   T

4) Item Type: True-False   Objective: 1   Section 26.1
   Level: A: Standard   Tag: None

   _____ At the equator, both days and nights are about 12 hours long in all seasons.

   T

5) Item Type: True-False   Objective: 1   Section 26.1
   Level: A: Standard   Tag: None

   _____ Latitude determines both the amount of solar energy an area receives and the
   prevailing wind patterns of the region.

   T

6) Item Type: True-False   Objective: 1   Section 26.1
   Level: A: Standard   Tag: None

   _____ The amount of solar radiation received by a region is determined primarily by the
   latitude of the region.

   T

7) Item Type: True-False   Objective: 2   Section 26.1
   Level: B: Enriched   Tag: None

   _____ El Niño causes a cooling of the waters off the west coast of South America.

   F

8) Item Type: True-False   Objective: 2   Section 26.1
    Level: A: Standard   Tag: None

    _____   Waves and other movements prevent the ocean's surface from cooling rapidly.

    F

9) Item Type: True-False   Objective: 2   Section 26.1
    Level: A: Standard   Tag: None

    _____   In certain regions, heat differences between the land and oceans cause winds to
            shift seasonally.

    T

10) Item Type: True-False   Objective: 2   Section 26.1
    Level: A: Standard   Tag: None

    _____   More heat is required to raise the temperature of 1 g of rock by 1°C than to raise
            the temperature of 1 g water by 1°C.

    F

11) Item Type: True-False   Objective: 2   Section 26.1
    Level: A: Standard   Tag: None

    _____   On lakes, waves and currents prevent surface temperatures from changing rapidly.

    F

12) Item Type: True-False   Objective: 2   Section 26.1
    Level: A: Standard   Tag: None

    _____   Water releases heat faster than land does.

    F

13) Item Type: True-False   Objective: 2   Section 26.1
    Level: A: Standard   Tag: None

    _____   The specific heat of land is higher than that of water.

    F

14) Item Type: True-False   Objective: 2   Section 26.1
    Level: A: Standard   Tag: None

    _____   The Gulf Stream has a moderating effect on the climate of the eastern coast of
            North America.

    F

15) Item Type: True-False   Objective: 2   Section 26.1
   Level: A: Standard   Tag: None

_____ Ocean currents have a stronger effect on land air masses when winds consistently blow away from the shore.

F

16) Item Type: True-False   Objective: 3   Section 26.1
   Level: A: Standard   Tag: None

_____ When air rises over a mountain, it tends to lose moisture through precipitation.

T

17) Item Type: True-False   Objective: 3   Section 26.1
   Level: A: Standard   Tag: None

_____ The bora is a warm wind that comes down from the mountains.

F

18) Item Type: True-False   Objective: 3   Section 26.1
   Level: A: Standard   Tag: None

_____ Winds that blow down mountain slopes may be either warm or cold.

T

19) Item Type: True-False   Objective: 4   Section 26.2
   Level: A: Standard   Tag: None

_____ Savannas are located in tropical areas.

T

20) Item Type: True-False   Objective: 4   Section 26.2
   Level: A: Standard   Tag: None

_____ Tropical climates are located near the equator.

T

21) Item Type: True-False   Objective: 4   Section 26.2
   Level: B: Enriched   Tag: None

_____ Tropical deserts coincide with belts of high-pressure areas.

T

22) Item Type: True-False   Objective: 4   Section 26.2
     Level: A: Standard   Tag: None

_____ Mass destruction of the earth's rain forests may be causing a rise in global temperatures.

T

23) Item Type: True-False   Objective: 5   Section 26.2
     Level: A: Standard   Tag: None

_____ Areas with a tundra climate receive less than 25 cm of precipitation each year.

T

24) Item Type: True-False   Objective: 5   Section 26.2
     Level: A: Standard   Tag: None

_____ The tundra climate has a larger yearly temperature range than the subarctic climate does.

F

25) Item Type: True-False   Objective: 5   Section 26.2
     Level: A: Standard   Tag: None

_____ Subarctic climates have small yearly temperature ranges.

F

26) Item Type: True-False   Objective: 5   Section 26.2
     Level: A: Standard   Tag: None

_____ Mosses, lichens, and small flowering plants are common in areas of tundra climate.

T

27) Item Type: True-False   Objective: 6   Section 26.2
     Level: B: Enriched   Tag: None

_____ The Great Plains of the United States are located in a steppe climate.

T

28) Item Type: True-False   Objective: 6   Section 26.2
     Level: A: Standard   Tag: None

_____ Most of the rainfall in a Mediterranean climate occurs in the summer.

F

29) Item Type: True-False Objective: 6 Section 26.2
  Level: A: Standard   Tag: None

_____ Most of the rainfall in middle-latitude areas results from wave cyclones.

T

30) Item Type: True-False Objective: 6 Section 26.2
  Level: A: Standard   Tag: None

_____ Middle-latitude deserts have seasonal temperature ranges similar to those of tropical deserts.

F

31) Item Type: True-False Objective: 7 Section 26.2
  Level: A: Standard   Tag: None

_____ Cities are warmed at night by radiation from buildings that have been heated by solar energy during the day.

T

32) Item Type: True-False Objective: 7 Section 26.2
  Level: A: Standard   Tag: None

_____ Radiational cooling at night is more prevalent in cities than in nearby rural areas.

F

33) Item Type: True-False Objective: 7 Section 26.2
  Level: A: Standard   Tag: None

_____ Lakes may have a moderating effect on local climates.

T

34) Item Type: True-False Objective: 7 Section 26.2
  Level: A: Standard   Tag: None

_____ Forests affect local climates by increasing the humidity.

T

35) Item Type: True-False Objective: 7 Section 26.2
  Level: A: Standard   Tag: None

_____ Cities are generally cooler than nearby rural areas.

F

36) Item Type: Multiple Choice   Objective: 1   Section 26.1
    Level: B: Enriched   Tag: None

Wave cyclones form in the region of

a) subpolar lows.   b) subtropical highs.

c) polar highs.   d) tropical lows.

a

37) Item Type: Multiple Choice   Objective: 1   Section 26.1
    Level: A: Standard   Tag: None

About how many hours of sunlight are received each day along the equator?

a) 8   b) 12   c) 16   d) 18

b

38) Item Type: Multiple Choice   Objective: 1   Section 26.1
    Level: A: Standard   Tag: None

Which region has a very large annual temperature range and a very small daily temperature range?

a) equatorial   b) middle-north latitudes

c) polar   d) middle-south latitudes

c

39) Item Type: Multiple Choice   Objective: 1   Section 26.1
    Level: A: Standard   Tag: None

A region that receives 8 hours of sunlight in December and 16 hours of light in June is located at about

a) the equator.   b) 45° north latitude.

c) the North Pole.   d) 45° south latitude.

b

40) Item Type: Multiple Choice   Objective: 1   Section 26.1
    Level: A: Standard   Tag: None

When average daily temperatures are plotted on a map, the boundaries of the zones formed coincide roughly with

a) coastlines.   b) longitudes.   c) mountain ranges. d) latitudes.

d

41) Item Type: Multiple Choice   Objective: 1   Section 26.1
Level: A: Standard   Tag: None

At the equator, the sun's rays strike the earth's surface at an angle of about

a) 10°.   b) 25°.   c) 45°.   d) 90°.

d

42) Item Type: Multiple Choice   Objective: 2   Section 26.1
Level: A: Standard   Tag: None

Specific heat is the amount of heat needed to raise the temperature of 1 g of a substance by

a) 1°C.   b) 5°C.   c) 10°C.   d) 50°C.

a

43) Item Type: Multiple Choice   Objective: 2   Section 26.1
Level: A: Standard   Tag: Diagram 2671

What is the name for the wind in the diagram?

a) bora   b) foehn   c) monsoon   d) mistral

c

44) Item Type: Multiple Choice   Objective: 3   Section 26.1
Level: A: Standard   Tag: None

Which wind flows down the eastern slopes of the Rocky Mountains?

a) mistral   b) bora   c) chinook   d) foehn

c

45) Item Type: Multiple Choice   Objective: 3   Section 26.1
Level: B: Enriched   Tag: None

Which of the following is a warm, dry wind that flows down the Alps?

a) mistral   b) bora   c) foehn   d) monsoon

c

46) Item Type: Multiple Choice   Objective: 3   Section 26.1
Level: B: Enriched   Tag: None

Which of these statements best describes what happens to an air mass moving over a mountain range?

a) Its temperature increases moving upslope.

b) It loses moisture moving upslope.

c) Its temperature decreases moving downslope.

d) It gains moisture moving downslope.

47) Item Type: Multiple Choice   Objective: 3   Section 26.1
Level: A: Standard   Tag: None

Which of these conditions describes a moving air mass that encounters a mountain range?

a) density decreases   b) pressure increases

c) moisture content decreases   d) temperature increases

48) Item Type: Multiple Choice   Objective: 3   Section 26.1
Level: A: Standard   Tag: None

Which of the following blows down from the Alps to the Mediterranean Sea?

a) chinook   b) mistral   c) foehn   d) bora

49) Item Type: Multiple Choice   Objective: 3   Section 26.1
Level: A: Standard   Tag: None

The warm winds called chinooks are found on the

a) eastern slopes of the Rocky Mountains.

b) eastern slopes of the Sierras.

c) southern slopes of the Himalayas.

d) southern slopes of the Alps.

Chapter 26

50) Item Type: Multiple Choice   Objective: 4   Section 26.2
    Level: B: Enriched   Tag: None

Where do maritime tropical air masses develop?

a) in a marine west-coast climate   b) close to the poles

c) in a humid subtropical climate   d) close to the equator

d

51) Item Type: Multiple Choice   Objective: 4   Section 26.2
    Level: A: Standard   Tag: None

Which of the following is typical of the savanna climate?

a) forests of hardwood and softwood trees

b) open areas of coarse grasses

c) forests of cone-bearing trees

d) large expanses of rocky land

b

52) Item Type: Multiple Choice   Objective: 4   Section 26.2
    Level: A: Standard   Tag: None

In which type of climate are the jungles of southern Asia located?

a) tropical savanna   b) rain forest

c) Mediterranean   d) humid subtropical

b

53) Item Type: Multiple Choice   Objective: 4   Section 26.2
    Level: B: Enriched   Tag: None

Which of these climates has wet summers and dry winters?

a) middle-latitude steppe   b) tropical savanna

c) Mediterranean   d) tundra

b

54) Item Type: Multiple Choice   Objective: 4   Section 26.2
    Level: A: Standard   Tag: None

Which type of climate does much of the interior of Australia have?

a) tropical desert   b) tropical savanna

c) middle-latitude desert   d) Mediterranean

a

55) Item Type: Multiple Choice   Objective: 5   Section 26.2
    Level: B: Enriched   Tag: None

The tundra climate has a lower yearly temperature range than the subarctic climate because the tundra

a) contains more vegetation.

b) is closer to the equator.

c) contains larger expanses of land.

d) is closer to the ocean.

d

56) Item Type: Multiple Choice   Objective: 5   Section 26.2
    Level: A: Standard   Tag: None

In which climate would the largest yearly temperature range usually be found?

a) humid continental

b) subarctic

c) middle-latitude steppe

d) tundra

b

57) Item Type: Multiple Choice   Objective: 5   Section 26.2
    Level: A: Standard   Tag: None

Areas poleward of subarctic climates are characterized by their

a) large daily temperature ranges.

b) small annual temperature ranges.

c) small amounts of precipitation.

d) vast areas of sand.

c

58) Item Type: Multiple Choice   Objective: 5   Section 26.2
    Level: A: Standard   Tag: None

The climate in which the vegetation consists of lichen, mosses, and small flowering plants is called

a) tundra.

b) middle-latitude steppe.

c) tropical savanna.

d) tropical desert.

a

59) Item Type: Multiple Choice   Objective: 5   Section 26.2
    Level: A: Standard   Tag: None

Which type of vegetation is characteristic of subarctic climates?

a) hardwood trees

b) drought-resistant shrubs

c) cone-bearing trees

d) dense grasses

c

60) Item Type: Multiple Choice   Objective: 6   Section 26.2
Level: A: Standard   Tag: None

Which middle-latitude climate receives the greatest average amount of yearly rainfall?

a) humid continental   b) marine west-coast

c) humid subtropical   d) Mediterranean

c

61) Item Type: Multiple Choice   Objective: 6   Section 26.2
Level: A: Standard   Tag: None

In which climate would dense forests of cone-bearing trees most likely grow?

a) tundra   b) marine west-coast

c) tropical savanna   d) middle-latitude desert

b

62) Item Type: Multiple Choice   Objective: 6   Section 26.2
Level: A: Standard   Tag: None

Which climate is characterized by mild winters and warm, humid summers?

a) humid continental   b) humid subtropical

c) Mediterranean   d) tropical rain forest

b

63) Item Type: Multiple Choice   Objective: 6   Section 26.2
Level: A: Standard   Tag: Diagram 2673

Which region in the map above has a middle-latitude steppe climate?

a) *1*   b) *2*   c) *3*   d) *5*

c

64) Item Type: Multiple Choice   Objective: 6   Section 26.2
Level: B: Enriched   Tag: Diagram 2673

Which region in the map above has the greatest average yearly rainfall?

a) *1*   b) *2*   c) *3*   d) *4*

a

65) Item Type: Multiple Choice   Objective: 6   Section 26.2
Level: A: Standard   Tag: None

A Mediterranean climate is characterized by winters that are

a) mild and wet.   b) cold and wet.   c) mild and dry.   d) cold and dry.

a

66) Item Type: Multiple Choice   Objective: 6   Section 26.2
Level: A: Standard   Tag: Diagram 2674

Which of the cities on the map above has a humid continental climate?

a) *1*   b) *3*   c) *5*   d) *6*

c

67) Item Type: Multiple Choice   Objective: 6   Section 26.2
Level: A: Standard   Tag: Diagram 2674

Which of the cities on the map has a humid subtropical climate?

a) *2*   b) *4*   c) *5*   d) *6*

d

68) Item Type: Multiple Choice   Objective: 7   Section 26.2
Level: A: Standard   Tag: None

Forests can modify local climate by reducing

a) humidity.   b) transpiration.   c) wind speed.   d) precipitation.

c

69) Item Type: Multiple Choice   Objective: 7   Section 26.2
Level: A: Standard   Tag: None

Rural areas usually have a lower average temperature than cities partly because of

a) large numbers of condensation nuclei.   b) more transpiration by vegetation.

c) high atmospheric carbon dioxide levels.   d) greater available water vapor.

b

70) Item Type: Multiple Choice   Objective: 7   Section 26.2
     Level: B: Enriched   Tag: None

Cities usually receive more rainfall than nearby rural areas as a result of having

a)   less radiational cooling at night.

b)   less vegetation to absorb solar radiation.

c)   increased energy usage in cities.

d)   increased numbers of condensation nuclei.

d

71) Item Type: Fill in the Blank   Objective: 1   Section 26.1
     Level: A: Standard   Tag: None

The sun's rays strike the earth's surface at almost a 90° angle at the

_____.

equator

72) Item Type: Fill in the Blank   Objective: 1   Section 26.1
     Level: A: Standard   Tag: None

The two weather factors most often used to describe climate are precipitation and average

_____.

temperature

73) Item Type: Fill in the Blank   Objective: 2   Section 26.1
     Level: B: Enriched   Tag: None

What is the name of the warm Pacific current that can affect up to 70 percent of the earth's

weather? _____

El Niño

74) Item Type: Fill in the Blank   Objective: 3   Section 26.1
     Level: A: Standard   Tag: None

What is the name of the cold, stormy winds that flow from the Alps toward the

Mediterranean Sea? _____

mistral

75) Item Type: Fill in the Blank   Objective: 3   Section 26.1
    Level: B: Enriched   Tag: None

What is the decrease in average temperature for every increase in altitude of 100 m?

_____

| 0.7°C |

76) Item Type: Fill in the Blank   Objective: 3   Section 26.1
    Level: A: Standard   Tag: None

When a moving air mass encounters a mountain range, it rises, expands, and

_____.

| cools |

77) Item Type: Fill in the Blank   Objective: 4   Section 26.2
    Level: B: Enriched   Tag: None

In which type of climate do approximately one-third of all species on the earth live?

_____

| rain forest |

78) Item Type: Fill in the Blank   Objective: 4   Section 26.2
    Level: A: Standard   Tag: None

Which tropical climate has annual rainfall of less than 25 cm? _____

| tropical desert |

79) Item Type: Fill in the Blank   Objective: 4   Section 26.2
    Level: A: Standard   Tag: None

Between the tropical and polar climate zones is the zone of temperate climates called

_____.

| middle-latitude climates |

80) Item Type: Fill in the Blank   Objective: 4   Section 26.2
    Level: A: Standard   Tag: None

Each principal climate zone is divided into several different types of climates because of

differences in the amount of _____.

| precipitation |

81) Item Type: Fill in the Blank   Objective: 4   Section 26.2
    Level: B: Enriched   Tag: None

What type of climate is characterized by a very small temperature range and more than

250 cm of rainfall per year? _____

rain forest ▮

82) Item Type: Fill in the Blank   Objective: 4   Section 26.2
    Level: A: Standard   Tag: None

The region roughly located between the Tropic of Cancer and the Tropic of Capricorn is

the type of climate zone called _____.

tropical ▮

83) Item Type: Fill in the Blank   Objective: 5   Section 26.2
    Level: A: Standard   Tag: None

What type of climate does the northern part of Alaska have? _____

subarctic ▮

84) Item Type: Fill in the Blank   Objective: 5   Section 26.2
    Level: B: Enriched   Tag: None

In what type of climate was the largest yearly temperature range recorded?

_____

subarctic ▮

85) Item Type: Fill in the Blank   Objective: 5   Section 26.2
    Level: A: Standard   Tag: None

Which polar climate contains the polar deserts? _____

tundra ▮

86) Item Type: Fill in the Blank   Objective: 6   Section 26.2
    Level: A: Standard   Tag: None

Which type of desert is characterized by cold winters and hot summers?

_____

middle-latitude ▮

87) Item Type: Fill in the Blank   Objective: 7   Section 26.2
Level: B: Enriched   Tag: None

What is the name of the process by which plants give off water vapor?

_____

| transpiration |

88) Item Type: Fill in the Blank   Objective: 7   Section 26.2
Level: B: Enriched   Tag: None

The most important factor affecting local weather conditions is the area's

_____.

| elevation |

89) Item Type: Essay   Objective: 2   Section 26.1
Level: B: Enriched   Tag: None

Explain how the wind patterns of southern and eastern Asia change with the seasons.

_____

_____

_____

_____

_____

| In the winter, the land loses heat more quickly than the water. A high-pressure system develops over the land and the cool air flows away from the land toward the ocean. Thus the wind moves seaward. In summer, low pressure develops over the land. Warm air over the land rises and is replaced by cool air flowing in from over the ocean. Thus the wind moves landward. |

90) Item Type: Essay   Objective: 3   Section 26.1
    Level: B: Enriched   Tag: None

Death Valley lies east of the Sierra Nevada of California and has a tropical desert climate, while about 300 km to the west in Los Angeles, the climate is Mediterranean. Explain the difference in climate.

_____

_____

_____

_____

| The Sierra Nevada influence the temperature and moisture content of the passing air masses. As the mild air rises from the Los Angeles area up the west side of the mountains, it cools, and through precipitation it loses most of its moisture. As the air descends into Death Valley on the east side, it is warmed. Therefore, the air moving into Death Valley is very warm and dry. |

91) Item Type: Essay   Objective: 7   Section 26.2
    Level: B: Enriched   Tag: None

How does a large lake, such as Lake Michigan, affect climate?

_____

_____

_____

_____

_____

| Large lakes can cause an increase in precipitation on the downwind shore. The eastern shore of Lake Michigan generally has more moderate temperatures, more cloudiness, and higher precipitation than the western shore does. |

1)   Tag Name: Diagram 2671

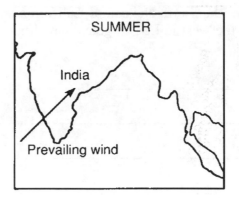

2)   Tag Name: Diagram 2673

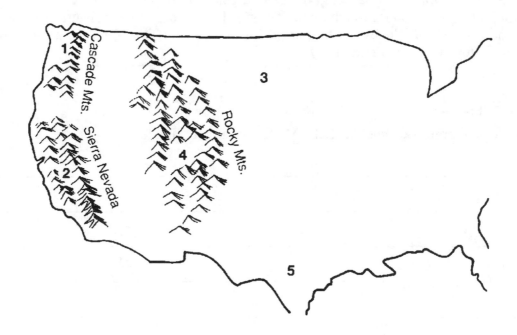

3) Tag Name: Diagram 2674

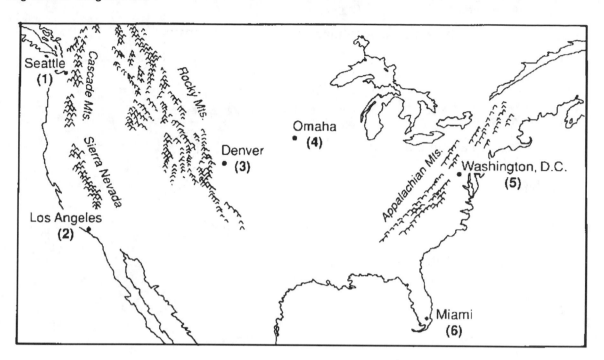

1) Item Type: True-False   Objective: 1   Section 27.1
   Level: A: Standard   Tag: None

_____ All stars have the same composition.

F

2) Item Type: True-False   Objective: 1   Section 27.1
   Level: A: Standard   Tag: None

_____ Stars vary in color.

T

3) Item Type: True-False   Objective: 1   Section 27.1
   Level: A: Standard   Tag: None

_____ The sun is the star nearest the earth.

T

4) Item Type: True-False   Objective: 2   Section 27.1
   Level: A: Standard   Tag: None

_____ A star's actual motion can be seen with the naked eye.

F

5) Item Type: True-False   Objective: 2   Section 27.1
   Level: A: Standard   Tag: None

_____ From the earth, stars appear to rise in the west and set in the east.

F

6) Item Type: True-False   Objective: 3   Section 27.1
   Level: A: Standard   Tag: None

_____ The Doppler effect is commonly used to measure the absolute magnitudes of stars far away from the earth.

F

7) Item Type: True-False   Objective: 3   Section 27.1
   Level: A: Standard   Tag: None

_____ A light-year is the distance that light travels in one year.

T

8) Item Type: True-False   Objective: 4   Section 27.1
    Level: A: Standard   Tag: None

_____ All cool stars are dim as seen from the earth.

F

9) Item Type: True-False   Objective: 4   Section 27.1
    Level: A: Standard   Tag: None

_____ Scientists can compare a star's apparent magnitude with its absolute magnitude to determine its distance from the earth.

T

10) Item Type: True-False   Objective: 5   Section 27.2
    Level: A: Standard   Tag: None

_____ A cloud of dust and gas from which a star forms is called a nebula.

T

11) Item Type: True-False   Objective: 5   Section 27.2
    Level: A: Standard   Tag: None

_____ Nebulae never produce more than one star.

F

12) Item Type: True-False   Objective: 5   Section 27.2
    Level: A: Standard   Tag: None

_____ The temperature inside a protostar increases because of particle collisions.

T

13) Item Type: True-False   Objective: 6   Section 27.2
    Level: B: Enriched   Tag: None

_____ A main-sequence star generates energy by fusing helium into hydrogen.

F

14) Item Type: True-False   Objective: 6   Section 27.2
    Level: A: Standard   Tag: None

_____ A main-sequence star maintains a stable size as long as there is enough hydrogen to fuse into helium.

T

15) Item Type: True-False   Objective: 7   Section 27.2
    Level: A: Standard   Tag: None

_____ A dying star can shed its outer atmosphere as a planetary nebula.

T

16) Item Type: True-False   Objective: 7   Section 27.2
    Level: B: Enriched   Tag: None

_____ A star with a mass 10 times greater than that of the sun can produce a supernova.

F

17) Item Type: True-False   Objective: 7   Section 27.2
    Level: B: Enriched   Tag: None

_____ White dwarfs can produce more than one nova.

T

18) Item Type: True-False   Objective: 7   Section 27.2
    Level: A: Standard   Tag: None

_____ As white dwarfs cool, they often explode as supernovas.

F

19) Item Type: True-False   Objective: 8   Section 27.3
    Level: A: Standard   Tag: None

_____ Most constellations look like the figures for which they were named.

F

20) Item Type: True-False   Objective: 8   Section 27.3
    Level: A: Standard   Tag: None

_____ A constellation is a type of galaxy.

F

21) Item Type: True-False   Objective: 8   Section 27.3
    Level: A: Standard   Tag: None

_____ Stars in a constellation are generally located close to one another.

F

22) Item Type: True-False   Objective: 8   Section 27.3
     Level: A: Standard   Tag: None

_____ Most visible stars are single stars.

F

23) Item Type: True-False   Objective: 8   Section 27.3
     Level: A: Standard   Tag: None

_____ The brightest star in a constellation is usually labeled with the Greek beta symbol.

F

24) Item Type: True-False   Objective: 9   Section 27.3
     Level: A: Standard   Tag: None

_____ The Milky Way Galaxy does not rotate.

F

25) Item Type: True-False   Objective: 9   Section 27.3
     Level: B: Enriched   Tag: None

_____ A globular star cluster has more stars than an open star cluster.

T

26) Item Type: True-False   Objective: 9   Section 27.3
     Level: B: Enriched   Tag: None

_____ Dark nebulae are dark because they absorb the light of more-distant stars behind them.

T

27) Item Type: True-False   Objective: 9   Section 27.3
     Level: A: Standard   Tag: None

_____ A globular cluster is a type of galaxy.

F

28) Item Type: True-False   Objective: 9   Section 27.3
     Level: A: Standard   Tag: None

_____ Elliptical galaxies contain very little dust and gas.

T

29) Item Type: True-False   Objective: 9   Section 27.3
    Level: A: Standard   Tag: None

_____ The Milky Way Galaxy is an example of an irregular galaxy.

F

30) Item Type: True-False   Objective: 9   Section 27.3
    Level: A: Standard   Tag: None

_____ Many galaxies contain nebulae as well as stars.

T

31) Item Type: True-False   Objective: 9   Section 27.3
    Level: A: Standard   Tag: None

_____ Stars are distributed unevenly in an irregular galaxy.

T

32) Item Type: True-False   Objective: 9   Section 27.3
    Level: B: Enriched   Tag: None

_____ Spiral galaxies are smaller and fainter than other types of galaxies.

F

33) Item Type: True-False   Objective: 9   Section 27.3
    Level: A: Standard   Tag: None

_____ The galaxy that includes the earth is called the Large Magellanic Cloud.

F

34) Item Type: True-False   Objective: 10   Section 27.3
    Level: B: Enriched   Tag: None

_____ According to the big bang theory, pulsars were among the first objects formed after the gigantic explosion that formed the universe.

F

35) Item Type: True-False   Objective: 10   Section 27.3
    Level: A: Standard   Tag: None

_____ Quasars have large red shifts.

T

36) Item Type: Multiple Choice   Objective: 1   Section 27.1
Level: B: Enriched   Tag: Diagram 2771

The color of a star in category O in the table above would most likely be

a) blue.   b) blue-white.   c) yellow.   d) red.

a

37) Item Type: Multiple Choice   Objective: 1   Section 27.1
Level: B: Enriched   Tag: Diagram 2771

A star in category G in the table above would most likely be

a) blue.   b) blue-white.   c) yellow.   d) red.

c

38) Item Type: Multiple Choice   Objective: 1   Section 27.1
Level: A: Standard   Tag: None

A device for separating light into different colors is known as a

a) bright-line emission.   b) spectroscope.

c) Cepheid variable.   d) H-R diagram.

b

39) Item Type: Multiple Choice   Objective: 1   Section 27.1
Level: A: Standard   Tag: None

Starlight passing through a device for separating light produces a display of colors and lines called a(n)

a) barred spiral.   b) shift.   c) open cluster.   d) spectrum.

d

40) Item Type: Multiple Choice   Objective: 2   Section 27.1
Level: A: Standard   Tag: None

An apparent shift in the wavelength of light emitted by a light source moving toward or away from an observer is known as

a) spectrum analysis.   b) parallax.

c) the Doppler effect.   d) a Cepheid variable.

c

41) Item Type: Multiple Choice   Objective: 2   Section 27.1
Level: B: Enriched   Tag: None

A star moving toward the earth would display a

a) red shift.   b) green shift.   c) yellow shift.   d) blue shift.

d

42) Item Type: Multiple Choice   Objective: 2   Section 27.1
Level: A: Standard   Tag: None

A star that has a blue-shifted spectrum is most likely moving

a) toward the earth.   b) away from the sun.

c) around the Milky Way Galaxy.   d) toward a black hole.

a

43) Item Type: Multiple Choice   Objective: 3   Section 27.1
Level: A: Standard   Tag: None

The apparent change in the position of an object resulting from a change in the angle or in the position from which it is viewed is called

a) stellar magnitude.   b) spectroscopy.

c) the Doppler effect.   d) parallax.

d

44) Item Type: Multiple Choice   Objective: 3   Section 27.1
Level: B: Enriched   Tag: None

How far is the sun from the earth?

a) 4.3 light-minutes   b) 8 light-minutes   c) 9 light-years   d) 680 light-years

b

45) Item Type: Multiple Choice   Objective: 3   Section 27.1
Level: A: Standard   Tag: None

A star that experiences regular changes in brightness over time is called a

a) Cepheid variable.   b) supernova.   c) binary star.   d) quasar.

a

46) Item Type: Multiple Choice   Objective: 4   Section 27.1
Level: A: Standard   Tag: Diagram 2772

In which region of the diagram does the sun lie?

a) **A**   b) **B**   c) **C**   d) **D**

b

47) Item Type: Multiple Choice   Objective: 4   Section 27.1
Level: A: Standard   Tag: Diagram 2772

What type of star is generally located in region **A** of this Hertzsprung-Russell diagram?

a) main-sequence   b) protostar   c) giant   d) white dwarf

d

48) Item Type: Multiple Choice   Objective: 4   Section 27.1
Level: B: Enriched   Tag: None

A star with which of the following apparent magnitudes would appear brightest?

a) 10   b) 5   c) 1   d) −5

d

49) Item Type: Multiple Choice   Objective: 4   Section 27.1
Level: A: Standard   Tag: None

The apparent magnitude of a star is a measure of its

a) surface temperature and composition.   b) distance from the earth.

c) brightness as it appears from the earth.   d) position on the H-R diagram.

c

50) Item Type: Multiple Choice   Objective: 4   Section 27.1
Level: B: Enriched   Tag: None

Which of the following apparent magnitudes is dimmest?

a) 4   b) 2   c) 0   d) −2

a

51) Item Type: Multiple Choice   Objective: 4   Section 27.1
Level: A: Standard   Tag: None

What stage of stellar evolution is the sun experiencing at this time?

a) giant   b) white dwarf   c) protostar   d) main-sequence

d

52) Item Type: Multiple Choice   Objective: 4   Section 27.1
     Level: A: Standard   Tag: None

A plot of surface temperatures of stars against their absolute magnitude is called

a) an H-R diagram.   b) a Cepheid variable.

c) a constellation map.   d) a spectrum.

a

53) Item Type: Multiple Choice   Objective: 5   Section 27.2
     Level: A: Standard   Tag: None

Which of the following stages is the earliest in the development of a star?

a) neutron star   b) protostar   c) dark nebula   d) giant

c

54) Item Type: Multiple Choice   Objective: 5   Section 27.2
     Level: B: Enriched   Tag: None

Nuclear fusion begins when temperatures within a protostar reach over

a) 30,000°C.   b) 300,000°C.   c) 10,000,000°C.   d) 150,000,000°C.

c

55) Item Type: Multiple Choice   Objective: 5   Section 27.2
     Level: A: Standard   Tag: None

The process in which smaller atomic nuclei combine into larger atomic nuclei is known as

a) gravitational attraction.   b) parallax.

c) nuclear fusion.   d) the Doppler effect.

c

56) Item Type: Multiple Choice   Objective: 6   Section 27.2
     Level: A: Standard   Tag: None

Which type of star maintains a stable size because the energy from fusion balances the force of gravity?

a) main-sequence   b) neutron star   c) pulsar   d) supergiant

a

57) Item Type: Multiple Choice   Objective: 6   Section 27.2
Level: A: Standard   Tag: None

Which stage of stellar evolution is characterized by the fusion of hydrogen atoms into helium atoms?

a) main-sequence   b) neutron star   c) nova   d) white dwarf

a

58) Item Type: Multiple Choice   Objective: 6   Section 27.2
Level: A: Standard   Tag: None

Which stage in stellar evolution is the longest?

a) supergiant   b) neutron star   c) black hole   d) main-sequence

d

59) Item Type: Multiple Choice   Objective: 6   Section 27.2
Level: A: Standard   Tag: None

A main-sequence star maintains a stable size as long as it has an ample supply of hydrogen to fuse into

a) helium.   b) magnesium.   c) protactinium.   d) iron.

a

60) Item Type: Multiple Choice   Objective: 7   Section 27.2
Level: A: Standard   Tag: None

A pulsar is a type of

a) protostar.   b) neutron star.   c) white dwarf.   d) supergiant.

b

61) Item Type: Multiple Choice   Objective: 7   Section 27.2
Level: B: Enriched   Tag: None

The end of which stage of stellar evolution is marked by the end of helium fusion?

a) protostar   b) neutron star   c) black dwarf   d) giant

d

62) Item Type: Multiple Choice   Objective: 7   Section 27.2
Level: A: Standard   Tag: None

Which of the following stages in a star's evolution follows fusion of heavy elements into iron?

a) white dwarf   b) supernova   c) black hole   d) main-sequence

b

63) Item Type: Multiple Choice   Objective: 7   Section 27.2
      Level: A: Standard   Tag: None

In the last stage of stellar evolution following a supernova, stars too massive to form neutron stars may form a

a) black dwarf.   b) supergiant.   c) white dwarf.   d) black hole.

d

64) Item Type: Multiple Choice   Objective: 7   Section 27.2
      Level: A: Standard   Tag: None

In which stage of stellar evolution does combined hydrogen fusion and helium fusion cause a star's outer shell to expand?

a) supernova   b) giant   c) main-sequence   d) protostar

b

65) Item Type: Multiple Choice   Objective: 7   Section 27.2
      Level: A: Standard   Tag: None

Stars that emit two beams of radiation that can be detected as radio waves are called

a) nebulae.   b) protostars.   c) pulsars.   d) supergiants.

c

66) Item Type: Multiple Choice   Objective: 7   Section 27.2
      Level: B: Enriched   Tag: None

How much of its hydrogen has the sun converted to helium?

a) 5%   b) 10%   c) 25%   d) 50%

a

67) Item Type: Multiple Choice   Objective: 7   Section 27.2
      Level: A: Standard   Tag: None

When a white dwarf star no longer emits energy, it may become a

a) black dwarf.   b) nova.   c) neutron star.   d) black hole.

a

68) Item Type: Multiple Choice   Objective: 7   Section 27.2
      Level: B: Enriched   Tag: Diagram 2773

What stage in the life of a star 10 times more massive than the sun is missing in this chart?

a) supernova   b) neutron star   c) black hole   d) giant

d

69) Item Type: Multiple Choice   Objective: 8   Section 27.3
    Level: B: Enriched   Tag: None

Astronomers label the stars within constellations according to

a) composition.   b) color.   c) surface temperature.   d) apparent magnitude.

d

70) Item Type: Multiple Choice   Objective: 8   Section 27.3
    Level: A: Standard   Tag: None

Astronomers label the stars in a constellation according to their

a) color.   b) apparent magnitude.   c) size.   d) absolute magnitude.

b

71) Item Type: Multiple Choice   Objective: 9   Section 27.3
    Level: A: Standard   Tag: None

Large-scale groups of stars bound together by gravitational attraction are known as

a) galaxies.   b) multiple-star systems.   c) Cepheids.   d) clusters.

a

72) Item Type: Multiple Choice   Objective: 10   Section 27.3
    Level: B: Enriched   Tag: None

Astronomers believe that Beta Pictoris, once thought to be a distant star, is actually a developing

a) solar system.   b) black hole.   c) quasar.   d) protostar.

a

73) Item Type: Fill in the Blank   Objective: 1   Section 27.1
    Level: A: Standard   Tag: None

Spherical star clusters that are distributed around the central core of the galaxy are called

_____.

globular clusters

74) Item Type: Fill in the Blank   Objective: 1   Section 27.1
    Level: A: Standard   Tag: None

The surface temperature of a star is indicated by what physical characteristic?

_____

color

75) Item Type: Fill in the Blank   Objective: 2   Section 27.1
    Level: B: Enriched   Tag: None

What term describes the wavelengths of light emitted from a star that appear to be longer

than they actually are? _____

red shifted ■

76) Item Type: Fill in the Blank   Objective: 2   Section 27.1
    Level: A: Standard   Tag: None

A star that is always visible in the night sky is called _____.

circumpolar ■

77) Item Type: Fill in the Blank   Objective: 2   Section 27.1
    Level: A: Standard   Tag: None

From the Northern Hemisphere, circumpolar stars appear to be circling

_____.

the North Star (Polaris) ■

78) Item Type: Fill in the Blank   Objective: 2   Section 27.1
    Level: A: Standard   Tag: None

The apparent shift in the spectrum of a star as it moves toward or away from the earth is

explained by the phenomenon known as the _____.

Doppler effect ■

79) Item Type: Fill in the Blank   Objective: 2   Section 27.1
    Level: A: Standard   Tag: None

The apparent motion of stars results from the movement of _____.

the earth ■

80) Item Type: Fill in the Blank   Objective: 2   Section 27.1
    Level: A: Standard   Tag: None

To observers in North America, most stars appear to move around the star called

_____.

Polaris (the North Star) ■

81) Item Type: Fill in the Blank   Objective: 3   Section 27.1
     Level: A: Standard   Tag: None

The only direct method scientists use to measure the distance from the earth to stars is

known as _____.

| parallax   ■ |

82) Item Type: Fill in the Blank   Objective: 3   Section 27.1
     Level: A: Standard   Tag: None

The distance between a star and the earth is measured in _____.

| light-years   ■ |

83) Item Type: Fill in the Blank   Objective: 4   Section 27.1
     Level: A: Standard   Tag: None

The H-R diagram plots the surface temperature of stars against their

_____.

| absolute magnitudes   ■ |

84) Item Type: Fill in the Blank   Objective: 4   Section 27.1
     Level: A: Standard   Tag: None

The distance from the earth to a star that has identical apparent and absolute magnitudes is

_____.

| 32.6 light-years   ■ |

85) Item Type: Fill in the Blank   Objective: 4   Section 27.1
     Level: A: Standard   Tag: None

The true brightness of a star is called its _____.

| absolute magnitude   ■ |

86) Item Type: Fill in the Blank   Objective: 5   Section 27.2
     Level: A: Standard   Tag: None

The process in which small atomic nuclei are combined into larger ones is known as

_____.

| nuclear fusion   ■ |

87) Item Type: Fill in the Blank   Objective: 5   Section 27.2
     Level: A: Standard   Tag: None

Planets are produced from clouds of gas and dust called _____.

nebulae

88) Item Type: Fill in the Blank   Objective: 6   Section 27.2
     Level: A: Standard   Tag: None

The phase of stellar evolution that is characterized by the fusion of hydrogen atoms into

helium atoms is called the _____.

main-sequence phase

89) Item Type: Fill in the Blank   Objective: 7   Section 27.2
     Level: A: Standard   Tag: None

An explosion 100 times brighter than a nova is called a(n) _____.

supernova

90) Item Type: Fill in the Blank   Objective: 7   Section 27.2
     Level: A: Standard   Tag: None

Massive objects whose dense cores collapse because of contraction leave areas in space

called _____.

black holes

91) Item Type: Fill in the Blank   Objective: 8   Section 27.3
     Level: A: Standard   Tag: None

Patterns of stars that appear relatively fixed in the night sky are known as

_____.

constellations

92) Item Type: Fill in the Blank   Objective: 8   Section 27.3
     Level: A: Standard   Tag: None

Scientists continue to divide the sky into sectors using star patterns called

_____.

constellations

Chapter 27

93) Item Type: Fill in the Blank   Objective: 9   Section 27.3
Level: B: Enriched   Tag: Diagram 2774

Galaxies such as the one shown in this diagram are known as _____.

spiral (or barred spiral) galaxies

94) Item Type: Fill in the Blank   Objective: 9   Section 27.3
Level: A: Standard   Tag: None

The Milky Way Galaxy and about 17 other galaxies within 3 million light-years of the

Milky Way Galaxy are collectively known as the _____.

Local Group

95) Item Type: Fill in the Blank   Objective: 9   Section 27.3
Level: A: Standard   Tag: None

Pairs of stars that revolve around each other are called _____.

binary stars

96) Item Type: Fill in the Blank   Objective: 10   Section 27.3
Level: A: Standard   Tag: None

The most distant objects that have been observed from the earth are called

_____.

quasars

97) Item Type: Fill in the Blank   Objective: 10   Section 27.3
Level: A: Standard   Tag: None

The true brightness of a star is known as its _____.

absolute magnitude

98) Item Type: Fill in the Blank   Objective: 10   Section 27.3
Level: A: Standard   Tag: None

The big bang theory is a widely accepted theory explaining the formation of

_____.

the universe

99) Item Type: Fill in the Blank   Objective: 10   Section 27.3
    Level: A: Standard   Tag: None

The starlike objects whose light must travel for 12 billion years to reach the earth are called

_____.

| quasars |

100) Item Type: Essay   Objective: 2   Section 27.1
     Level: B: Enriched   Tag: None

A photograph of the motion of stars in the northern sky during an extended period shows
a pattern of curved trails. Explain what causes this apparent motion of the stars.

_____

_____

_____

_____

_____

The curved trails left by most stars suggest that they are moving in a circle around Polaris,
the North Star. This circular pattern actually results from the rotation of the earth on its
axis. Polaris is located almost directly above the North Pole and does not appear to move.

101) Item Type: Essay   Objective: 3   Section 27.1
     Level: B: Enriched   Tag: None

Explain how scientists use a Cepheid variable star to measure the distance to the star's
galaxy.

_____

_____

_____

_____

_____

A Cepheid variable is a type of star that brightens and fades in a regular pattern caused by
the swelling and shrinking of the star. Astronomers measure the cycle of brightness
changes and estimate the star's true brightness. Then they compare the Cepheid's apparent
brightness with its true brightness and calculate the distance to the galaxy where the
Cepheid is located.

Chapter 27

102) Item Type: Essay   Objective: 5   Section 27.2
     Level: B: Enriched   Tag: None

Explain how parallax can be used to measure the distance between the earth and a star.

_____

_____

_____

_____

_____

Parallax is used to determine the distance to a nearby star by observing it from slightly different angles. During a six-month period, a nearby star will appear to shift slightly relative to more-distant stars. The closer a nearby star is, the greater will be the amount of shift. From the amount of shift, astronomers can calculate the distance to any star within 1,000 light-years.

103) Item Type: Essay   Objective: 7   Section 27.2
     Level: B: Enriched   Tag: None

How do astronomers try to locate black holes?

_____

_____

_____

_____

_____

Astronomers theorize that a black hole can be located by observing its effect on a companion star. Just before matter from a companion star is pulled into a black hole, it gives off X rays. Astronomers try to locate the black holes by detecting these X rays from the earth.

104) Item Type: Essay   Objective: 7   Section 27.2
     Level: B: Enriched   Tag: None

The core of a star is five times more massive than the sun and is composed mostly of helium and carbon. Which stage of stellar evolution is the star most likely in? Why?

_____

_____

_____

_____

_____

> The star has most likely become a giant. Giants are 10 times bigger than the sun. In the giant stage, hydrogen is no longer a source of fuel, and the star's core contracts under the force of its own gravity. This contraction causes increased pressure in the core, causing the helium atoms there to fuse into carbon atoms.

105) Item Type: Essay   Objective: 7   Section 27.2
     Level: B: Enriched   Tag: None

Explain how a nova occurs.

_____

_____

_____

_____

_____

> Astronomers think novas occur in white dwarfs that revolve around a giant or main-sequence star. The white dwarf has a greater surface gravitational attraction because it is denser, and gases from the companion star accumulate on the white dwarf. This causes the pressure in the white dwarf to increase until it explodes, creating a nova.

106) Item Type: Essay   Objective: 10   Section 27.3
Level: A: Standard   Tag: None

Explain the big bang theory of the formation of the universe.

_____

_____

_____

_____

_____

> According to the big bang theory, all matter and energy of the universe was once concentrated in an extremely small volume. About 17 billion years ago, the big bang occurred, sending this matter and energy out in all directions. As matter and energy moved outward with the expanding universe, gravity began to affect matter, causing it to condense and form galaxies.

1)   Tag Name: Diagram 2771

CATEGORY	COLOR	SURFACE TEMPERATURE (°C)
M		less than 3,500
K	orange	3,500–5,000
G		5,000–6,000
F	yellow-white	6,000–7,500
A		7,500–10,000
B	blue-white	10,000–30,000
O		above 30,000

2)   Tag Name: Diagram 2772

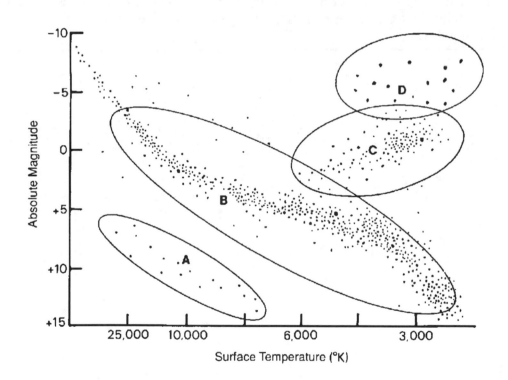

3)   Tag Name: Diagram 2773

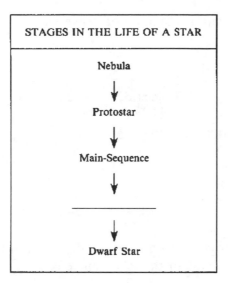

4)   Tag Name: Diagram 2774

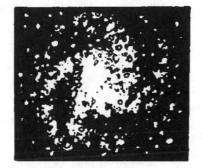

1) Item Type: True-False   Objective: 1   Section 28.1
    Level: A: Standard   Tag: None

_____ The core of the sun is made up of dense gas.

T

2) Item Type: True-False   Objective: 1   Section 28.1
    Level: A: Standard   Tag: None

_____ Nuclear fusion occurs in the sun's core.

T

3) Item Type: True-False   Objective: 1   Section 28.1
    Level: A: Standard   Tag: None

_____ During the hydrogen fusion reaction, the mass of the two original hydrogen nuclei is more than the mass of the fused nucleus.

T

4) Item Type: True-False   Objective: 1   Section 28.1
    Level: A: Standard   Tag: None

_____ The sun's energy output remains constant over time.

F

5) Item Type: True-False   Objective: 1   Section 28.1
    Level: A: Standard   Tag: None

_____ A decrease in mass occurs in the hydrogen-nuclei-to-helium-nucleus reaction in the sun.

T

6) Item Type: True-False   Objective: 1   Section 28.1
    Level: A: Standard   Tag: None

_____ Nuclear fusion occurs in the sun's corona.

F

7) Item Type: True-False   Objective: 1   Section 28.1
    Level: A: Standard   Tag: None

_____ The most common nuclear reaction occurring inside the sun is nuclear fusion.

T

8) Item Type: True-False   Objective: 2   Section 28.1
    Level: A: Standard   Tag: None

    _____   The layer of the sun immediately surrounding the core is called the convective zone.

    F

9) Item Type: True-False   Objective: 2   Section 28.1
    Level: B: Enriched   Tag: None

    _____   Cool gases are denser than hot gases.

    T

10) Item Type: True-False   Objective: 2   Section 28.1
    Level: A: Standard   Tag: None

    _____   The radiative zone is the hottest region of the sun.

    F

11) Item Type: True-False   Objective: 3   Section 28.1
    Level: A: Standard   Tag: None

    _____   The corona is part of the sun's magnetosphere.

    F

12) Item Type: True-False   Objective: 3   Section 28.1
    Level: A: Standard   Tag: None

    _____   The photosphere is considered to be the surface of the sun.

    T

13) Item Type: True-False   Objective: 3   Section 28.1
    Level: A: Standard   Tag: None

    _____   The chromosphere and corona are usually not visible from earth.

    T

14) Item Type: True-False   Objective: 4   Section 28.2
    Level: A: Standard   Tag: None

    _____   The number of sunspots increases at the beginning of a sunspot cycle.

    T

15) Item Type: True-False   Objective: 4   Section 28.2
      Level: A: Standard   Tag: None

_____ The sunspot cycle lasts an average of 20 years.

F

16) Item Type: True-False   Objective: 4   Section 28.2
      Level: B: Enriched   Tag: None

_____ The movement of sunspots was one of the first indications to astronomers that the sun rotates on its axis.

T

17) Item Type: True-False   Objective: 4   Section 28.2
      Level: A: Standard   Tag: None

_____ Sunspots are located in the sun's chromosphere.

F

18) Item Type: True-False   Objective: 5   Section 28.2
      Level: A: Standard   Tag: None

_____ Solar flares usually occur near sunspots.

T

19) Item Type: True-False   Objective: 5   Section 28.2
      Level: A: Standard   Tag: None

_____ Prominences form arches from one sunspot area to another.

T

20) Item Type: True-False   Objective: 5   Section 28.2
      Level: A: Standard   Tag: None

_____ Prominences are the most violent of all solar disturbances.

F

21) Item Type: True-False   Objective: 5   Section 28.2
      Level: A: Standard   Tag: None

_____ Magnetic storms can disrupt radio communications on the earth.

T

22) Item Type: True-False   Objective: 5   Section 28.2
    Level: A: Standard   Tag: None

_____ Prominences follow curved lines of magnetic force.

T

23) Item Type: True-False   Objective: 6   Section 28.2
    Level: A: Standard   Tag: None

_____ Bands of colored light that result from a magnetic storm are called solar flares.

F

24) Item Type: True-False   Objective: 6   Section 28.2
    Level: A: Standard   Tag: None

_____ Auroras are usually seen close to the earth's magnetic poles.

T

25) Item Type: True-False   Objective: 6   Section 28.2
    Level: A: Standard   Tag: None

_____ Auroras usually occur near sunspots.

F

26) Item Type: True-False   Objective: 6   Section 28.2
    Level: A: Standard   Tag: None

_____ The northern and southern lights are effects of magnetic storms.

T

27) Item Type: True-False   Objective: 6   Section 28.2
    Level: A: Standard   Tag: None

_____ The northern and southern lights are solar flares.

F

28) Item Type: True-False   Objective: 7   Section 28.3
    Level: B: Enriched   Tag: None

_____ The nebular theory states that the sun formed before the planets.

F

29) Item Type: True-False   Objective: 7   Section 28.3
     Level: B: Enriched   Tag: None

_____ The nebular theory was developed from Laplace's hypothesis.

T

30) Item Type: True-False   Objective: 8   Section 28.3
     Level: A: Standard   Tag: None

_____ Planetesimals coalesced into the protoplanets.

T

31) Item Type: True-False   Objective: 8   Section 28.3
     Level: A: Standard   Tag: None

_____ During the formation of the solar system, the distance between the developing sun and a protoplanet influenced the composition of the planets that formed.

T

32) Item Type: True-False   Objective: 8   Section 28.3
     Level: A: Standard   Tag: None

_____ Uranus is a gas giant.

T

33) Item Type: True-False   Objective: 8   Section 28.3
     Level: A: Standard   Tag: None

_____ Planetesimals developed from protoplanets.

F

34) Item Type: True-False   Objective: 9   Section 28.3
     Level: B: Enriched   Tag: None

_____ High-energy particles from radioactive materials were a source of heat when the earth first formed.

T

35) Item Type: True-False   Objective: 9   Section 28.3
     Level: A: Standard   Tag: None

_____ Water absorbs carbon dioxide from the atmosphere.

T

36) Item Type: True-False   Objective: 9   Section 28.3
     Level: A: Standard   Tag: None

    ____ The earth's crust is more dense than its core.

F

37) Item Type: True-False   Objective: 9   Section 28.3
     Level: A: Standard   Tag: None

    ____ As the solid earth developed, the less-dense materials flowed to its center.

F

38) Item Type: Multiple Choice   Objective: 1   Section 28.1
     Level: A: Standard   Tag: None

The fusion of hydrogen nuclei in the sun produces a nucleus of

a) hydrogen.   b) oxygen.   c) helium.   d) carbon.

c

39) Item Type: Multiple Choice   Objective: 1   Section 28.1
     Level: A: Standard   Tag: None

Einstein's equation $E = mc^2$ can be used to calculate the amount of energy produced from a given amount of

a) heat.   b) volume.   c) light.   d) mass.

d

40) Item Type: Multiple Choice   Objective: 1   Section 28.1
     Level: A: Standard   Tag: Diagram 2871

In the diagram above, which of the following types of nuclei is substance **X**?

a) nitrogen   b) iron   c) helium   d) carbon

c

41) Item Type: Multiple Choice   Objective: 1   Section 28.1
     Level: B: Enriched   Tag: None

The transfer of energy in moving liquids and gases occurs by the process called

a) convection.   b) granulation.   c) fusion.   d) radiation.

a

42) Item Type: Multiple Choice   Objective: 2   Section 28.1
Level: A: Standard   Tag: None

The convective zone of the sun is surrounded by the

a) corona.   b) photosphere.   c) core.   d) chromosphere.

b

43) Item Type: Multiple Choice   Objective: 2   Section 28.1
Level: A: Standard   Tag: None

The layer of the sun in which the transfer of energy occurs by moving liquids or gases is the

a) convective zone.   b) corona.   c) radiative zone.   d) core.

a

44) Item Type: Multiple Choice   Objective: 3   Section 28.1
Level: B: Enriched   Tag: None

Which layer of the sun gives off the visible light normally seen from the earth?

a) core   b) magnetosphere   c) corona   d) photosphere

d

45) Item Type: Multiple Choice   Objective: 3   Section 28.1
Level: A: Standard   Tag: None

Ions that stream out into space through holes in the corona create the

a) ozone layer.   b) solar wind.   c) chromosphere.   d) solar nebula.

b

46) Item Type: Multiple Choice   Objective: 3   Section 28.1
Level: A: Standard   Tag: None

The layer that prevents most of the atomic particles from the sun's surface from escaping into space is the

a) ozone.   b) corona.   c) mantle.   d) crust.

b

47) Item Type: Multiple Choice   Objective: 3   Section 28.1
Level: A: Standard   Tag: None

In which layer of the sun does granulation occur?

a) core   b) chromosphere   c) corona   d) photosphere

d

Chapter 28

48) Item Type: Multiple Choice   Objective: 3   Section 28.1
Level: A: Standard   Tag: None

The innermost layer of the sun's atmosphere is the

a) prominence.   b) photosphere.   c) corona   d) chromosphere.

b

49) Item Type: Multiple Choice   Objective: 3   Section 28.1
Level: A: Standard   Tag: None

The region of the sun in which much of the energy given off is in the form of visible light is called the

a) radiative zone.   b) photosphere.   c) convective zone.   d) chromosphere.

b

50) Item Type: Multiple Choice   Objective: 3   Section 28.1
Level: A: Standard   Tag: None

The outermost layer of the sun's atmosphere is the

a) aurora.   b) chromosphere.   c) corona.   d) radiative zone.

c

51) Item Type: Multiple Choice   Objective: 3   Section 28.1
Level: B: Enriched   Tag: Diagram 2872

Which part of the diagram represents the color sphere?

a) *1*   b) *2*   c) *3*   d) *4*

c

52) Item Type: Multiple Choice   Objective: 3   Section 28.1
Level: A: Standard   Tag: None

The surface of the sun is called the

a) troposphere.   b) corona.   c) photosphere.   d) aurora.

c

53) Item Type: Multiple Choice   Objective: 4   Section 28.2
Level: B: Enriched   Tag: None

Compared to surrounding areas, regions of the photosphere near strong magnetic fields are

a) more reactive.   b) cooler.   c) less dense.   d) brighter.

b

54) Item Type: Multiple Choice   Objective: 4   Section 28.2
    Level: A: Standard   Tag: None

The sunspot cycle lasts an average of

a) 2 years.   b) 11 years.   c) 20 years.   d) 17 years.

b

55) Item Type: Multiple Choice   Objective: 4   Section 28.2
    Level: A: Standard   Tag: None

Cooler regions of the photosphere near strong magnetic fields are called

a) solar flares.   b) prominences.   c) solar winds.   d) sunspots.

d

56) Item Type: Multiple Choice   Objective: 5   Section 28.2
    Level: A: Standard   Tag: None

One of the most violent of all solar disturbances is called a

a) solar flare.   b) prominence.   c) solar wind.   d) sunspot.

a

57) Item Type: Multiple Choice   Objective: 5   Section 28.2
    Level: A: Standard   Tag: None

A sudden outward eruption of electrically charged atomic particles from the sun is called

a) a solar flare.   b) a sunspot.   c) a solar wind.   d) an aurora.

a

58) Item Type: Multiple Choice   Objective: 6   Section 28.2
    Level: A: Standard   Tag: None

Auroras are an effect caused by

a) nuclear fusion.   b) radio waves.   c) magnetic storms.   d) solar nebulae.

c

59) Item Type: Multiple Choice   Objective: 7   Section 28.3
    Level: A: Standard   Tag: None

It is generally believed that the entire solar system originated from a

a) nebula.   b) protoplanet.   c) star.   d) planetesimal.

a

60) Item Type: Multiple Choice   Objective: 7   Section 28.3
    Level: B: Enriched   Tag: None

Which of the following is the approximate percentage of matter in the solar nebula that became part of the sun?

a) 38%   b) 55%   c) 73%   d) 99%

d

61) Item Type: Multiple Choice   Objective: 7   Section 28.3
    Level: A: Standard   Tag: None

Which of the following is best described as a cloud of gas and dust?

a) a nebula   b) a gas giant   c) an aurora   d) a prominence

a

62) Item Type: Multiple Choice   Objective: 8   Section 28.3
    Level: A: Standard   Tag: None

The inner planets may have lost their original atmospheres of lighter gases because of an intense solar

a) wind.   b) flare.   c) prominence.   d) nebula.

a

63) Item Type: Multiple Choice   Objective: 8   Section 28.3
    Level: A: Standard   Tag: None

Which of the following planets is a gas giant?

a) Uranus   b) Venus   c) Mars   d) Mercury

a

64) Item Type: Multiple Choice   Objective: 8   Section 28.3
    Level: A: Standard   Tag: None

Which of the following planets consists of water and frozen gases?

a) Venus   b) Mars   c) Mercury   d) Saturn

d

65) Item Type: Multiple Choice   Objective: 8   Section 28.3
    Level: A: Standard   Tag: None

Which of the following planets may have lost its original atmosphere of lighter gases because of an intense solar wind?

a) Mars   b) Neptune   c) Jupiter   d) Uranus

a

66) Item Type: Multiple Choice   Objective: 8   Section 28.3
Level: A: Standard   Tag: None

The small bodies of matter that were present in the solar nebula while the sun was forming are called

a) auroras.   b) planetesimals.   c) moons.   d) protoplanets.

b

67) Item Type: Multiple Choice   Objective: 8   Section 28.3
Level: A: Standard   Tag: None

Which of the following planets originated from one of the four protoplanets closest to the sun?

a) Uranus   b) Mars   c) Saturn   d) Pluto

b

68) Item Type: Multiple Choice   Objective: 9   Section 28.3
Level: B: Enriched   Tag: None

The earth's second atmosphere was composed of large amounts of gaseous

a) oxygen.   b) methane.   c) nitrogen.   d) helium.

b

69) Item Type: Multiple Choice   Objective: 9   Section 28.3
Level: B: Enriched   Tag: None

As a result of photosynthesis, there is an increase in the atmospheric amount of the element

a) carbon.   b) nitrogen.   c) hydrogen.   d) oxygen.

d

70) Item Type: Multiple Choice   Objective: 9   Section 28.3
Level: A: Standard   Tag: None

The process of photosynthesis increases the amount of which gas in the earth's atmosphere?

a) nitrogen   b) carbon dioxide   c) oxygen   d) water vapor

c

71) Item Type: Multiple Choice   Objective: 9   Section 28.3
Level: A: Standard   Tag: Diagram 2873

Which of the following gases is indicated by the blank line in this table?

a) helium   b) ammonia   c) methane   d) carbon dioxide

a

72) Item Type: Multiple Choice   Objective: 9   Section 28.3
    Level: A: Standard   Tag: None

Which of the following shields the earth from the harmful ultraviolet radiation of the sun?

a) crust   b) ozone   c) mantle   d) magnetosphere

b

73) Item Type: Multiple Choice   Objective: 9   Section 28.3
    Level: A: Standard   Tag: None

The core of the earth consists primarily of nickel and

a) helium.   b) carbon.   c) iron.   d) magnesium.

c

74) Item Type: Fill in the Blank   Objective: 1   Section 28.1
    Level: B: Enriched   Tag: None

The primary element fueling the sun's nuclear fusion reactions is _____.

hydrogen

75) Item Type: Fill in the Blank   Objective: 1   Section 28.1
    Level: B: Enriched   Tag: None

The amount of energy that hits each square centimeter of the earth each second is called

the _____.

solar constant

76) Item Type: Fill in the Blank   Objective: 2   Section 28.1
    Level: A: Standard   Tag: None

The layer of the sun just outside the core is the _____.

radiative zone

77) Item Type: Fill in the Blank   Objective: 2   Section 28.1
    Level: A: Standard   Tag: None

The region of the sun in which energy is transferred between atoms in the form of

electromagnetic waves is the _____.

radiative zone

78) Item Type: Fill in the Blank   Objective: 2   Section 28.1
Level: A: Standard   Tag: None

In the radiative zone, heat energy moves from atom to atom in the form of

_____.

electromagnetic waves

79) Item Type: Fill in the Blank   Objective: 4   Section 28.2
Level: A: Standard   Tag: None

One of the first indications to astronomers that the sun rotates on its axis was the

movement of _____.

sunspots

80) Item Type: Fill in the Blank   Objective: 4   Section 28.2
Level: A: Standard   Tag: None

Cooler areas of the sun that appear darker than the surrounding areas are called

_____.

sunspots

81) Item Type: Fill in the Blank   Objective: 5   Section 28.2
Level: A: Standard   Tag: None

A prominence arches from one sunspot area to another following curved lines of

_____.

magnetic force

82) Item Type: Fill in the Blank   Objective: 5   Section 28.2
Level: A: Standard   Tag: None

As solar wind particles enter the earth's atmosphere, they can generate disturbances called

_____.

magnetic storms

83) Item Type: Fill in the Blank   Objective: 5   Section 28.2
Level: A: Standard   Tag: None

One of the most violent of all solar disturbances is called a solar _____.

flare

Chapter 28

84) Item Type: Fill in the Blank   Objective: 6   Section 28.2
Level: A: Standard   Tag: None

The northern and southern lights are also called _____.

| auroras |

85) Item Type: Fill in the Blank   Objective: 6   Section 28.2
Level: A: Standard   Tag: None

Electrically charged particles of the solar wind that strike the earth's upper atmosphere

create sheets of light called _____.

| auroras |

86) Item Type: Fill in the Blank   Objective: 6   Section 28.2
Level: A: Standard   Tag: None

Charged particles of the solar wind are guided to the earth's magnetic poles by the

_____.

| magnetosphere |

87) Item Type: Fill in the Blank   Objective: 7   Section 28.3
Level: B: Enriched   Tag: None

The hypothesis that the sun and planets formed out of the same cloud of dust and gas was

proposed by a mathematician named _____.

| Laplace |

88) Item Type: Fill in the Blank   Objective: 7   Section 28.3
Level: A: Standard   Tag: None

The small bodies of matter that eventually formed protoplanets are called

_____.

| planetesimals |

89) Item Type: Fill in the Blank   Objective: 7   Section 28.3
Level: A: Standard   Tag: None.

Laplace's hypothesis on the formation of the solar system developed into what is known

as the _____.

| nebular theory |

90) Item Type: Fill in the Blank   Objective: 7   Section 28.3
    Level: A: Standard   Tag: None

The solar system developed from a cloud of gas and dust called a(n)

_____.

nebula

91) Item Type: Fill in the Blank   Objective: 9   Section 28.3
    Level: B: Enriched   Tag: None

The earth's core is composed mostly of the two elements _____.

iron and nickel

92) Item Type: Fill in the Blank   Objective: 9   Section 28.3
    Level: B: Enriched   Tag: None

The earth's second atmosphere resulted from gases emitted by _____.

volcanoes

93) Item Type: Fill in the Blank   Objective: 9   Section 28.3
    Level: A: Standard   Tag: None

As the earth's atmosphere developed, much of the carbon dioxide that was present in the

atmosphere was absorbed by the _____.

ocean water

94) Item Type: Fill in the Blank   Objective: 9   Section 28.3
    Level: A: Standard   Tag: None

When the earth cooled, oceans formed as the water vapor _____.

condensed

95) Item Type: Essay   Objective: 2   Section 28.1
     Level: B: Enriched   Tag: None

Explain how heat energy is transferred to the sun's surface by convection.

_____

_____

_____

_____

_____

In the convective zone, as hot gas atoms move outward and expand, they radiate and lose
heat. These cooler gases become denser and sink to the bottom of the convective zone,
where they are heated by energy from the radiative zone and rise again.

96) Item Type: Essay   Objective: 4   Section 28.2
     Level: A: Standard   Tag: None

Describe the way in which the sun rotates.

_____

_____

_____

_____

_____

Because the sun is a ball of hot gases rather than a solid sphere, parts of the sun rotate at
different speeds. Places closest to the sun's equator rotate faster than points near the poles.

97) Item Type: Essay   Objective: 4   Section 28.2
    Level: B: Enriched   Tag: None

Explain how sunspots are formed.

_____

_____

_____

_____

_____

The up-and-down movement of gases in the sun's convective zone and the movement of
the sun's rotation produce magnetic fields that slow down activity in the convective zone.
Slower convection means less gas is transferring heat from the core to the photosphere.
Therefore, regions of the photosphere near these magnetic fields are cooler and appear
darker than surrounding areas. These darker areas are sunspots.

98) Item Type: Essay   Objective: 4   Section 28.2
    Level: B: Enriched   Tag: None

Explain how sunspots are caused by powerful magnetic fields.

_____

_____

_____

_____

_____

Magnetic fields slow down activity in the sun's convective zone, so less heat is transferred
from the core to the photosphere. Areas of the photosphere near these magnetic fields that
are cooler and appear darker than surrounding areas are called sunspots.

99) Item Type: Essay   Objective: 7   Section 28.3
Level: A: Standard   Tag: None

Explain how the sun formed.

_____

_____

_____

_____

_____

> The solar nebula contracted and became denser because of shock waves from a nearby supernova or some other force. The center became denser and hotter because of heat from collisions and pressure caused by the force of gravity. When the temperature in the nebula's center became great enough, hydrogen fusion occurred and the sun formed.

100) Item Type: Essay   Objective: 7   Section 28.3
Level: B: Enriched   Tag: None

What two proposals were stated in Laplace's hypothesis on the origin of the solar system?

_____

_____

_____

_____

_____

> It stated that the sun and the planets condensed out of the same spinning nebula. It also stated that the entire solar system formed at approximately the same time.

101) Item Type: Essay   Objective: 9   Section 28.3
      Level: B: Enriched   Tag: None

If the earth were much more massive than it is today and had no green plants on its surface, what changes in the earth's atmosphere might be expected?

_____

_____

_____

_____

If the earth were more massive, its gravitational pull would be greater and it would have been able to retain most of the hydrogen that originally was present but escaped into space. Also, if no green plants existed, there would be no photosynthesis. Thus the amount of hydrogen would increase and the amount of oxygen would decrease.

102) Item Type: Essay   Objective: 9   Section 28.3
      Level: B: Enriched   Tag: None

What three sources of heat contributed to the high temperature on the earth as it formed?

_____

_____

_____

_____

First, the earth retained much of the heat produced when it collided with planetesimals. Second, the increasing weight of the outer layers compressed the inner layers and generated heat. Third, high-energy particles emitted from radioactive materials were absorbed by rocks. Some of the energy of motion of the particles was converted to heat.

1) Tag Name: Diagram 2871

Nuclear Fusion Reaction
(p=proton, n=neutron)

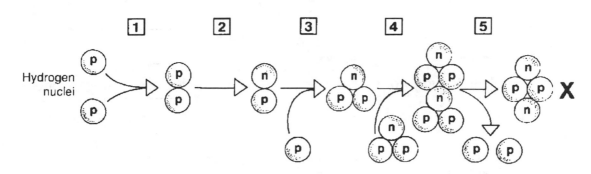

2) Tag Name: Diagram 2872

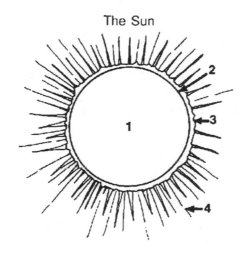

The Sun

3) Tag Name: Diagram 2873

Major Components of the
Earth's Atmospheres

First Atmosphere	Present Atmosphere
hydrogen _____	nitrogen oxygen

1)   Item Type: True-False   Objective: 1   Section 29.1
            Level: B: Enriched   Tag: None

_____ Ptolemy developed a heliocentric model of the universe.

F

2)   Item Type: True-False   Objective: 1   Section 29.1
            Level: A: Standard   Tag: None

_____ Ptolemy's model of the universe proposed that each planet has two motions.

T

3)   Item Type: True-False   Objective: 1   Section 29.1
            Level: A: Standard   Tag: None

_____ The geocentric model of the solar system states that the planets revolve around the sun.

F

4)   Item Type: True-False   Objective: 1   Section 29.1
            Level: A: Standard   Tag: None

_____ Galileo's observations confirmed the validity of a heliocentric solar system model.

T

5)   Item Type: True-False   Objective: 1   Section 29.1
            Level: A: Standard   Tag: None

_____ Aristotle developed the first accurate model of the solar system.

F

6)   Item Type: True-False   Objective: 1   Section 29.1
            Level: A: Standard   Tag: None

_____ The geocentric model of the solar system was proposed by Copernicus.

F

7)   Item Type: True-False   Objective: 1   Section 29.1
            Level: A: Standard   Tag: None

_____ Aristotle was the first astronomer to explain the retrograde motion of planets.

F

8) Item Type: True-False   Objective: 1   Section 29.1
    Level: A: Standard   Tag: None

    _____ Galileo's observations with a telescope helped confirm Copernicus's model of the
    solar system.

    T

9) Item Type: True-False   Objective: 2   Section 29.1
    Level: B: Enriched   Tag: None

    _____ A circular orbit has neither a perihelion nor an aphelion.

    T

10) Item Type: True-False   Objective: 2   Section 29.1
    Level: A: Standard   Tag: None

    _____ Kepler suggested that planets move at different speeds at different points in their
    orbit.

    T

11) Item Type: True-False   Objective: 2   Section 29.1
    Level: B: Enriched   Tag: None

    _____ According to Kepler's third law, the cube of the average distance of a planet from
    the sun is always proportional to the square of the planet's orbit period.

    T

12) Item Type: True-False   Objective: 2   Section 29.1
    Level: B: Enriched   Tag: None

    _____ The square of a planet's orbit period measured in earth years is equal to the cube of
    the planet's distance from the sun measured in astronomical units (AU).

    T

13) Item Type: True-False   Objective: 2   Section 29.1
    Level: B: Enriched   Tag: None

    _____ A planet's orbit period measured in years is equal to its distance from the sun
    measured in astronomical units (AU).

    F

14) Item Type: True-False   Objective: 2   Section 29.1
    Level: A: Standard   Tag: None

    _____ The point at which an orbit comes closest to the sun is called perihelion.

    T

15) Item Type: True-False   Objective: 3   Section 29.2
    Level: B: Enriched   Tag: None

_____ Basalt and granite are found on the surface of Venus.

T

16) Item Type: True-False   Objective: 3   Section 29.2
    Level: A: Standard   Tag: None

_____ Hardened lava in some craters on Mercury suggests that the planet was once volcanic.

T

17) Item Type: True-False   Objective: 3   Section 29.2
    Level: A: Standard   Tag: None

_____ The surface of Venus is probably covered with oceans.

F

18) Item Type: True-False   Objective: 3   Section 29.2
    Level: A: Standard   Tag: None

_____ Mercury is a terrestrial planet.

T

19) Item Type: True-False   Objective: 4   Section 29.2
    Level: A: Standard   Tag: None

_____ Mars has polar ice caps.

T

20) Item Type: True-False   Objective: 4   Section 29.2
    Level: A: Standard   Tag: None

_____ Mars has a shorter period of revolution than the earth.

F

21) Item Type: True-False   Objective: 4   Section 29.2
    Level: A: Standard   Tag: None

_____ The earth completes one rotation in about 365 days.

F

22) Item Type: True-False   Objective: 5   Section 29.3
    Level: A: Standard   Tag: None

_____ Astronomers believe that the Great Red Spot is a giant atmospheric storm.

T

23) Item Type: True-False   Objective: 5   Section 29.3
    Level: A: Standard   Tag: None

_____ Jupiter rotates faster than any other planet in the solar system.

T

24) Item Type: True-False   Objective: 5   Section 29.3
    Level: A: Standard   Tag: None

_____ Jupiter's Great Red Spot may be the result of volcanic activity.

T

25) Item Type: True-False   Objective: 5   Section 29.3
    Level: B: Enriched   Tag: None

_____ Because of high temperature and pressure, the surface of Jupiter is partially molten rock.

F

26) Item Type: True-False   Objective: 5   Section 29.3
    Level: A: Standard   Tag: None

_____ Venus is one of the Jovian planets.

F

27) Item Type: True-False   Objective: 6   Section 29.3
    Level: A: Standard   Tag: None

_____ Many scientists believe that Pluto was a moon of Uranus.

F

28) Item Type: True-False   Objective: 6   Section 29.3
    Level: A: Standard   Tag: None

_____ The clouds covering Uranus are a greenish color.

T

29) Item Type: True-False   Objective: 6   Section 29.3
      Level: A: Standard   Tag: None

_____ Data from the *Voyager* spacecraft indicate that the planet Neptune is mostly solid rock.

F

30) Item Type: True-False   Objective: 6   Section 29.3
      Level: A: Standard   Tag: None

_____ Neptune is composed largely of ice.

F

31) Item Type: True-False   Objective: 6   Section 29.3
      Level: A: Standard   Tag: None

_____ Data from the *Voyager* spacecraft indicate that Uranus is made up primarily of ice.

F

32) Item Type: True-False   Objective: 6   Section 29.3
      Level: A: Standard   Tag: None

_____ Pluto is presently inside Neptune's orbit.

T

33) Item Type: True-False   Objective: 6   Section 29.3
      Level: A: Standard   Tag: None

_____ Neptune's upper atmosphere is composed of frozen methane.

T

34) Item Type: True-False   Objective: 7   Section 29.4
      Level: A: Standard   Tag: None

_____ Regardless of the direction a comet is traveling, its tail points away from the sun.

T

35) Item Type: True-False   Objective: 7   Section 29.4
      Level: A: Standard   Tag: None

_____ All asteroids have the same composition.

F

36) Item Type: True-False   Objective: 7   Section 29.4
    Level: B: Enriched   Tag: None

_____ The tail of a comet can extend millions of kilometers behind it.

T

37) Item Type: True-False   Objective: 7   Section 29.4
    Level: A: Standard   Tag: None

_____ The tail of a comet usually points toward the sun.

F

38) Item Type: True-False   Objective: 7   Section 29.4
    Level: A: Standard   Tag: None

_____ The asteroid belt is located between the orbits of Mars and Jupiter.

T

39) Item Type: True-False   Objective: 8   Section 29.4
    Level: A: Standard   Tag: None

_____ A fireball indicates that a meteorite is about to land somewhere on the earth.

F

40) Item Type: True-False   Objective: 8   Section 29.4
    Level: A: Standard   Tag: None

_____ Meteor showers occur when the earth's orbit intersects the orbits of comets.

T

41) Item Type: True-False   Objective: 8   Section 29.4
    Level: A: Standard   Tag: None

_____ Fireballs are produced by meteors.

T

42) Item Type: Multiple Choice   Objective: 1·   Section 29.1
    Level: A: Standard   Tag: None

Kepler's laws were developed from observations made by

a) Copernicus.   b) Galileo.   c) Ptolemy.   d) Brahe.

d

43) Item Type: Multiple Choice   Objective: 1   Section 29.1
Level: A: Standard   Tag: Diagram 2971

In the diagram above, perihelion is located at the point labeled

a) *1.*   b) *2.*   c) *3.*   d) *4.*

c

44) Item Type: Multiple Choice   Objective: 1   Section 29.1
Level: A: Standard   Tag: None

Who first proposed a heliocentric model of the solar system?

a) Ptolemy   b) Copernicus   c) Brahe   d) Galileo

b

45) Item Type: Multiple Choice   Objective: 2   Section 29.1
Level: B: Enriched   Tag: None

The presence of a magnetic field indicates that a planet may have

a) an iron core.   b) an atmosphere.

c) oceans.   d) a high surface temperature.

a

46) Item Type: Multiple Choice   Objective: 2   Section 29.1
Level: B: Enriched   Tag: Diagram 2974

According to Kepler's third law, which planet in the table above has the longest orbit period?

a) *1*   b) *2*   c) *5*   d) *9*

d

47) Item Type: Multiple Choice   Objective: 2   Section 29.1
Level: B: Enriched   Tag: None

Based on Kepler's third law, the orbit period of an object four astronomical units from the sun would be

a) 4 years.   b) 8 years.   c) 16 years.   d) 32 years.

b

48) Item Type: Multiple Choice   Objective: 2   Section 29.1
Level: A: Standard   Tag: None

Kepler's first law states that planets orbit the sun in paths called

a) circles.   b) periods.   c) epicycles.   d) ellipses.

d

49) Item Type: Multiple Choice   Objective: 2   Section 29.1
Level: A: Standard   Tag: None

The tendency of an object to remain at rest until acted upon by an outside force is called

a) inertia.   b) magnetism.   c) gravity.   d) velocity.

a

50) Item Type: Multiple Choice   Objective: 2   Section 29.1
Level: B: Enriched   Tag: None

One astronomical unit is equal to the distance between the earth and

a) Mercury.   b) the moon.   c) Pluto.   d) the sun.

d

51) Item Type: Multiple Choice   Objective: 2   Section 29.1
Level: A: Standard   Tag: None

The first astronomer to mathematically model most aspects of planetary motion was

a) Galileo.   b) Brahe.   c) Copernicus.   d) Kepler.

d

52) Item Type: Multiple Choice   Objective: 2   Section 29.1
Level: A: Standard   Tag: None

The planets are kept in orbit by the sun's

a) inertia.   b) magnetism.   c) gravity.   d) period.

c

53) Item Type: Multiple Choice   Objective: 2   Section 29.1
Level: A: Standard   Tag: None

Kepler's law describing the shape of planetary orbits is called the law of

a) equal areas.   b) periods.   c) epicycles.   d) ellipses.

d

54) Item Type: Multiple Choice   Objective: 3   Section 29.2
Level: A: Standard   Tag: None

The atmosphere of Venus is 96 percent

a) sulfuric acid.   b) oxygen.   c) carbon dioxide.   d) nitrogen.

c

55) Item Type: Multiple Choice   Objective: 3   Section 29.2
Level: A: Standard   Tag: None

Clouds in the atmosphere of Venus are composed of droplets of

a) liquid hydrogen.   b) water vapor.   c) carbon dioxide.   d) sulfuric acid.

d

56) Item Type: Multiple Choice   Objective: 3   Section 29.2
Level: A: Standard   Tag: None

Heat is trapped on the surface of Venus by high atmospheric levels of

a) water vapor.   b) sulfuric acid.   c) ammonia.   d) carbon dioxide.

d

57) Item Type: Multiple Choice   Objective: 3   Section 29.2
Level: A: Standard   Tag: None

The terrestrial planets are composed mostly of

a) liquid metal.   b) ice.   c) solid rock.   d) gases.

c

58) Item Type: Multiple Choice   Objective: 4   Section 29.2
Level: A: Standard   Tag: None

The earth is the only planet in the solar system that has

a) clouds.   b) oceans of water.

c) a core.   d) an atmosphere.

b

59) Item Type: Multiple Choice   Objective: 4   Section 29.2
Level: A: Standard   Tag: None

Great cracks in the surface of Mars were probably formed by

a) erosion.   b) ice caps.   c) meteors.   d) fault zones.

d

60) Item Type: Multiple Choice   Objective: 4   Section 29.2
Level: A: Standard   Tag: Diagram 2972

In the photograph, the circular features on the planet's surface are

a) ocean basins.   b) valleys.   c) impact craters.   d) volcanoes.

c

61) Item Type: Multiple Choice   Objective: 4   Section 29.2
Level: A: Standard   Tag: None

Which of the following is a terrestrial planet?

a) Pluto   b) Uranus   c) Mars   d) Jupiter

c

62) Item Type: Multiple Choice   Objective: 4   Section 29.2
Level: A: Standard   Tag: None

The largest volcanoes in the solar system are most likely on

a) Mars.   b) Jupiter.   c) the earth.   d) Pluto.

a

63) Item Type: Multiple Choice   Objective: 4   Section 29.2
Level: A: Standard   Tag: None

Olympus Mons is a large volcano on

a) the moon.   b) the earth.   c) Mars.   d) Jupiter.

c

64) Item Type: Multiple Choice   Objective: 5   Section 29.3
Level: A: Standard   Tag: None

Which of the following is one of the giant planets?

a) Mars   b) Uranus   c) Venus   d) Pluto

b

65) Item Type: Multiple Choice   Objective: 5   Section 29.3
Level: B: Enriched   Tag: Diagram 2974

Which planet in the table above represents Jupiter?

a) *4*   b) *5*   c) *6*   d) *7*

b

66) Item Type: Multiple Choice   Objective: 5   Section 29.3
Level: B: Enriched   Tag: Diagram 2973

Which planet in the diagram above has the lowest density?

a) *2*   b) *4*   c) *6*   d) *8*

c

67) Item Type: Multiple Choice   Objective: 5   Section 29.3
    Level: A: Standard   Tag: Diagram 2973

The bands on planet **5** in the diagram above are caused by

a) volcanic activity.   b) rotating moons.   c) swirling gases.   d) nuclear fusion.

c

68) Item Type: Multiple Choice   Objective: 6   Section 29.3
    Level: A: Standard   Tag: Diagram 2973

What is the name of planet **7** in the diagram above?

a) Pluto   b) Venus   c) Neptune   d) Uranus

d

69) Item Type: Multiple Choice   Objective: 6   Section 29.3
    Level: B: Enriched   Tag: None

The first planet whose existence was predicted before it was discovered was

a) Venus.   b) Neptune.   c) Uranus.   d) Pluto.

b

70) Item Type: Multiple Choice   Objective: 6   Section 29.3
    Level: A: Standard   Tag: None

The planet discovered in 1930 by Clyde Tombaugh is called

a) Pluto.   b) Uranus.   c) Neptune.   d) Charon.

a

71) Item Type: Multiple Choice   Objective: 7   Section 29.4
    Level: A: Standard   Tag: None

Which of the following are hypothesized to originate in the Oort cloud?

a) comets   b) fireballs   c) meteors   d) asteroids

a

72) Item Type: Fill in the Blank   Objective: 2   Section 29.1
    Level: A: Standard   Tag: None

What is the shape of the path followed by each planet as it orbits the sun?

_____

an ellipse

73) Item Type: Fill in the Blank   Objective: 2   Section 29.1
    Level: A: Standard   Tag: None

One astronomical unit is the average distance between the sun and

_____.

the earth

74) Item Type: Fill in the Blank   Objective: 2   Section 29.1
    Level: A: Standard   Tag: None

Planets are kept in their orbits by the gravitational attraction of the

_____.

sun

75) Item Type: Fill in the Blank   Objective: 3   Section 29.2
    Level: A: Standard   Tag: None

The inner planets are sometimes called _____.

terrestrial planets

76) Item Type: Fill in the Blank   Objective: 3   Section 29.2
    Level: A: Standard   Tag: None

The name of the planet closest to the sun is _____.

Mercury

77) Item Type: Fill in the Blank   Objective: 3   Section 29.2
    Level: A: Standard   Tag: None

What are the bowl-shaped depressions that result from the collision of a planet with

meteors? _____

impact craters

78) Item Type: Fill in the Blank   Objective: 4   Section 29.2
    Level: B: Enriched   Tag: None

The major evidence of geologic activity on Mars is the presence of

_____.

volcanoes

79) Item Type: Fill in the Blank   Objective: 4   Section 29.2
Level: B: Enriched   Tag: None

The *Valles Marineris* is a large canyon on the surface of _____.

Mars

80) Item Type: Fill in the Blank   Objective: 4   Section 29.2
Level: A: Standard   Tag: None

Almost all of the visible water on Mars appears to be trapped in _____.

polar ice caps

81) Item Type: Fill in the Blank   Objective: 5   Section 29.3
Level: B: Enriched   Tag: None

The least dense planet in the solar system is _____.

Saturn

82) Item Type: Fill in the Blank   Objective: 5   Section 29.3
Level: A: Standard   Tag: None

The most distinguishing feature of Saturn is its complex system of _____.

rings

83) Item Type: Fill in the Blank   Objective: 5   Section 29.3
Level: A: Standard   Tag: None

What is another name for the outer planets? _____

Jovian planets

84) Item Type: Fill in the Blank   Objective: 6   Section 29.3
Level: B: Enriched   Tag: None

What United States spacecraft transmitted the first pictures of Uranus?

_____

Voyager 2

85) Item Type: Fill in the Blank   Objective: 6   Section 29.3
Level: A: Standard   Tag: Diagram 2975

The name of the planet labeled **X** in the diagram is _____.

Pluto

86) Item Type: Fill in the Blank   Objective: 7   Section 29.4
    Level: B: Enriched   Tag: None

The famous short-period comet that last appeared in 1986 is called

_____.

| Halley's comet   ▪ |

87) Item Type: Fill in the Blank   Objective: 7   Section 29.4
    Level: A: Standard   Tag: None

The spherical cloud of gas and dust surrounding the nucleus of a comet is called the

_____.

| coma   ▪ |

88) Item Type: Fill in the Blank   Objective: 7   Section 29.4
    Level: B: Enriched   Tag: Diagram 2976

In the diagram above, the line labeled **Y** represents the orbit of a type of asteroid called

_____.

| an earth-grazer   ▪ |

89) Item Type: Fill in the Blank   Objective: 7   Section 29.4
    Level: A: Standard   Tag: Diagram 2976

The area labeled **X** in the diagram above represents the location of the

_____.

| asteroid belt   ▪ |

90) Item Type: Fill in the Blank   Objective: 7   Section 29.4
    Level: A: Standard   Tag: None

What is the name for the asteroids that are concentrated ahead of and behind Jupiter?

_____

| Trojan asteroids   ▪ |

91) Item Type: Fill in the Blank   Objective: 7   Section 29.4
    Level: A: Standard   Tag: None

What type of comet completes its orbit in less than 100 years? _____

| short-period comet   ▪ |

92) Item Type: Fill in the Blank   Objective: 7   Section 29.4
    Level: A: Standard   Tag: None

The Oort cloud is the probable source of _____.

comets

93) Item Type: Fill in the Blank   Objective: 8   Section 29.4
    Level: A: Standard   Tag: None

The rarest of the three basic types of meteorites is called _____.

stony-iron

94) Item Type: Fill in the Blank   Objective: 8   Section 29.4
    Level: A: Standard   Tag: None

Small bits of rock or metal moving through the solar system are called

_____.

meteoroids

95) Item Type: Fill in the Blank   Objective: 8   Section 29.4
    Level: A: Standard   Tag: None

Generally, the type of meteorite that is easiest to find because of its metallic appearance is

_____.

iron

96) Item Type: Fill in the Blank   Objective: 8   Section 29.4
    Level: A: Standard   Tag: None

What is visible when large numbers of small meteoroids enter the earth's atmosphere

during a short time? _____

meteor shower

97) Item Type: Fill in the Blank   Objective: 8   Section 29.4
    Level: A: Standard   Tag: None

What is a meteor called after it hits the earth? _____

meteorite

98) Item Type: Fill in the Blank   Objective: 8   Section 29.4
     Level: A: Standard   Tag: None

The most abundant type of meteorite is called _____.

> stony

99) Item Type: Essay   Objective: 2   Section 29.1
     Level: B: Enriched   Tag: None

Describe the law of equal areas.

_____

_____

_____

_____

_____

> Since planetary orbits are ellipses, the distance from the center of a planet to the center of the sun is not always the same. Because a planet moves fastest when it is closest to the sun, an imaginary line along the orbit's moving radius creates, over a given period of time, a wide triangular sector. When the planet is farther from the sun, the sector created in the same time period is longer and more narrow. Both sectors, however, will have equal areas.

100) Item Type: Essay   Objective: 3   Section 29.2
      Level: B: Enriched   Tag: None

Give two explanations for Mercury's lack of an atmosphere.

_____

_____

_____

_____

_____

> Since Mercury is so close to the sun, solar heat causes the gas molecules near the surface of the planet to move rapidly. Because the planet is so small, its gravitational pull is too weak to hold enough rapidly moving gas molecules to form a dense atmosphere.

101) Item Type: Essay   Objective: 3   Section 29.2
     Level: B: Enriched   Tag: None

Describe the effects of inertia on a moving object and on a stationary object.

_____

_____

_____

_____

_____

> A moving object will continue to move in a straight line unless something causes it to change direction or stop. A stationary object will remain at rest until an outside force acts on it.

102) Item Type: Essay   Objective: 7   Section 29.4
     Level: B: Enriched   Tag: None

Identify the three parts of a comet, and describe their composition.

_____

_____

_____

_____

> The core nucleus is made up of rock or metals and ice. The coma surrounding the nucleus is a cloud of gas and dust. The tail is the gas and dust that stream out from the head.

1) Tag Name: Diagram 2971

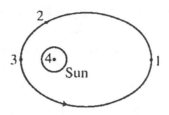

2) Tag Name: Diagram 2972

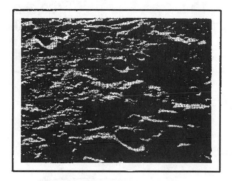

3) Tag Name: Diagram 2973

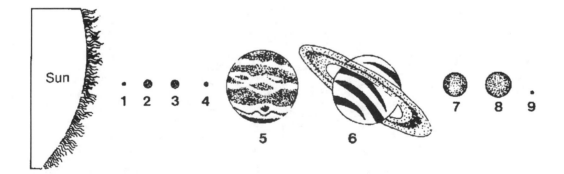

4)   Tag Name: Diagram 2974

Planetary Data

Planet	Average distance from sun (10^6 km)	Diameter (km)	Rate of rotation (Earth time)
1	57.9	4,878	59 d
2	108.2	12,104	−243 d
3	149.6	12,756	23 hr 56 min
4	227.9	6,794	24 hr 37 min
5	778.3	142,796	9 hr 50 min
6	1,427	120,660	10 hr 40 min
7	2,871	52,400	17 hr 14 min
8	4,497	50,450	16 hr 06 min
9	5,914	2,445	6.4 d

5)   Tag Name: Diagram 2975

Orbits of the Outer Planets

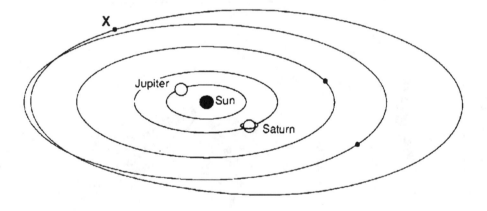

6)   Tag Name: Diagram 2976

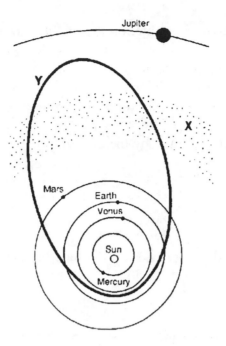

1) Item Type: True-False   Objective: 1   Section 30.1
    Level: A: Standard   Tag: None

    _____ The force of gravity is greater on the moon than on the earth.

    | F |

2) Item Type: True-False   Objective: 1   Section 30.1
    Level: A: Standard   Tag: None

    _____ The first satellite launched by the United States was *Voyager 1*.

    | F |

3) Item Type: True-False   Objective: 1   Section 30.1
    Level: A: Standard   Tag: None

    _____ Rilles look like dry river beds on the lunar surface.

    | T |

4) Item Type: True-False   Objective: 1   Section 30.1
    Level: A: Standard   Tag: None

    _____ Maria are long, deep channels found on the lunar surface.

    | F |

5) Item Type: True-False   Objective: 2   Section 30.1
    Level: A: Standard   Tag: None

    _____ Seismographs have been placed on the moon by astronauts.

    | T |

6) Item Type: True-False   Objective: 2   Section 30.1
    Level: A: Standard   Tag: None

    _____ A promising method for mining moon rocks uses the process of hydrolysis.

    | F |

7) Item Type: True-False   Objective: 2   Section 30.1
    Level: A: Standard   Tag: None

    _____ The moon's mantle is rich in silica, magnesium, and iron.

    | T |

8) Item Type: True-False   Objective: 2   Section 30.1
    Level: A: Standard   Tag: None

_____ The side of the moon facing the earth always receives the same amount of sunlight as the side facing away from the earth.

T

9) Item Type: True-False   Objective: 3   Section 30.1
    Level: A: Standard   Tag: None

_____ During the fourth and final stage of lunar development, large numbers of meteorites punctured the moon's solid crust.

F

10) Item Type: True-False   Objective: 3   Section 30.1
    Level: A: Standard   Tag: None

_____ The moon may have formed when the earth was struck by a Mars-sized body.

T

11) Item Type: True-False   Objective: 3   Section 30.1
    Level: B: Enriched   Tag: None

_____ Lunar surface features have undergone significant modification since their formation.

F

12) Item Type: True-False   Objective: 4   Section 30.2
    Level: A: Standard   Tag: None

_____ The moon revolves around the earth in a circular orbit.

F

13) Item Type: True-False   Objective: 4   Section 30.2
    Level: A: Standard   Tag: None

_____ The center of mass of the earth-moon system follows a smooth orbit around the sun.

T

14) Item Type: True-False   Objective: 4   Section 30.2
    Level: A: Standard   Tag: None

_____ The moon passes closest to the earth at apogee.

F

15) Item Type: True-False   Objective: 4   Section 30.2
Level: A: Standard   Tag: None

_____ The same side of the moon always faces the earth.

T

16) Item Type: True-False   Objective: 4   Section 30.2
Level: A: Standard   Tag: None

_____ When the moon is closest to the earth, it is at apogee.

F

17) Item Type: True-False   Objective: 4   Section 30.2
Level: B: Enriched   Tag: None

_____ Each rotation of the earth takes slightly longer than the previous one.

T

18) Item Type: True-False   Objective: 5   Section 30.2
Level: A: Standard   Tag: None

_____ An eclipse occurs when one planetary body passes through the shadow of another.

T

19) Item Type: True-False   Objective: 5   Section 30.2
Level: A: Standard   Tag: None

_____ In the umbra, sunlight is partially blocked.

F

20) Item Type: True-False   Objective: 5   Section 30.2
Level: A: Standard   Tag: None

_____ Lunar eclipses are visible from any location on the dark side of the earth.

T

21) Item Type: True-False   Objective: 5   Section 30.2
Level: A: Standard   Tag: None

_____ Eclipses are the result of one planetary body casting a shadow on another.

T

22) Item Type: True-False   Objective: 6   Section 30.3
    Level: A: Standard   Tag: None

_____ A new moon occurs when the lighted surface of the moon faces toward the earth.

F

23) Item Type: True-False   Objective: 6   Section 30.3
    Level: A: Standard   Tag: None

_____ When the moon is waxing, the size of the visible portion is increasing.

T

24) Item Type: True-False   Objective: 7   Section 30.3
    Level: A: Standard   Tag: None

_____ A planetary body makes one rotation on its axis each day.

T

25) Item Type: True-False   Objective: 7   Section 30.3
    Level: B: Enriched   Tag: None

_____ A solar year is about 304 days long.

F

26) Item Type: True-False   Objective: 8   Section 30.4
    Level: A: Standard   Tag: None

_____ The moons of Mars are heavily cratered.

T

27) Item Type: True-False   Objective: 8   Section 30.4
    Level: A: Standard   Tag: None

_____ Each of the moons of Mars is much larger than the earth's moon.

F

28) Item Type: True-False   Objective: 8   Section 30.4
    Level: A: Standard   Tag: None

_____ The two moons of Mars are believed to be fairly young because their surfaces are characterized by very few craters.

F

29) Item Type: True-False   Objective: 9   Section 30.4
     Level: A: Standard   Tag: None

     _____ Jupiter's ring is made of particles of dark rock.

T

30) Item Type: True-False   Objective: 9   Section 30.4
     Level: A: Standard   Tag: None

     _____ Galileo discovered the four moons closest to Jupiter and what later were discovered to be the rings of Saturn.

T

31) Item Type: True-False   Objective: 9   Section 30.4
     Level: A: Standard   Tag: None

     _____ The moon Charon is always above the same spot on the surface of Pluto because the period of Charon's orbit is equal to one day on Pluto.

T

32) Item Type: True-False   Objective: 9   Section 30.4
     Level: A: Standard   Tag: None

     _____ Jupiter's moon Ganymede is the smallest and densest moon in the solar system.

F

33) Item Type: True-False   Objective: 9   Section 30.4
     Level: A: Standard   Tag: None

     _____ Jupiter's moon Io has active volcanoes.

T

34) Item Type: True-False   Objective: 9   Section 30.4
     Level: A: Standard   Tag: None

     _____ The planet Jupiter has only four moons.

F

35) Item Type: True-False   Objective: 9   Section 30.4
     Level: B: Enriched   Tag: None

     _____ Uranus's rings are made of tiny ice particles.

F

Chapter 30

36) Item Type: True-False   Objective: 9   Section 30.4
Level: B: Enriched   Tag: None

_____ One hypothesis about the formation of Saturn's rings states that the rings formed when a body orbiting Saturn was torn apart by the gravity of the planet.

T

37) Item Type: Multiple Choice   Objective: 9   Section 30.4
Level: A: Standard   Tag: None

The streaks that form when surface material has been splashed around lunar craters are called

a) rilles.   b) maria.   c) regolith.   d) rays.

d

38) Item Type: Multiple Choice   Objective: 1   Section 30.1
Level: A: Standard   Tag: None

A body that orbits a larger body is called a

a) penumbra.   b) satellite.   c) ray.   d) regolith.

b

39) Item Type: Multiple Choice   Objective: 1   Section 30.1
Level: A: Standard   Tag: None

The earth's first artificial satellite was called

a) *Sputnik 1.*   b) *Explorer 1.*   c) *Voyager 1.*   d) *Apollo 13.*

a

40) Item Type: Multiple Choice   Objective: 1   Section 30.1
Level: A: Standard   Tag: None

A long, deep channel on the moon is called a

a) mare.   b) ray.   c) streak.   d) rille.

d

41) Item Type: Multiple Choice   Objective: 1   Section 30.1
Level: A: Standard   Tag: None

What are bowl-shaped lunar depressions called?

a) breccia   b) craters   c) anorthosites   d) basalts

b

42) Item Type: Multiple Choice Objective: 1 Section 30.1
Level: A: Standard Tag: None

The outer layer of dust and small rock fragments on the moon is called

a) maria. b) umbra. c) rille. d) regolith.

d

43) Item Type: Multiple Choice Objective: 2 Section 30.1
Level: A: Standard Tag: None

The moon may have a small liquid core made of

a) silicon. b) magnesium. c) iron. d) aluminum.

c

44) Item Type: Multiple Choice Objective: 2 Section 30.1
Level: B: Enriched Tag: Diagram 3071

In the diagram, the moon's mantle is the structure labeled

a) *1.* b) *2.* c) *3.* d) *4.*

b

45) Item Type: Multiple Choice Objective: 2 Section 30.1
Level: B: Enriched Tag: None

Scientists believe that the only part of the moon that could possibly be liquid is the

a) highlands. b) mantle. c) core. d) crust.

c

46) Item Type: Multiple Choice Objective: 3 Section 30.1
Level: A: Standard Tag: None

During which stage of the moon's development was the surface covered with hot molten rock?

a) first b) second c) third d) last

b

47) Item Type: Multiple Choice Objective: 3 Section 30.1
Level: A: Standard Tag: None

According to the giant-impact hypothesis of the moon's origin, the moon was once part of

a) Mercury. b) Venus. c) the earth. d) Mars.

c

Chapter 30

48) Item Type: Multiple Choice   Objective: 3   Section 30.1
     Level: A: Standard   Tag: None

How many stages of lunar development have scientists described?

a) 1   b) 2   c) 3   d) 4

d

49) Item Type: Multiple Choice   Objective: 3   Section 30.1
     Level: B: Enriched   Tag: None

Approximately how many billion years ago did the moon take on relatively the same appearance that it has today?

a) 1   b) 2   c) 3   d) 4

c

50) Item Type: Multiple Choice   Objective: 4   Section 30.2
     Level: A: Standard   Tag: None

As the moon rotates, one side of the moon always faces the earth because

a) the moon rotates and revolves at the same rate.

b) the earth's gravity attracts the near side of the moon.

c) the moon has more mass than the earth.

d) gravity on the moon is greater than on the earth.

a

51) Item Type: Multiple Choice   Objective: 4   Section 30.2
     Level: A: Standard   Tag: None

The orbit of the moon around the earth forms

a) a sphere.   b) a cone.   c) a circle.   d) an ellipse.

d

52) Item Type: Multiple Choice   Objective: 4   Section 30.2
     Level: A: Standard   Tag: None

What is the difference, in minutes, in the rising time of the moon each day?

a) 10   b) 25   c) 50   d) 90

c

53) Item Type: Multiple Choice   Objective: 4   Section 30.2
    Level: B: Enriched   Tag: None

The center of mass of the earth-moon system is at a balance point, which is located

a) within the earth's interior.

b) less than half the distance from the earth to the moon.

c) within the moon's interior.

d) more than half the distance from the earth to the moon.

a

54) Item Type: Multiple Choice   Objective: 5   Section 30.2
    Level: A: Standard   Tag: None

Which of the following will occur if the moon crosses the plane of the earth's orbit when the moon is between the earth and the sun?

a) solar eclipse   b) full-moon phase

c) lunar eclipse   d) last-quarter phase

a

55) Item Type: Multiple Choice   Objective: 5   Section 30.2
    Level: A: Standard   Tag: None

What type of eclipse is seen by the most people?

a) annular   b) solar   c) waning   d) lunar

d

56) Item Type: Multiple Choice   Objective: 5   Section 30.2
    Level: A: Standard   Tag: None

Which of the following occurs when one planetary body passes through the shadow of another?

a) eclipse   b) umbra   c) apogee   d) perigee

a

57) Item Type: Multiple Choice   Objective: 5   Section 30.2
    Level: B: Enriched   Tag: Diagram 3072

According to the diagram above, a solar eclipse is most likely to occur when the moon is in position

a) *1.*   b) *2.*   c) *3.*   d) *4.*

a

58) Item Type: Multiple Choice   Objective: 5   Section 30.2
Level: B: Enriched   Tag: Diagram 3072

According to the diagram above, observers on the earth see a full moon when the moon is in position

a) *1.*   b) *2.*   c) *3.*   d) *4.*

c

59) Item Type: Multiple Choice   Objective: 5   Section 30.2
Level: B: Enriched   Tag: Diagram 3073

Which type of eclipse is pictured in the diagram above?

a) total solar eclipse   b) annular eclipse

c) penumbral eclipse   d) lunar eclipse

b

60) Item Type: Multiple Choice   Objective: 5   Section 30.2
Level: B: Enriched   Tag: Diagram 3074

A lunar eclipse is most likely to occur when the moon is in which position in the diagram above?

a) *1*   b) *2*   c) *3*   d) *4*

c

61) Item Type: Multiple Choice   Objective: 6   Section 30.3
Level: A: Standard   Tag: None

When the visible portion of the moon is increasing, the moon is

a) waxing.   b) waning.   c) full.   d) waning crescent.

a

62) Item Type: Multiple Choice   Objective: 6   Section 30.3
Level: A: Standard   Tag: None

When only a small part of the moon is visible, the moon may be in its

a) first-quarter phase.   b) new-moon phase

c) waning-crescent phase.   d) last-quarter phase.

c

63) Item Type: Multiple Choice   Objective: 6   Section 30.3
Level: A: Standard   Tag: None

Approximately how many days does it take the moon to go through a complete cycle of phases?

a) 7   b) 11   c) 27   d) 29

d

64) Item Type: Multiple Choice   Objective: 6   Section 30.3
Level: A: Standard   Tag: None

The changing shapes of the moon, as viewed from the earth, are called

a) gibbous.   b) phases.   c) maria.   d) umbras.

b

65) Item Type: Multiple Choice   Objective: 6   Section 30.3
Level: A: Standard   Tag: None

In which phase is the moon when the unlighted side faces the earth?

a) crescent   b) gibbous   c) full   d) new

d

66) Item Type: Multiple Choice   Objective: 6   Section 30.3
Level: B: Enriched   Tag: None

How many days does it take the moon to revolve around the earth?

a) 27.3   b) 29.5   c) 30.1   d) 35.6

a

67) Item Type: Multiple Choice   Objective: 6   Section 30.3
Level: B: Enriched   Tag: None

Which diagram below shows the moon in a gibbous phase?

a)   b)   c)   d)

c

68) Item Type: Multiple Choice   Objective: 7   Section 30.3
Level: A: Standard   Tag: None

What is the name of the calendar that was introduced in 46 B.C. to make 365.25 days the average length of a year?

a) World   b) Sol   c) Gregorian   d) Julian

d

69) Item Type: Multiple Choice   Objective: 7   Section 30.3
Level: A: Standard   Tag: None

Systems that measure the passage of time are called

a) phases.   b) calendars.   c) years.   d) sols.

b

70) Item Type: Multiple Choice   Objective: 7   Section 30.3
Level: A: Standard   Tag: None

How many rotations does the earth make on its axis in a day?

a) 1   b) 2   c) 3   d) 4

a

71) Item Type: Multiple Choice   Objective: 7   Section 30.3
Level: B: Enriched   Tag: None

What is the name of the calendar currently in use in the United States?

a) Julian   b) Gregorian   c) Sol   d) World

b

72) Item Type: Multiple Choice   Objective: 7   Section 30.3
Level: B: Enriched   Tag: None

Prior to the introduction of the Julian calendar, how many days made up a Roman year?

a) 106   b) 197   c) 265   d) 304

d

73) Item Type: Multiple Choice   Objective: 8   Section 30.4
Level: A: Standard   Tag: None

Phobos is a moon of

a) Venus.   b) Mars.   c) Jupiter.   d) Saturn.

b

74) Item Type: Multiple Choice   Objective: 9   Section 30.4
Level: A: Standard   Tag: None

Which of the following is a moon of Jupiter?

a) Io   b) Titan   c) Oberon   d) Triton

a

75) Item Type: Multiple Choice   Objective: 9   Section 30.4
    Level: B: Enriched   Tag: None

Which of the following is a moon of Pluto?

a) Charon   b) Ariel   c) Ganymede   d) Titan

a

76) Item Type: Fill in the Blank   Objective: 1   Section 30.1
    Level: B: Enriched   Tag: None

Light-colored, coarse-grained rocks called anorthosites make up much of the lunar surface

called the _____.

highlands

77) Item Type: Fill in the Blank   Objective: 1   Section 30.1
    Level: B: Enriched   Tag: None

What metal may be extracted from moon rocks using the process of electrolysis?

_____

iron

78) Item Type: Fill in the Blank   Objective: 2   Section 30.1
    Level: A: Standard   Tag: None

The layer beneath the moon's crust is called the _____.

mantle

79) Item Type: Fill in the Blank   Objective: 2   Section 30.1
    Level: A: Standard   Tag: None

What instruments were left on the moon's surface by Apollo astronauts to help study the

moon's interior? _____

seismographs

80) Item Type: Fill in the Blank   Objective: 4   Section 30.2
    Level: A: Standard   Tag: None

When the moon is farthest from the earth, it is said to be at what position?

_____

apogee

81) Item Type: Fill in the Blank   Objective: 4   Section 30.2
    Level: A: Standard   Tag: None

How long does it take for the moon to make one revolution around the earth?

_____

27.3 days

82) Item Type: Fill in the Blank   Objective: 4   Section 30.2
    Level: B: Enriched   Tag: None

What is the average distance of the moon from the earth? _____

384,000 km

83) Item Type: Fill in the Blank   Objective: 5   Section 30.2
    Level: A: Standard   Tag: None

What part of a planetary body's shadow is cone-shaped and completely blocks the

sunlight? _____

umbra

84) Item Type: Fill in the Blank   Objective: 6   Section 30.3
    Level: A: Standard   Tag: None

What phase is the moon in when the side of the moon facing the earth is unlighted?

_____

new moon phase

85) Item Type: Fill in the Blank   Objective: 6   Section 30.3
    Level: A: Standard   Tag: None

What is sunlight reflected off the earth called? _____

earthshine

86) Item Type: Fill in the Blank   Objective: 6   Section 30.3
    Level: A: Standard   Tag: None

What is the major source of the light that reflects off the moon? _____

the sun

87) Item Type: Fill in the Blank   Objective: 7   Section 30.3
    Level: A: Standard   Tag: None

In what period of time does the moon go through a complete cycle of phases?

_____

| one month (29.5 days) |

88) Item Type: Fill in the Blank   Objective: 7   Section 30.3
    Level: A: Standard   Tag: None

What is the time required for the earth to make one complete rotation on its axis?

_____

| one day (about 24 hrs.) |

89) Item Type: Fill in the Blank   Objective: 7   Section 30.3
    Level: A: Standard   Tag: None

What is the name of the proposed calendar that would add a day at the end of June every

four years? _____

| World calendar |

90) Item Type: Fill in the Blank   Objective: 7   Section 30.3
    Level: A: Standard   Tag: None

How many months did Julius Caesar divide the year into? _____

| 12 |

91) Item Type: Fill in the Blank   Objective: 7   Section 30.3
    Level: A: Standard   Tag: None

What are years with an extra day called? _____

| leap years |

92) Item Type: Fill in the Blank   Objective: 8   Section 30.4
    Level: A: Standard   Tag: None

Which planet has two, small, irregularly-shaped moons? _____

| Mars |

93) Item Type: Fill in the Blank   Objective: 9   Section 30.4
Level: A: Standard   Tag: None

Which moon of Neptune is unusual because it travels from east to west as it revolves

around the planet? _____

Triton

94) Item Type: Fill in the Blank   Objective: 9   Section 30.4
Level: A: Standard   Tag: None

Which planet has a moon called Triton, which is known to revolve from east to west

around the planet? _____

Neptune

95) Item Type: Fill in the Blank   Objective: 9   Section 30.4
Level: A: Standard   Tag: None

What materials compose most of Saturn's rings and moons? _____

ice and rock

96) Item Type: Fill in the Blank   Objective: 9   Section 30.4
Level: B: Enriched   Tag: None

Which of Jupiter's moons may be the most densely cratered moon in the solar system?

_____

Callisto

97) Item Type: Essay   Objective: 1   Section 30.1
     Level: B: Enriched   Tag: None

Would an individual on the moon weigh the same, more, or less than he or she weighs on the earth? Explain your answer.

_____

_____

_____

_____

_____

> An individual on the moon weighs much less than he or she does on the earth. Since the moon has less mass than the earth does, its gravity is weaker. An individual's weight is determined by the force of gravity acting on the individual. The moon's surface gravity is about one-sixth the surface gravity of the earth, so an individual on the moon would weigh about one-sixth of what he or she weighs on the earth.

98) Item Type: Essay   Objective: 2   Section 30.1
     Level: B: Enriched   Tag: None

Describe the structure of the moon.

_____

_____

_____

_____

_____

> The outer layer of crust is 60–100 km thick. Below the crust is a dense mantle of rocks rich in silica, magnesium, and iron. The mantle extends to a depth of 1,000 km. At the center may be a small iron core. The lower portion of the mantle and the core may be slightly molten, and the core may even be liquid.

99) Item Type: Essay   Objective: 3   Section 30.1
Level: B: Enriched   Tag: None

Describe the giant-impact hypothesis of the moon's origin.

_____

_____

_____

_____

_____

The moon formed when a Mars-sized body struck the earth, causing an ejection of fragments into orbit around the earth. These fragments eventually joined to form the moon. Most of the ejected materials were from the silicate-rich mantle of the earth and the colliding body, while dense metallic core materials remained in the earth.

100) Item Type: Essay   Objective: 3   Section 30.1
Level: B: Enriched   Tag: None

Describe the stage of the moon's development that began when its surface cooled.

_____

_____

_____

_____

_____

The outer surface of the moon cooled and formed a solid crust over the molten rock. Some meteorites punctured this crust, and molten rock flowed out onto the surface through the breaks. This molten rock partially filled the craters and formed the maria.

101) Item Type: Essay   Objective: 4   Section 30.2
     Level: B: Enriched   Tag: None

How does the rate of lunar rotation affect our view of the moon?

_____

_____

_____

_____

_____

> The moon rotates on its axis very slowly, completing a rotation only once during each orbit around the earth. Because the moon rotates at the same rate that it revolves around the earth, the same side of the moon always faces the earth. From earth, we can only see the near side of the moon.

102) Item Type: Essay   Objective: 5   Section 30.2
     Level: A: Standard   Tag: None

Describe an annular solar eclipse.

_____

_____

_____

_____

_____

> During an annular solar eclipse, the moon comes between the earth and sun, but its umbra does not reach the earth. The sun is not completely blocked out and appears as a bright ring around the moon.

103) Item Type: Essay   Objective: 7   Section 30.3
      Level: B: Enriched   Tag: None

Explain the revision introduced by the use of the Gregorian calendar.

_____

_____

_____

_____

The Gregorian calendar has three fewer days every 400 years than the Julian calendar. Years ending in "00" are not leap years unless they are divisible by 400.

104) Item Type: Essay   Objective: 9   Section 30.4
      Level: B: Enriched   Tag: None

What are the two theories used to explain the formation of Saturn's rings?

_____

_____

_____

_____

One theory hypothesizes that Saturn's rings formed from the breakup of a satellite of Saturn; the other theory hypothesizes that the rings are made of material that was unable to condense into a moon.

1)   Tag Name: Diagram 3071

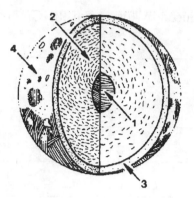

2)   Tag Name: Diagram 3072

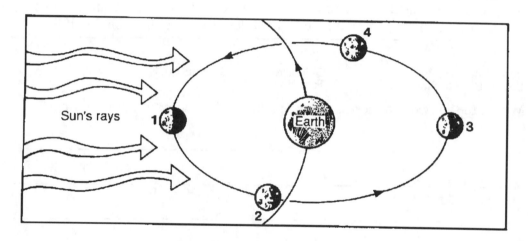

3)   Tag Name: Diagram 3073

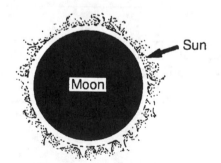

4) Tag Name: Diagram 3074

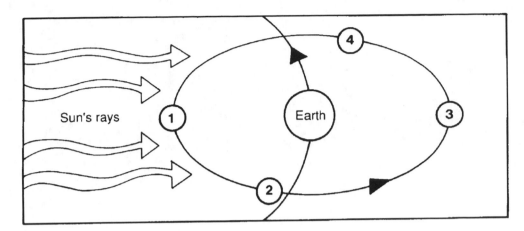